AGRICULTURE ISSUES AND POLICIES

CROP ROTATIONS

FARMING PRACTICES, MONITORING AND ENVIRONMENTAL BENEFITS

AGRICULTURE ISSUES AND POLICIES

Additional books in this series can be found on Nova's website
under the Series tab.

Additional e-books in this series can be found on Nova's website
under the e-book tab.

AGRICULTURE ISSUES AND POLICIES

CROP ROTATIONS

FARMING PRACTICES, MONITORING AND ENVIRONMENTAL BENEFITS

BAO-LUO MA
EDITOR

nova publishers
New York

NOTICE TO THE READER

The Publisher has taken reasonable care in the preparation of this book, but makes no expressed or implied warranty of any kind and assumes no responsibility for any errors or omissions. No liability is assumed for incidental or consequential damages in connection with or arising out of information contained in this book. The Publisher shall not be liable for any special, consequential, or exemplary damages resulting, in whole or in part, from the readers' use of, or reliance upon, this material. Any parts of this book based on government reports are so indicated and copyright is claimed for those parts to the extent applicable to compilations of such works.

Independent verification should be sought for any data, advice or recommendations contained in this book. In addition, no responsibility is assumed by the publisher for any injury and/or damage to persons or property arising from any methods, products, instructions, ideas or otherwise contained in this publication.

This publication is designed to provide accurate and authoritative information with regard to the subject matter covered herein. It is sold with the clear understanding that the Publisher is not engaged in rendering legal or any other professional services. If legal or any other expert assistance is required, the services of a competent person should be sought. FROM A DECLARATION OF PARTICIPANTS JOINTLY ADOPTED BY A COMMITTEE OF THE AMERICAN BAR ASSOCIATION AND A COMMITTEE OF PUBLISHERS.

Additional color graphics may be available in the e-book version of this book.

Library of Congress Cataloging-in-Publication Data

Names: Ma, Bao-Luo, editor.
Title: Crop rotations : farming practices, monitoring and environmental benefits / editor: Bao-Luo Ma.
Other titles: Agriculture issues and policies series.
Description: Hauppauge, New York : Nova Science Publishers, [2016] | Series: Agriculture issues and policies | Includes index.
Identifiers: LCCN 2015050199 (print) | LCCN 2016000159 (ebook) | ISBN 9781634844963 (hardcover) | ISBN 9781634845281 ()
Subjects: LCSH: Crop rotation.
Classification: LCC S603 .C767 2016 (print) | LCC S603 (ebook) | DDC 631.5/82--dc23
LC record available at http://lccn.loc.gov/2015050199

Published by Nova Science Publishers, Inc. † New York

CONTENTS

PREFACE

The global population is projected to reach 9 billion by mid-century. Questions continue to arise concerning the ability of the agriculture sector to keep pace with the demands for food, feed, fibre and fuel of an increasing population in the near future, in a way of sustaining both the production system and the environment. Crop rotations, an ancient practice that has gained renewed interest in recent years, involve growing different crop species/varieties on the same piece of land in consecutive growing seasons (years). The direct and indirect benefits of this practice to the production system and the environment have been recognized for millennia. In recent years, it has been documented that crop rotations, coupled with conservation tillage, enhances the physical, chemical and biological properties of soil, improves seasonal nitrogen availability, and provides nitrogen inputs through the symbiotic nitrogen fixation by legumes. This strategy can also lead to a better balance of plant nutritional requirements and a shift in soil mycorrhizal populations, interrupt insect populations, increase root activity, reduce disease severity, enhance environmentally-friendly biodiversity, and lower per-area greenhouse gas emissions or per-yield carbon footprints. This book presents the latest innovations and integrated knowledge from sciences as diverse as agronomy, soil science, ecology, economy, and social sciences, in this dynamic field from around the world.

Chapter 1 – presents a thorough review on crop productivity and environmental impact in a maize-legume rotation system. Maize-legume rotation, a popular cereal-based cropping system, has been practiced for thousands of years and has recently been adopted by numerous small-holder farms worldwide. Maize, grown in rotation with legume crops, such as alfalfa or soybean, often yields more and requires less synthetic nitrogen (N) than continuous monoculture. Meta-data analysis indicates that the maize yield advantage is approximately 9.6% in a maize-soybean rotation, and up to 40% in a maize-green manure (or legume forage) rotation, compared to continuous maize monoculture. In addition, soil N amendment, such as the addition of farm manure, green manure or composted municipal waste to maize not only provides the crop with N and other necessary nutrients, but perhaps more importantly, increases soil N release through seasonal N mineralization and N availability to the current and succeeding crops. Thus, maize-legume rotations with soil amendments have multifaceted potential to improve maize yields and resource-use efficiency on one hand, while protecting the environment on the other. This chapter examines the rationale, merits, recent research trends and future opportunities of maize-legume rotations with soil amendments.

Chapter 2 – focuses on crop rotation trends: past, present and future benefits and drivers in temperate production regions. Before 1950, complex crop rotations provided such benefits as weed, pest and insect management, nutrient supply and labour distribution. More recently however, technological advancements in nitrogen fertilizers, pesticides, plant genetics and equipment have reduced the apparent need for crop rotation complexity in favour of more "simple" rotations. These simple rotations have the perception as the "most profitable" when intensively managed, and may consist of only two crops or the continuous planting of one crop. Long-term rotation trials have, however, demonstrated that simple rotations are associated with reduced yields and resiliency of a system, along with negative environmental impacts such as reduced soil organic matter, reduced nutrient use efficiency and increased nutrient loss to air and water. The costs of these negative impacts are often not borne by the producer but by other segments of society. The future effects of a changing climate, emerging biomass industries, and intensification of production systems could increase the overall costs associated with simple rotations, thereby compromising long-term profitability and leading to the need for the development of more rotation diversity with associated environmental benefits.

Chapter 3 – gathers recent research progress on the legume-cereal rotation cropping system in China. The science and rationale of legume-cereal crop rotation system in China can be traced back to the Chinese book – 'Qi Min Yao Shu', published during the Western Han Dynasty, sometime 2000 years ago. This chapter presents the history, and the recent understanding of legume-cereal crop rotation systems in China. Considering the long-term benefits that legumes-cereals can have on maintaining crop yield, improvement of soil properties and environmental conditions, the authors emphasized the role of legume-cereal rotation in sustaining the agriculture sector, and the society and environment as a whole. They also proposed research priorities in the future, including (1) optimize and improve legume-cereal systems, taking into consideration diverse soils, climate, crops, and cropping systems; (2) establish legume-cereal research networks and links involving multidisciplinary teams; (3) identify suitable rotation patterns for the small land holders and diverse farming areas; and (4) link food security with environmental protection, sustainable soil management, and climate change.

Chapter 4 – introduces a case study on the rotation of peanut and cotton with bahiagrass to improve soil quality and crop productivity in the USA. Bahiagrass (*Paspalum notatum*), a perennial grass, has been rotated with other row crops in the southeast United States. The research focused on a short-term rotation system that keeps bahiagrass in the rotation with row crops and has been found to be economically and environmentally advantageous. This system has been found to increase soil organic matter content and water infiltration along with improving growth, yields, and profits of peanut (*Arachis hypogea* L.) and cotton (*Gossypium hirsutum* L.). One of the main contributing factors to the improved profit potential of the sod-based rotation is reduction in input costs compared to the conventional rotation. The system incorporates a short term bahiagrass-bahiagrass-peanut-cotton rotation (sod-based rotation) system as compared with the conventional peanut-cotton-cotton rotation in the region. In both the conventional and sod-based rotations, reduced tillage techniques have been utilized as an added benefit to water and soil conservation. In this chapter, the authors review and update recent research and provide information about rotation of row crops (peanut and cotton) with bahiagrass to improve soil quality, crop physiology, growth, and productivity.

Chapter 5 – presents land use practices, cropping systems and climate change vulnerability to mountain agro-ecosystems of Nepal. Characterized by fragile geo-ecology, marginality, inaccessibility, and subsistence livelihoods, the land resources in the Middle Mountain region of Nepal are intensively cultivated beyond their carrying capacity. The mountain cropping systems have to face numerous natural and human-induced challenges, including land degradation and loss of agro-biodiversity, leading to food insecurity and unsustainable livelihoods. This chapter outlines the typical characteristics of the mountain farming systems and discusses how they are coping with the ongoing natural and socio-economic changes. The mountain agriculture seems to be highly vulnerable to climate change particularly related to erratic rainfall events and droughts, and land degradation due to soil erosion, landslides, flash floods, and siltation leading to loss of productive lands and crops failure. Effective measures to cope with such impacts on the mountain agro-ecosystem are suggested. Increased public awareness about the climate change effects, building adaptive capacity to cope with such effects, sustainable soil and water conservation practices, and resilient cropping practices (drought tolerant crop/varieties, change in crop rotations, and water use efficiency) seem to be the key strategies to be adopted to cope with the climate change in the fragile mountain agro-ecosystems.

Chapter 6 – deciphers soil moisture and crop productivity in grass-crop rotation systems under the semiarid Loess Plateau conditions in northwest China. This chapter provides a comprehensive review of the forefront of the ridge-furrow planting with plastic film mulching system. During the past two decades, adoption of this technology has led to the introduction of corn as a new major crop to the region and has revolutionized the crop/forage production systems in the poorer northwestern arid and semi-arid regions, leading to increases of 36-50% in wheat yields, of 17-80% in corn yields, and a doubling of potato yields. In recent years, the introduction of alfalfa and a range of native shrub species to the region, has led to the rehabilitation of eroded and abandoned farmland on the Loess Plateau. This chapter reviews the theoretical understanding of the improved overall resource use efficiency, including water saving and water use efficiency and crop productivity, the cumulative benefits to small-hold farm community and the impact on the agroecosystem and the overall environment.

Chapter 7 – reviews the benefits and challenges of crop rotation and cover crop to soil biodiversity, pest and disease management in advancing sustainable agriculture. Only recently has it been realized that the ecological and environmental problems associated with the monoculture and excessive use of chemical pesticides and fertilizers in the Green Revolution can be mitigated through well-planned crop rotation and cover crop. One of the newly found benefits of the cultural practices such as crop rotation and cover crop is the increased diversity of flura and fauna within the agro-system, which tends to be healthy, and resilient to pests and diseases. Successful crop rotations and cover crops for pest and disease controls start with an understanding of the pests and diseases. This chapter reviews the latest scientific knowledge by studying the current cropping systems, and cover crops in reducing pest and disease populations in the context of integrated pest management (IPM) systems. The shift of adoption from selecting a specific (non-host) to a general (only taxonomical distant) rotating crop or cover crop for pest and disease control is discussed.

Chapter 8 – offers a contextual view of the science of carbon footprints in agricultural crop production systems. Global climate is rapidly changing due to anthropogenic greenhouse gas (GHG) emissions, causing substantial risks to agricultural production systems associated with frequent occurrence of catastrophic weather events. Since agriculture itself is one of the major contributors to GHG emissions, producers, researchers and policy makers strive to develop effective crop management practices to minimize GHGs while maximizing farmer's net returns. Thus, quantification of GHGs with diverse cropping systems is essential to mitigate GHGs from agriculture and in turn to develop more sustainable practices. Carbon footprint is a measure of the intensity of GHGs and productivity of different agricultural practices. Because of the easy conveyance of information to the general public about the GHG intensity of a variety of products and diverse activities, carbon footprint, as a new quantitative indicator, has attracted the attention of scientists and policy-makers and gained public acceptance. Although scientific literature on carbon footprint targeting GHGs from farming practices is still sparse, accumulated convincing evidence indicates that a significant part of the GHGs related to agriculture can be mitigated through improved agronomic practices, including the adoption of diversified cropping systems with well-defined crop sequences including cereal, oilseed and legume crops. Effective crop rotation systems have been shown to increase crop productivity with an efficient use of resources by individual crops, as well as improved soil carbon storage and reduced carbon footprints. This chapter comprises an overview of GHGs in relation to different crop management practices, followed by the concept and general principle of estimating carbon footprints of agricultural products. It also reviews available scientific literature on calculations of carbon footprint, its application, boundaries and challenges, and effective measures to reduce GHGs in agriculture, particularly focusing on diverse crop rotation systems.

Dr. Bao-Luo Ma
Editor
Senior Research Scientist, Crop Physiology,
Ottawa Research and Development Centre,
Agriculture and Agri-Food Canada,
Ottawa, ON, Canada

ACKNOWLEDGMENTS

I sincerely thank Dr. Chang Liang, a leading expertise in sustainable agriculture and a senior agricultural scientist at Environment Canada, who has reviewed and provided all the authors with thorough and constructive comments and suggestions. I am thankful for my family. My parents, who both were illiterate, lived in hardship most of their lives, encouraged and supported me to take education to the highest level I could reach, and taught me to work hard no matter what I am doing, which becomes the lifetime gift to me. The hardship in my childhood makes me appreciate what we have every day. I thank my family for their unconditional support to my effort on scientific paper writing.

In: Crop Rotations
Editor: Bao-Luo Ma
ISBN: 978-1-63484-496-3

Chapter 1

CROP PRODUCTIVITY AND ENVIRONMENT IMPACT IN A MAIZE-LEGUME ROTATION SYSTEM: A REVIEW

***Bao-Luo Ma*[1,*] *and Wei Wu*[1,2]**
[1]Ottawa Research and Development Centre,
Agriculture and Agri–Food Canada, Ottawa, ON, Canada
[2]College of Agronomy, Northwest A and F University, Yangling, Shaanxi, P. R. China

ABSTRACT

The challenge of securing the global food supply and environmental health is increasingly recognized by various levels of policy makers and scientific communities. The overuse of inorganic fertilizer to increase crop yields comes with a high environmental cost, and has exacerbated concerns over environmental sustainability. Alternatively, the use of leguminous rotation for biological nitrogen fixation (BNF) and soil nitrogen (N) amendment with organic N input, instead of relying entirely on N from commercial fertilizer, can be regarded as an environmentally-friendly strategy for sustainable agriculture development with enhanced crop productivity. Crop rotation is an agronomic practice of growing a series of dissimilar types of crop plants on the same piece of land in sequential growing seasons. Maize (or corn)-legume rotation, a popular cereal-based cropping system, has been practiced for thousands of years and has recently been adopted by hundreds of smallholder farms worldwide. Maize, grown in rotation with legume crops, such as alfalfa or soybean, often yields more and requires less synthetic N than continuous monoculture. In addition, soil N amendment, such as the addition of farm manure, green manure or composted municipal waste to maize, not only provides the crop with N and other necessary nutrients, but perhaps more importantly, increases soil N release through seasonal N mineralization and N availability to the current and succeeding crops. Thus, maize-legume rotations with soil amendments have multifaceted potential to improve maize yields and resource-use efficiency on one hand, while protecting the environment on the other. This chapter examines the rationale, progress, merits and implications of maize-legume rotations with soil amendments.

[*] Corresponding author: baoluo.ma@agr.gc.ca or baoluo.ma@canada.ca.

Keywords: biological nitrogen fixation, crop rotation, legume, maize, meta-analysis, sustainable agricultural development

1. Introduction

1.1. Global Food Demand and Environment Impact

Grain cereals such as maize (*Zea mays* L.) are the major constituents of our food and are responsible for fulfilling most caloric requirements. Questions continue to arise concerning the ability of agriculture to sustainably keep pace with the food demands of an increasing population in the near future (Tilman et al., 2011). To address this challenge, most practices of modern agriculture, such as intensive monocultures, genetic improvements, and the heavy use of agrochemicals for fertilization and pest/disease control, have led to a simplification of the components of agricultural systems and to worldwide increases in crop productivity from farming systems. However, it is widely recognized that modern agriculture has developed at the price of agricultural sustainability (Lichtfouse et al., 2009; Wu and Ma, 2015). Many farmers, researchers, and policy makers worldwide have now recognized the importance of the development of self-sustaining, low-input, energy-efficient and highly diversified agricultural systems.

In recent decades, the injudicious use of mineral fertilizers, such as N fertilizer, is one of the fundamental causes of environmental pollution, such as eutrophication and greenhouse gas (GHG) emissions (Davidson et al., 2014; IPCC, 2014). Now is the time to search for innovative cropping systems that can guarantee higher crop yields while minimizing further deterioration of our environment (Wu and Ma, 2015). Legume-cereal rotation has been practiced for thousands of years and has been recently adopted by hundreds of smallholder farms worldwide. As the most popular crop grown with alternate practices, maize when grown in rotation with legume crops such as alfalfa (*Medicago sativa* L.) or soybean (*Glycine max* L.) often yields more products and requires less external N fertilizer than continuous monoculture (Advient-Borbe et al., 2007; Edwards et al., 1988; Ma et al., 2003). The increased yield has been attributed to the beneficial effects of crop rotation, which include a better balance of plant nutritional factors, improved soil physical properties, interrupted insect populations, increased root activity, a shift in soil mycorrhizal populations, reduced disease severity, and enhanced seasonal N mineralization (Howard et al., 1998; Ma et al., 2003; McGrath, 2007; Smith and McSorley, 2000; Stoner, 2007).

More importantly, legume-cereal rotation systems not only maximize cereal and legume productivity, but also optimize agricultural sustainability across a far more complex landscape of production, which has recently caused unprecedented changes in our environment (Davis et al., 2012; IPCC, 2014). Thus, this practice has been regarded as a promising strategy for reducing GHG emissions, and thus lowering the negative environmental impact (Davis et al., 2012; Ma et al., 2012).

1.2. Main Legume Crops

In the legume-maize cropping system, legumes are important in achieving the direct "rotational effect" as these plants easily access atmospheric N_2 through symbiosis with a group of soil bacteria that are collectively called rhizobia, thereby requiring less input of chemical N fertilizers (Dwivedi et al., 2015; Giller and Cadisch, 1995; Nadar and Faught, 1984; Peoples et al., 1995). When part of this 'free' N is made available to a subsequent crop, the use of legumes in a rotation can lead to a reduction in the N fertilizer used by the non-legume crop. A large effort has been devoted to the study of legume effects on subsequent cereal crops grown in rotation systems, especially for the legume-maize rotation system (Kessel and Hartley, 2000).

A legume is a plant in the family Fabaceae (Leguminosae). In agriculture production, legume crops are grown primarily for food, livestock forage, silage fodder, and as soil-improving green manure. Legumes are notable for their biological N fixation (BNF) with symbiotic bacteria, known as rhizobia, which live inside the plant root nodules (Graham and Vance, 2000; Mikić et al., 2015). Well-known legumes include alfalfa, clover (*Trifolium spp.*; mainly used as forage or green manure), soybean, peanut (*Arachis hypogaea*), pea (*Pisum sativum*), beans (*Phaseolus spp.*), lentil (*Lens culinaris*), lupin (*Lupinus spp.*), mesquite (*Prosopis spp.*), carob (*Ceratonia siliqua*), and tamarind (*Tamarindus indica*; used as dual-purpose for food and feed). Legumes have been used as important crop plants for centuries. They are important sources of protein-rich food and feed, oil, fibre, minerals and vitamins; and direct contribution of N to soil fertility. Production systems including these crops also improve soil structure and increase soil organic carbon status, reduce the incidence of pest and diseases in rotated crops, and increase the overall productivity and economic benefits of the production system (Liebman and Davis, 2000; Ma et al., 2003; Ndakidemi et al., 2006; Rao and Mathuva, 2000; Smith and McSorley, 2000).

Alfalfa, also called lucerne, a perennial forage plant that could grow for more than 20 years, depending on the variety and climate, is one of the most cultivated legumes worldwide (Wang et al., 2008, 2009). Alfalfa has a deep root system, which makes it very resilient, especially important in response to global warming, i.e., drought and heat (Singh et al., 2009). However, this plant exhibits auto-toxicity, which makes it impossible to grow in existing stands of alfalfa. Therefore, alfalfa fields are recommended for rotation with other crops, such as maize and wheat (*Triticum aestivum* L.) (Levine et al., 2002).

The global harvested area of alfalfa is approximately 30 million ha, down from approximately 33 million ha in the 1970s. Of the current production, North America produces 41% (approximately 12 million ha), Europe accounts for 25%, South America produces 23%, Asia produces 8%, and Africa and Oceania produce the remainder (http://www.fao.org/ag/agp/ AGPC/doc/ ningxia_guide/chapter1.pdf). The United States was the largest alfalfa producer in the world with 9 million ha in 2009, followed by Canada, Russia, Italy, and China (http://faostat.fao.org/site/626/default.aspx).

Currently, grain legumes, well known as dual-purpose legumes (such as soybean, cowpea, groundnut, and pigeon pea) have become increasingly important in supplying protein for animals and food for mankind (Singh and Singh, 1992). The production area of these grain legumes accounted for 20% of the world's arable lands in 2013, increasing from only 5% in 1961. Dual-purpose legumes are particularly attractive to small-scale farmers who practice a mixed crop/livestock intensification system because the grain legume not only provides oil and proteins for human consumption needs, but also provides fodder for farm livestock, making it an important legume in intensive cropping systems worldwide (Ndakidemi et al., 2006; Ojiem et al., 2014). In addition to increasing cash income to farmers through the sale of grain and/or livestock products (such as milk, meat and manure), preceding grain legumes often result in an improved yield of subsequent cereal crops, even with a reduction in the use of inorganic N fertilizer. This indirect rotation effect can be attributed to the improved chemical, physical and biological properties of the soil by the legume crop production (Franke et al., 2008; Yusuf et al., 2009).

Globally, the average harvested area of grain legumes (including soybean, beans, groundnuts, chick pea, cow pea, field peas, pigeon peas, lentils, green peas, broad horse beans, green beans, lupins, bambara bean and green string beans) in 2013 was 215 million ha, or twice as much as in 1961. During the past 50 years, both annual production and yield have more than doubled, with total annual production of 70 million tonnes and yield of 0.8 t ha^{-1} in 1961 (except for the green legumes) to the current 390 million tonnes of production and 1.84 t ha^{-1} of yield in 2013 (Figure 1). Asia accounts for more than 60% of the global grain legume production, followed by Africa, the American Continent, Europe, and Oceania (http://www.faostat.fao.org/).

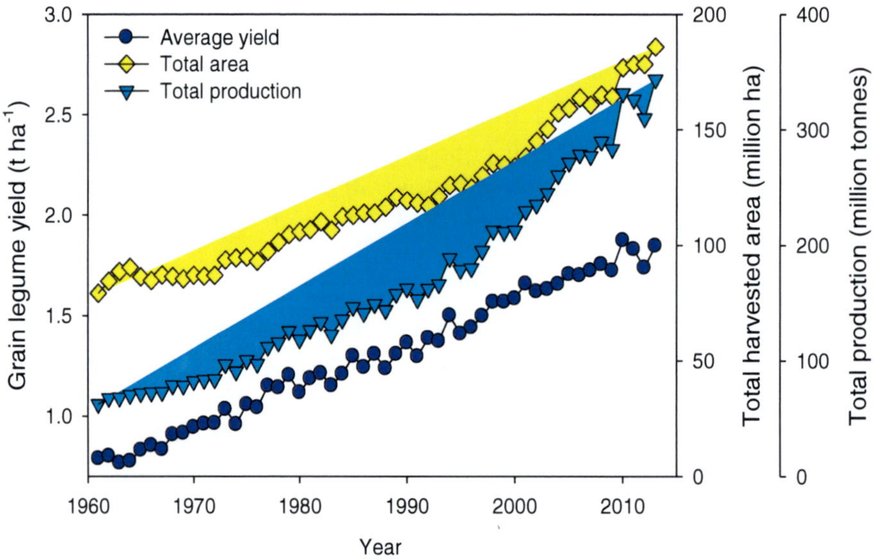

Figure 1. The dynamic changes in average yield, total area and total production of grain legume from 1961 to 2013. All data are the average for all grain legumes, including soybeans, beans, groundnuts, chick peas, cowpeas, peas, pigeon peas, lentils, broad horse beans, lupins and bambara beans. Adapted from: http://faostat.fao.org/site/626/default.aspx.

The main types of grain legumes and their respective rotation systems are shown in Figure 2. Soybean, beans and groundnuts are the three major leguminous crops, accounting for 77% of the total area of harvested legumes. The global acreage of soybean production increased fivefold from 23 million ha in 1961 to 112 million ha in 2013, while the average productivity increased from 1.13 t ha^{-1} to 2.47 t ha^{-1}. These data indicate the potential for more rotation benefits from soybean as the legume due to its large harvested area and increasing demand. Soybeans have been produced predominantly on the American Continent and Asia, representing roughly 86% and 12%, respectively, of the global total of 276 million tonnes. The global acreage of beans in 2013 was 29.1 million ha, with a total annual harvest of 22.8 million tonnes and average yield of 0.78 t ha^{-1}. The global groundnut acreage in 2013 was 25.4 million ha, with 45.6 million tonnes of production and an average yield of 1.8 t ha^{-1}. From 1961 to 2013, the annual global production of beans and groundnuts increased, by one and twofold, respectively.

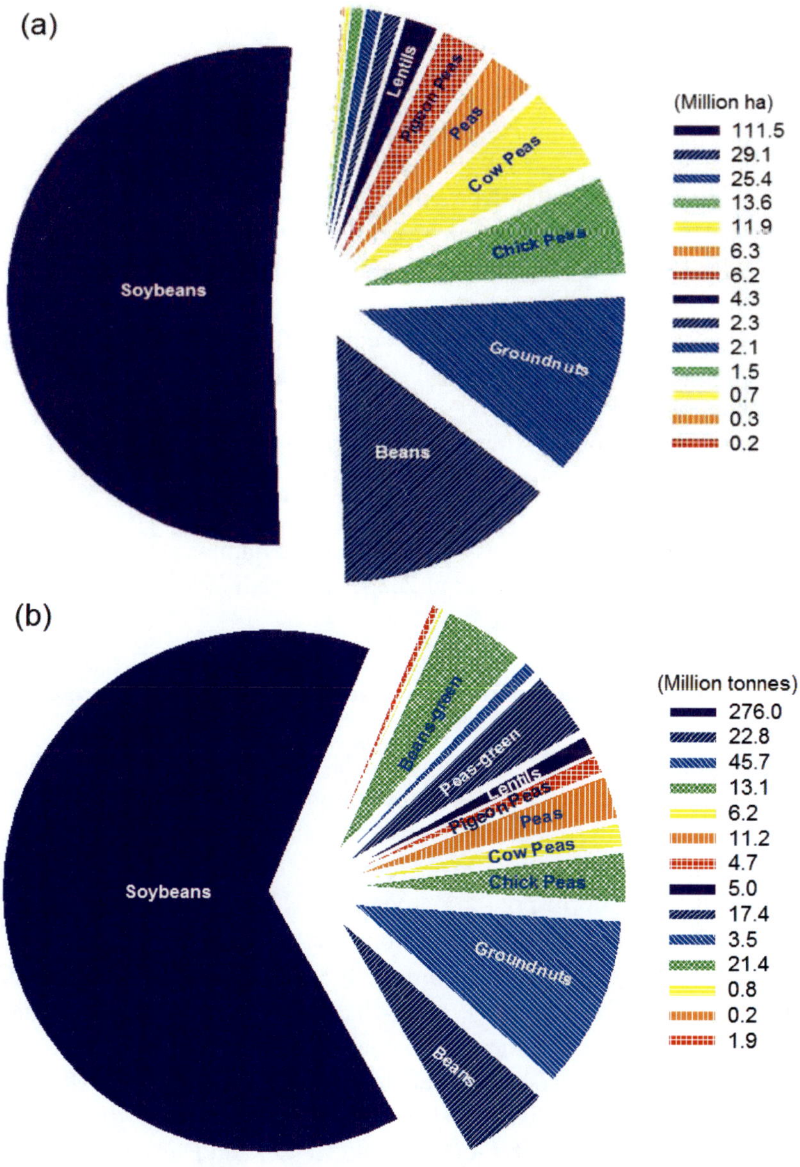

Figure 2. The globally harvested area (a) and production (b) of pulses (including soybeans, beans, groundnuts, chick peas, cowpeas, peas, pigeon peas, lentils, green peas, broad horse beans, green beans, lupins, bambara beans, and green string beans, in order according to harvested area) in 2013. Adapted from: http://faostat.fao.org/site/626/default.aspx.

This chapter will focus on the rationale and assessment methodology of a maize-legume rotation system. It highlights the main advantages of maize-legume rotation with special emphasis on the superiority of maize yield; the suppression of pests, diseases and weeds; and the potential benefits to the environment.

2. Rationale and Methodology of Legume-Cereal Rotation

2.1. The Concept of Biological N Fixation

Biological N fixation (BNF) is the process by which leguminous plants or other species having similar functions assimilate N_2 from the atmosphere, incorporate the molecules into their tissues and, subsequently, into the ground, thus improving their own growth as well as improving soil health and the overall productivity of the farming systems (Peoples et al., 1995). BNF is seen as a sustainable source of N for replacement or for use as a natural complement to chemical N fertilizer for the subsequent cereal crop (Giller and Cadisch, 1995).

In a cereal-legume rotation system, a fundamental function of legume plants is the formation of a symbiotic relationship with root-nodule bacteria (rhizobia). The rhizobia are gram-negative bacteria from a limited set of clades, belonging mainly to the Alphaproteobacteria and grouped into distinct genera (*Azorhizobium, Bradyrhizobium, Mesorhizobium, Rhizobium, and Sinorhizobium*), species, and symbiovars (Gyaneshwar et al., 2011; Rogel et al., 2011). Other rhizobia such as *Betaproteobacteria, Cupriavidus, and Pseudomonas* (*Ralstonia*) of the subclass Betaproteobacteria, are also able to form symbiotic associations with legumes (Balachandar et al., 2007; Dwivedi et al., 2015).

2.2. Estimation of N_2 Fixation Efficiency and Its Related Factors

Preliminary trials in the early 1980s indicated higher maize yields in rotation with single crop legumes (cowpea and bean) than in intercropping with legumes or continuous single cropping of maize, due to the great BNF produced in the leguminous season (Nadar and Faught, 1984). The N removed by the maize could be as much as 25 to 50 kg N ha^{-1} per season, which means that a corresponding amount of N needs to be supplied for the maize crop for the long-term sustainability of production (Ma et al., 2003).

Since those preliminary trials, some commonly used methods have been developed for estimating the efficiency of N_2 fixation: (1) N balance based on the difference in total N accumulation between a grain legume and a non-N_2-fixing reference crop; (2) ^{15}N-isotope dilution; (3) enriched ^{15}N-fertilizers; and (4) natural ^{15}N abundance level (Rennie and Kemp, 1983; Shearer and Kohl, 1986; Ndiyae et al., 2000). A more in-depth discussion on methods for the estimation of N_2 fixation can be found in van Kessel and Hartley (2000).

Based on the various methods available for estimating the amount of N_2 fixation by a legume crop, large differences were noted in the proportion of N_2 fixation occurring by the different legume crops in leguminous rotation systems. For example, 75% of the total N in the plants was derived from BNF in faba bean; 62-94% in cowpeas, chick peas and pigeon peas;

and only 39% in the common bean (Dwivedi et al., 2015). Pilbeamet al. (1995) indicated that N fixation by beans is notoriously inconsistent whether in inoculated or non-inoculated conditions, but that cowpeas are well nodulated by the ubiquitous *Bradyrhizobium* spp. and fix up to 197 kg N ha^{-1}, according to assessment with the ^{15}N isotope dilution technique.

Regional variation in the amount of N fixation has been observed among legume crops. For instance, soybean has been suggested to fix 193 kg N ha^{-1} in Africa and 300 kg N ha^{-1} in South America; common bean produced 75 kg N ha^{-1} in North America; groundnut, approximately 110 kg N ha^{-1} in South and Southeast Asia; pea, approximately 130 kg N ha^{-1} in Europe; cowpea, approximately 70 kg N ha^{-1} in South Asia and Africa; 122 kg N ha^{-1} by lentil in West Asia; and 58 kg N ha^{-1} via pigeon pea in South Asia (for more details, see the review by Dwivedi et al. (2015)).

Perennial tree legumes may also have a greater scope in replenishing soil fertility than annual grain legumes by their strong ability to exploit the water and subsoil nutrients that crops cannot utilize, withstand drought, and hence produce higher biomass (Rao and Mathuva, 2000). Their year-round growth habit may contribute to higher BNF (Dommergues, 1995). Other advantages of perennial legumes include an absence of recurring establishment costs, the opportunity to grow intercropping crops simultaneously without sacrificing land (Kang et al., 1990) and improved soil fertility and water infiltration because of their great root activity, as suggested by Rao et al. (1998).

Several factors contribute to the differences in the amount and efficiency of biological N_2 fixation (van Kessel and Hartley, 2000; Weisany et al., 2013; Dwivedi et al., 2015). Most importantly, factors that directly influence legume growth, such as water, nutrient availability, soil texture and tillage, pathogens and pests, crop husbandry practices and natural resource management, which either limit the presence of effective rhizobia in the soil or enhance competition for soil mineral N, are critical to the amount of atmospheric N_2 fixation that occurs from the legume-rhizobium symbiosis (Weisany et al., 2013).

2.2.1. Traits Related to Host-Rhizobium Association

Photosynthesis is strongly related to BNF in legumes. Some studies have reported that the availability of photosynthetic products can result in a corresponding improvement in BNF efficiency from better root nodule growth due to the appropriate net carbon exchange rate (Bethlenfalvar and Phillips, 1977). Possibly, carbohydrates from photosynthesis are imported into the nodules and then used as carbon skeletons in ammonia assimilation (Larrainzar et al., 2009). In addition, Ben Salah et al. (2009) suggested that photosynthesis always occurs as a positive feedback to BNF, whereas it may decrease due to the partial or complete blockage of BNF.

A recent study (Rodrigues et al., 2013) showed that the efficiency of BNF was positively correlated with the sucrose content in a nodule, rather than with the content of total soluble carbohydrates, sugars, and starch. Interestingly, their research reported that increased BNF under a triple inoculation treatment was not significantly correlated with sucrose synthase activity, but was related to soluble acid invertase activity in nodules at the beginning of senescence. Glutamine synthase, glutamate synthase, and glutamate dehydrogenase were stimulated by double or triple inoculation, in comparison to inoculation with *Rhizobium* alone.

2.2.2. Mineral Nutrition Status of the Host Plant

Mineral nutrition of the host plant is of great importance to plant growth and root nodule development; both of which can subsequently affect BNF. The necessary mineral nutrients required for legume crop growth serve as the basis for normal establishment and physiological functions. These include the six macronutrients: nitrogen (N), phosphorus (P), potassium (K), sulfur (S), magnesium (Mg), and calcium (Ca), which are present in relatively high concentrations in plant tissues, and some other essential micronutrients, including, boron (B), zinc (Zn), manganese (Mn), iron (Fe), copper (Cu), molybdenum (Mo), etc. Each nutrient is necessary for specific physiological and biochemical functions and is required in optimum concentrations for the establishment of symbiosis between the host plant and the rhizobium (Dwivedi et al., 2015; Weisany et al., 2013; Zahran, 1999).

Among those nutrients, P is essential for host plant growth and for nodulation and N_2 fixation. Generally, legumes are grown in conditions of insufficient soil P, and an adequate supply of available P is a prerequisite for high N fixation efficiency (Bünemann et al., 2004; Sahrawat et al., 2001). There is a strong relationship between soil pH and available concentration of P or other nutrients, which makes pH an important index of nutrient availability. Soil pH in the neutral range mobilizes all nutrients. However, in acidic soils, the availability of P becomes limited, whereas in the alkaline soils, the toxicity of sodium can affect the physiological function of the host plant and root nodulation development, thereby decreasing the N fixation efficiency (Brockwell et al., 1991; Dwivedi et al., 2015).

2.2.3. Soil Fertility and Starter N

It is widely accepted that under conditions of high soil fertility, especially those soils rich in organic matter, the application of starter N is not necessary (Dwivedi et al., 2015). Furthermore, the addition of chemical fertilizer will decrease the proportion of atmospheric N_2 fixed (Hungria et al., 2006). Hungria et al. (2006) also reported that the application of N at later stages to host plants does not increase N fixation or soybean yield. However, in conditions of low soil fertility, the application of N at a rate of 20-30 kg ha^{-1} as a base fertilizer is generally beneficial to the growth and yield of several grain legumes (Erman et al., 2009; Sogutet al., 2013). Thus, the recommended amount of starter N application for legumes depends on the fertility status of the soil condition and the N requirements for plant growth. Anyanzwa et al. (2010) showed that application of inorganic fertilizers (starter fertilizer of 60 kg P ha^{-1} and 60 kg N ha^{-1}) resulted in a higher maize yield of 5.2 t ha^{-1} in comparison with control plots whose yields were as low as 2 t ha^{-1} during the third season in a maize-legume cropping system. Ma and Wu (2008) suggested that the rate of N fertilizer for sidedress application to maize can be recommended based on a pre-plant soil nitrate test.

2.2.4. Abiotic Stress from Drought, Heat and Salinity

There are several environmental stressors that are regarded as limiting factors to BNF efficiency. BNF is highly sensitive to soil drought, which has been well documented in studies that show legume species exhibiting a great reduction in N fixation when exposed to soil water deficiency (Karmakar et al., 2014; Zahran et al., 1999). Soil water availability has a significant influence on N fixation because BNF processes in term of nodule initiation, growth, and activity are all positively regulated by or sensitive to soil water content (Albrecht et al., 1994; Bueckert et al., 2015; Zahran and Sprent, 1986). In addition, the response of BNF to water stress depends on the growth stage of the plant and the age of the nodule. Soil drought that occurs during the vegetative growth is possibly more detrimental to nodule growth and N fixation efficiency than that imposed during the late reproductive stage (Pena-Cabriales and Castellanos, 1993). It was estimated that there was an approximately 26% reduction in N derived from N_2 fixation when leguminous tree species was exposed to water stress (Sellstedt et al., 1993). Several mechanisms in legume plants are involved in the various physiological responses to drought stress, such as osmotic adjustment, accumulation of specific soluble solutes and potassium ions (for more details, see the review by Zahran et al. (1999)).

Similarly, high temperatures in tropical and subtropical regions are major problems for BNF in legume crops (Michiels et al., 1994; Aranjuelo et al., 2015). High-temperature environments strongly affect nodule formation, growth and N_2 fixation efficiency in many legume species, especially for some grain legume species, such as soybean (Munevar and Wollum, 1982), peanut (Kishinevsky et al., 1992), cowpea (Rainbird et al., 1983), bean (Hungria and Franco, 1993) and pea (Bueckert et al., 2015). Michiels et al. (1994) estimated that the critical temperature for N fixation is totally different among legume species, i.e., 30 °C for clover and pea, and 35-40 °C for soybean, peanut and cowpea. High temperatures affect root hair infection, bacteroid differentiation, nodule structure and growth, and result in low N_2 fixation efficiency (Roughley, 1970; Andrés et al., 2012). Some studies have found that heat stress is related to heat shock proteins in *Rhizobium*. These proteins have been detected in both heat-tolerant and heat-sensitive strains (Michiels et al., 1994).

Moreover, there are other environment-induced abiotic stressors that could influence the population and efficacy of rhizobia, such as soil acidity and salinity (please see a review paper by Zahran et al. (1999) for more detail). In addition, agronomic management practices, such as tillage, nutrient management, water management in relation to crop production and crop protection practices significantly influence the N_2 fixation ability of rhizobia in agroecosystems (Dwivedi et al., 2015).

3. Key Findings and Interpretation

3.1. Yield Advantage of Maize-Legume Rotation

Studies during the past 50 years have clearly identified the yield advantage of a maize crop when it is rotated with a legume crop, compared with a continuous maize monoculture system (CC) (Gentry et al., 2013; Howard et al., 1998; Katsvairo and Cox 2000; Ma et al., 2003; Varvel and Wilhelm 2003). We conducted a comprehensive literature search from the internet database of the Web of Knowledge and scholar google search (http://scholar.google.ca/) with "maize/corn-legume rotation" entered as keywords, and performed a meta-analysis. The criterion for including an article in the meta-analysis database was whether it contained detailed data on maize yield, comparing maize-grain legume rotation (mainly including maize-soybean/cowpea rotation and maize-alfalfa/green manure rotation) with CC from field experiments, which had been conducted in an acceptable scientific manner. Using the criterion of having comparable and detailed data that would permit quantitative analysis and comparison, a total of 9 articles (see Figure 3) were selected and used for the meta-analysis, of which most papers originated from long-term experiments. Due to fewer publications available for this meta-analysis, we extracted all the paired data from each selected article among different soil amendments, sites and years. Thus, 148 pairs of data were available for comparative analysis.

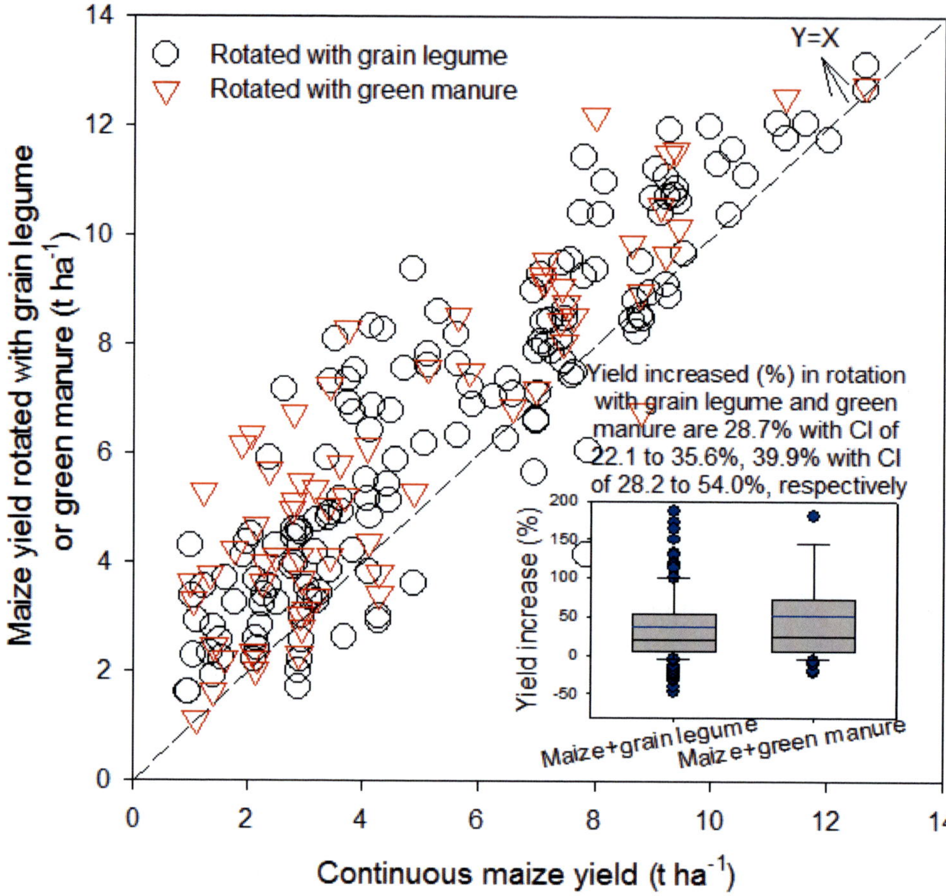

Figure 3. A distributed plot showing the paired maize yields of annual crop rotation with grain legume (Y axis, circle point), or non-grain legume (Y axis, triangle point) and continuous maize (X axis) from the selected articles that met the selection criteria for meta-analysis. Scatter points above the diagonal line (Y = X) indicate yield advantage of crop rotation over continuous monoculture. The inserted box-plot shows the yield advantage of crop rotation over continuous monoculture. Sources of data are from Rao and Mathuva (2000); Ma et al. (2003); Yusuf et al. (2009); Ojiem et al. (2014); Franke et al. (2008); Varvel and Wilhelm (2003); Mallarino et al. (2002); Lee (2003); and Gentry et al. (2013), among various soil amendments, sites and years.

The meta-analysis was conducted using a meta-analytical software package (Metawin 2.1, Sinauer Associates, Inc., Sunderland, MA, USA). To estimate the maize-legume rotation (MLR) effect compared with CC in term of maize yield, the natural log of the response ratio (R = variable in MLR/CC) was used as the metric for analysis (Hedges et al., 1999), and it is reported as the percentage changes from control as (R-1) * 100% (Ainsworth et al., 2002). A positive percentage change indicates a yield increase from the MLR treatment compared with the yield from CC, while a negative value indicates a yield reduction due to MLR.

This meta-analysis used an un-weighted approach in which the variance of the effect size was calculated using resampling techniques after 9999 iterations (Feng et al., 2008). Confidence limits around the effect size were calculated using a bias method. Estimates of the effect size were assumed to be significant if the 95% confidence intervals (CI) did not overlap zero (Curtis and Wang, 1998).

We used the pooled data to determine the average yield increase percentage with MLR compared with CC. The MLR was categorized into two classes: maize-grain legume (including soybean) rotation (MGLR) and maize-green manure rotation (including maize-forage rotation, MGMR). Overall, compared with CC, maize yield in the MGLR rotation significantly increased by 28.7% with CI of 22.1-35.6%, whereas in the MGMR system, maize yield increased by an average of 39.9% with CI of 28.2-54.0% (Figure 3).

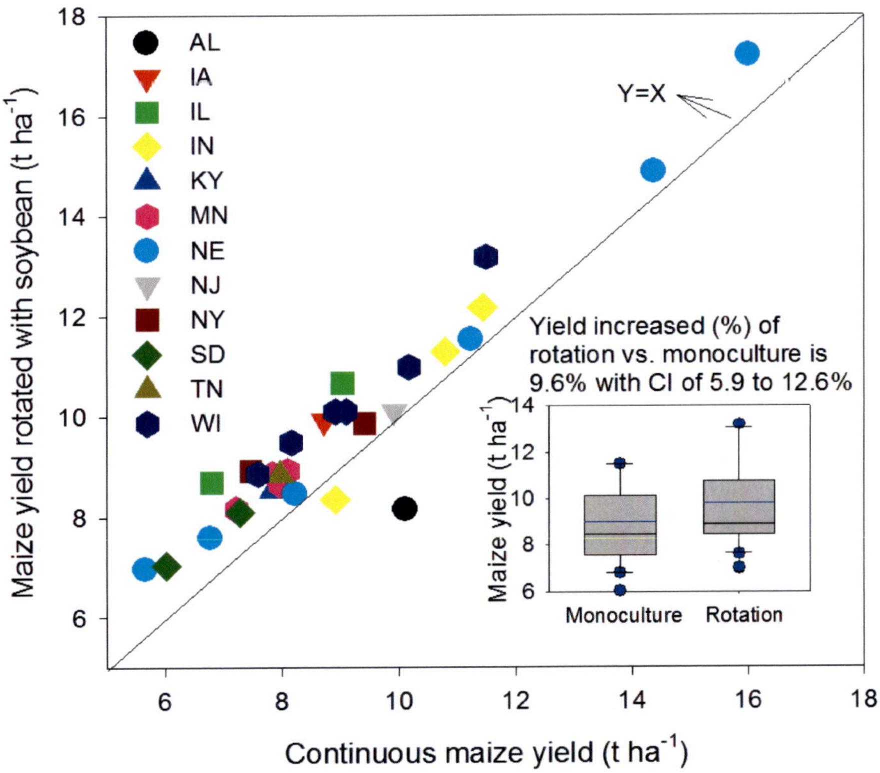

Figure 4. A distributed plot showing the maize yields of annual crop rotation with soybean (Y axis) in comparison with continuous monoculture (X axis) from the selected studies in USA. Scatter points above the diagonal line (Y = X) indicate yield advantage of maize-soybean rotation over continuous monoculture. The inserted box-plot shows the average yield of rotational and continuous monoculture maize. Data was adapted and updated from Erickson (2008). Sources: Peterson and Varvel (1989); Meese et al. (1991); Lund et al. (1993); Lauer et al. (1997); Porter et al. (1997); Crookston et al. (1998); Edwards et al. (1988); Griffith et al. (1988); Howard et al. (1988); Riedell et al. (1998); Singer and Cox (1998); Katsvairo and Cox (2000); Mallarino et al. (2002); Pedersen and Joseph (2002); Univ. of IL (2002); Pedersen and Joseph (2003); Singer et al. (2003); Varvel and Wilhelm (2003); Lee and John (2003); Walters et al. (2005); Wilhelm et al. (2004); Pikul et al. (2005); Vyn (2006); Adviento-Borbe et al. (2007); Stanger et al. (2008); and Gentry et al. (2013).

Table 1. Yield performances of maize rotated with soybeans in comparison with continuous maize in the USA. Adapted from Erickson (2008)

References	Location	Year	Maize yield in rotation with soybeans	Continuous maize	Yield increase
			t ha^{-1}	t ha^{-1}	%
Edwards et al. (1988)	Crossville, AL	1981-1984	8.2	10.1	-19.3
Griffith et al. (1988)	Butlerville, IN	1980-1986	8.3	8.9	-6.3
Peterson and Varvel(1989)	Mead, NE	1983-1986	7.6	6.8	12.0
Crookston et al. (1991)	Lamberton and Waseca, MN	1981-1989	8.7	8.0	8.7
Meese et al. (1991)	Arlington, WI	1987-1989	8.9	7.6	16.5
Lund et al. (1993)	Arlington, WI	1989-1991	11.0	10.2	8.0
Lauer et al. (1997)	Lamberton and waseca, MN, Arlington, WI	1981-1996	8.9	7.8	12.8
Porter et al. (1997)	Lamberton, MN	1985-1995	8.2	7.2	13.0
Porter et al. (1997)	Waseca, MN	1986-1995	8.9	8.1	10.1
Porter et al. (1997)	Arington, WI	1987-1995	9.5	8.2	16.2
Howard et al. (1988)	Grand Jct., TN	1986-1992	8.9	8.0	11.0
Riedell et al. (1998)	Brookings, SD	1994-1995	8.1	7.3	11.2
Singer and Cox (1998)	Aurora, NY	1993-1994	9.9	9.4	4.7
Katsvairo and Cox (2000)	Aurora, NY	1993-1997	8.9	7.5	19.3
Mallarino et al. (2002)	Lexington, KY	1979-2004	9.9	8.7	13.7
Pedersen and Joseph (2002)	Arlington,WI	1995-1997	10.1	8.9	13.4
Univ. of IL (2002)	IL	17 site-years	10.7	9.0	18.1
Pedersen and Joseph (2003)	Arlington, WI	1998-2001	13.2	11.5	14.8
Singer et al. (2003)	Pittstown,NJ	2000-2001	10.1	9.9	1.9
Varvel and Wilhelm (2003)	Shelton, NE	1993-2003	11.6	11.2	2.8
Varvel and Wilhelm (2003)	Mead, NE	1983-2003	8.5	8.2	3.1
Lee and John (2003)	Lexington, KY	1984-1997	8.5	7.8	8.8
Walters et al. (2005)	Mead, NE	1999-2004	14.9	14.4	3.5
Wilhelm et al. (2004)	Mead,NE	1986-2001	7.0	5.7	23.3
Pikul et al. (2005)	Brookings, SD	1992-2003	7.0	6.0	16.7
Vyn (2006)	Wanatah, IN	1997-2006	12.2	11.4	6.6
Vyn (2006)	West Lafayette, IN	1975-2006	11.3	10.8	4.7
Adviento-Borbe et al. (2007)	Lincoln, NE	1999-2005	17.2	16.0	7.5
Stanger et al. (2008)	Lancaster, WI	1990-2004	10.1	9.1	11.0
Gentry et al. (2013)	Urbana, IL	2005-2010	8.7	6.8	27.9

Similarly, a summary of 29 studies comparing continuous maize with maize-soybean rotation across various states of the USA (updated from Erickson, 2008) illustrated the overall yield superiority in the maize-soybean rotation (MSR), with only two earlier studies that displayed a yield reduction (Edwards et al., 1998; Griffith et al., 1988). In this analysis, only one pair of data was extracted from each reference, and averaged across all treatments and years or even the sites within the same State of U.S. (Table 1). These paired data are then used for the determination of the average percent yield increase with MSR compared with CC, using the same meta-analytical method as described above. Overall, MSR significantly increased the maize yield by 9.6% compared with CC, with 95% confidence intervals (CI) of 5.9% to 12.6% (Figure 4). On average, maize yield in the MSR rotation system was 9.6 t ha^{-1}, compared to 8.8 t ha^{-1} in the CC system (shown in the inset of Figure 4). This estimated yield increase of MSR over CC in U.S. is apparently lower than the meta-data analysis as illustrated above.

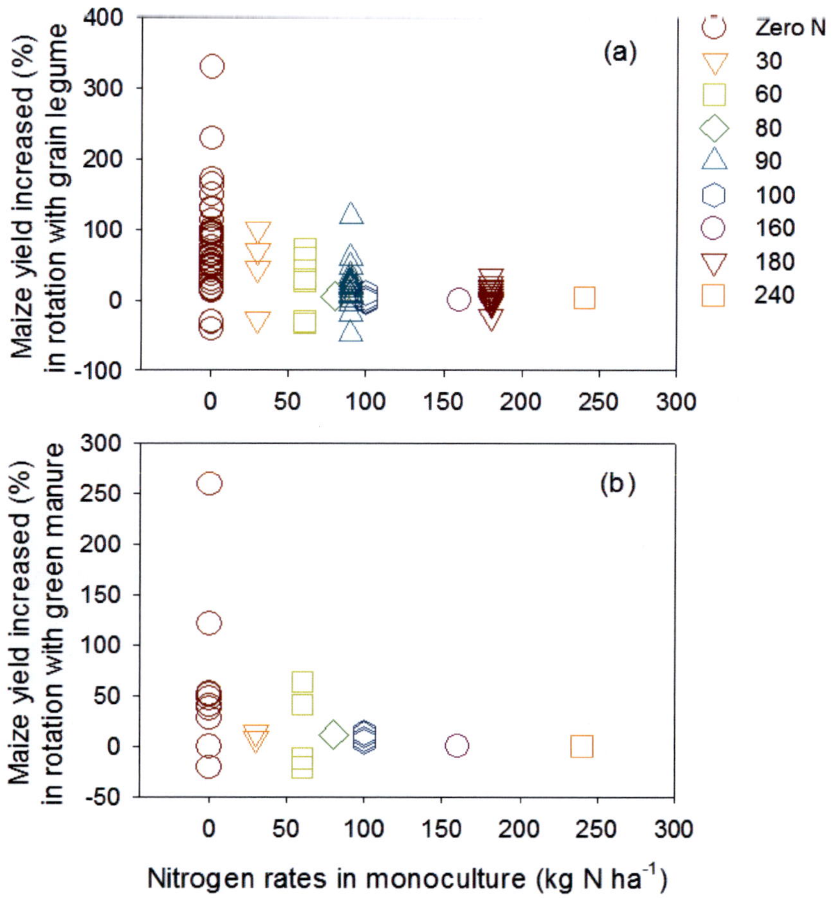

Figure 5. Maize yields increased in rotation with grain legume (a) and green manure (b) in response to various nitrogen rates under continuous monoculture. Scatter points indicate yield advantage of crop rotation over continuous monoculture decreases with increasing nitrogen input. Sources of data are from Ma et al. (2003); Yusuf et al. (2009); Ojiem et al. (2014); Franke et al. (2008); Varvel and Wilhelm (2003); Mallarino et al. (2004); and Gentry et al. (2013), among various soil amendments, sites and years.

The above meta-analysis data of MGLR vs. MGMR are further categorized into different classes according to the respective fertilizer N application rates.

The results clearly displayed that with increasing N rates, the difference in maize yields between CC and MGLR (Figure 5a) or MGMR (Figure 5b) becomes smaller.

This implies that there is a greater N credit available for maize in rotation with grain legume or green manure to achieve a higher grain yield at low than at ample N fertilizer input. In other words, CC cropping does need more inorganic N input to reach its attainable yield. In the Corn Belt of USA, Varvel et al. (2003) suggested that the recommended N applications should be reduced for maize grown in rotation with soybean to reduce N loss and unnecessary input costs. A similar recommendation for eastern Ontario and southwest Quebec was drawn from our previous study (Ma et al., 2003).

3.2. Improvement of Soil Fertility and Soil Quality

Crop rotations influence not only the crop growth but also the resource uptake and utilization, and therefore the resource-use efficiency (mainly N and P) (Ding et al., 1998; Van Kessel and Hartley, 2000). Rotation promotes changes in various nutrient sources and, subsequently, affects availability of these nutrients to the plant. The availability of N and P is often thought to play a dominant role in explaining the yield advantage of crop rotation (Rao and Mathuva, 2000). There is better synchronization between N/P mineralization and crop uptake in rotation cropping systems than in monoculture (Ma et al., 1999b, 2003; Varvel and Wilhelm, 2003; Wu et al., 2008). A long-term rotation study conducted since 1992 in Canada showed that total N uptake, N uptake efficiency and nitrogen use efficiency (NUE) of maize, especially during the grain-filling stage, were significantly higher under maize-legume rotations than in continuous maize cropping systems (Ma et al., 2003), resulting in superior yield performance (Franke et al., 2008; Gentry et al., 2013). Figure 6 is adopted from Ma et al. (2003) and illustrates clearly that the greater accumulation of N or more N credit resulting from the alfalfa preceding crop in the rotation is the main contribution to a superior maize yield in a maize-alfalfa rotation than in continuous maize or even a maize-soybean rotation.

Because there is a high N availability in maize cropping as a result of the rotation effect, Ma et al. (2003) suggested that fertilizer application for maize following soybean should be reduced by 60-70 kg ha^{-1} to attain a better economic performance. Furthermore, their results also showed that soil N amendments (utilization of stockpiled manure and rotted manure) had significantly positive effects on seasonal N mineralization, N supply to the crop and maize yield (Ma et al., 1999a, 1999b; Wu et al., 2008). With repeated application in succeeding years, dairy manure can provide the maize crop with up to 100% of the N required to reach maximum-attainable yield. Thus, their study implied that an appropriate combination of manure amendment and legume rotation is the most economic and environmentally-friendly approach in maize production.

Root exudates of some legume varieties can help release unavailable phosphorous into the soil for recovery by the legumes themselves or the rotated crops (Liu et al., 2005; Sinclair and Vadez, 2012). Furthermore, legume crop residues have better nutrient balance and nutritional qualities than cereal straws for use as biological fertilizer or for use as fodder for farm livestock (Blümmel et al., 2012).

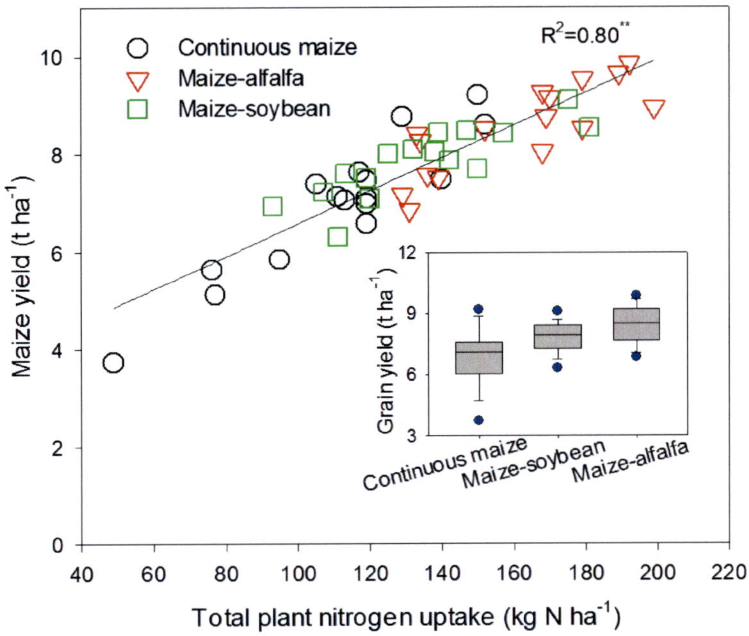

Figure 6. The low productivity of continuous monoculture in comparison with maize-soybean rotation and maize-alfalfa rotation and its association with inferior nitrogen accumulation. Data was adapted from Ma et al. (2003).

The N fertilizer replacement value (FRV) has been used frequently and is defined as the amount of inorganic N fertilizer required to achieve the same yield in continuous maize as that attained by non-N-fertilized maize that followed a preceding legume crop (such as forage legume or grain legume crops) (Hesterman et al., 1987; Ma et al., 2003). In contrast to the FRV method, the ^{15}N-trace method is accurate, but far more resource-intensive (LaRue and Patterson, 1981), in which specialized and costly fertilizer that is ^{15}N-tagged and analytical instrument must be used to obtain the data required for the calculation of N contributions from the legume. Thus, very few papers with ^{15}N data are available for use in constructing a consensus on the N flow in crop rotation, especially from long-term experiments conducted under various environmental conditions.

Conversely, FRV can be computed with relatively easy-to-gather data such as yield of a non-legume crop over a series of N fertilizer rates in both monoculture and rotation with a legume (Varvel and Wilhelm, 2003). These data are frequently collected in cropping system experiments. Therefore, estimates using the FRV can be made over a wide range of environmental conditions and soil amendments to arrive at a consensus estimate for the N contribution by the legume to the maize in a rotation system.

In Pennsylvania, USA, Fox and Piekielek (1988) estimated the FRV of three forage legumes (alfalfa, birdsfoot trefoil and red clover), with FRV values ranging from 146 to 187 kg N ha^{-1}. Similarly, Blevins et al. (1990) showed that 75 and 65 kg N ha^{-1} are provided from hairy vetch and big flower vetch preceding legume crop respectively. Vanotti and Bundy (1995) reported an FRV of 153 kg N ha^{-1} for alfalfa in Lancaster from a long-term experiment. On the other hand, a relative low FRV of 7-37 kg N ha^{-1} for soybean and cowpea was reported in the northern Guinea Savanna of Nigeria (Yusuf et al., 2009). Similarly, a relatively low FRV for grain legume compared with forage legume or manure legume has been reported (Ding et al., 1998; Vanotti and Bundy, 1995; Ma et al., 2003; Varvel and Wilhelm, 2003; Yusuf et al., 2009), as shown by a summary of FRV that compares grain legume with green manure in rotation with a maize crop (Table 2). In general, the average FRV for grain legume is approximately 62 kg N ha^{-1}, whereas a significantly higher FRV (149 kg N ha^{-1}) is estimated for green manure or forage legume from the available data (Figure 7). Taking full advantage of N credit from the preceding legume to the maize crop will help farm operators to consider the amount of N needed to avoid excess fertilizer for optimum yield. This reduction in N application to maize following a legume crop will prevent the inefficient use of N, which results in N losses through leaching or denitrification.

Bundy et al. (1993) estimated that the FRV of soybean differed markedly among locations and years and ranged from -22 to 210 kg N ha^{-1}. Soybean provided little N to subsequent crops on sandy soils due to the probable loss of residue N through leaching prior to use by the following crop. These authors also suggest that the fixed amount of N credits based on N response data combined over the years seldom accurately predict actual soybean N contributions. Thus, site-specific diagnostic tests are needed to improve the crediting of the N supplied by legume plant in crop sequences.

Table 2. A summary report on the fertilizer N replacement value (kg N ha^{-1}) in maize-legume rotation systems

References	Fertilizer N replacement value	Cropping system	Year	Site
Varvel and Wilhelm (2003)	65	Soybean-maize	1983-2002	Eastern Nebraska, Central Nebraska, USA
Blevins et al. (1990)	75 and 65 for hairy vetch and bigflower vetch, respectively.	Vetch-maize	1980-1983	Kentucky, USA
Ding et al. (1998)	41-59	Soybean-maize	1993-1994	Elora, Ontario, Canada
Fox and Piekielek (1988)	187, 169 and 147 for alfalfa, birdsfoot trefoil and red clover	Three forage legumes-maize	Two years	Pennsylvania, USA
Vanotti and Bundy (1995)	153 and 75 for alfalfa and soybean, respectively	Alfalfa/soybean-maize	1977-1991	Lancaster, Pennsylvania, USA
Ma et al. (2003)	43-113 for soybean; 56-187 for alfalfa	Alfalfa/soybean-maize	1993-1996	Ottawa, ON, Canada
Yusuf et al. (2009)	7-37 for two genotypes each of soybean and cowpea	Soybean/cowpea-maize	2003-2004	Samaru, Guinea, Nigeria

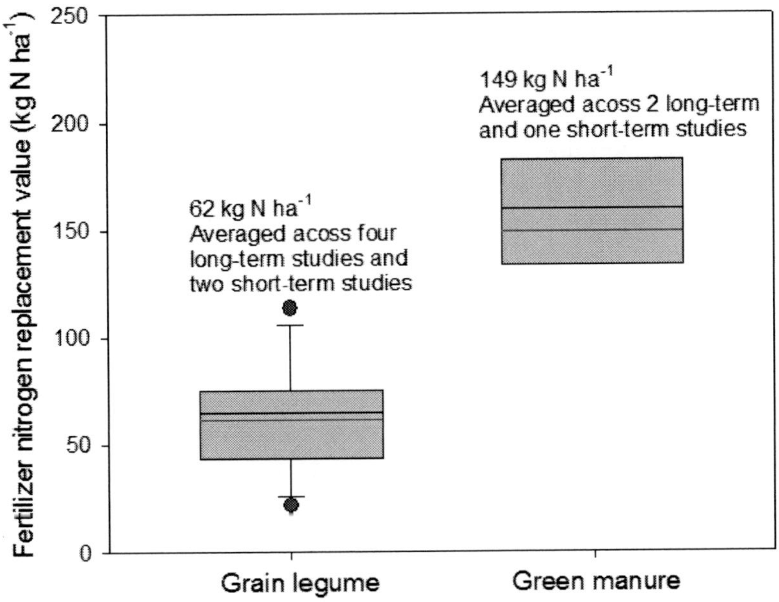

Figure 7. A summary of fertilizer nitrogen replacement value of grain legume in comparison with green manure. Sources of data are from Ma et al. (2003); Vanotti and Bundy (1995); Fox and Piekielek (1988); Varvel and Wilhelm (2003); Ding et al. (1998); Blevins et al. (1990); and Yusuf et al. (2009). For more detail, see Table 1.

The N contribution by legumes to maize also depends on the leguminous species and its BNF efficiency and is influenced by climate, soil, management of crop residues, etc. (Hesterman et al., 1987; Fox and Piekielek, 1988; Bundy et al., 1993; Ding et al., 1998). In general, grain legumes contribute less N credit than green manure/forage legume to the subsequent maize crop in rotation because most of the BNF from grain legumes is translocated to grain, and both the grain and the residues are invariably removed from the fields for human and livestock use (Giller et al., 1997; Figure 7). Hence, the N requirement of maize crops cannot be met from the residual effects of grain legumes, particularly in favorable seasons with high-yielding potential when large amount of N is needed by the maize plants (Ma et al., 1999a; Ma et al., 2006). Additional N from fertilizers and/or other organic sources with appropriate soil amendment strategies is required to exploit the potential of such seasons (Ma et al., 1999a; Ma et al., 2003).

3.3. Reduction in Incidences of Pests and Diseases and Suppression of Weeds

There are many cultural practices available for farmers to control diseases, insects and weeds in their crops, such as tillage practices, fertilization management, crop rotation, adjusting sowing date, plant population, row spacing and irrigation management (Munkwold, 2003). Most of these practices take advantage of reducing the amount of the pathogens (including weed seed banks) present in the field soil to keep the disease incidence or severity under control. Crop rotation may be the one efficient and practical method that farmers are well aware of and frequently use (Mora and Moreno, 1984; Munkwold, 2003).

The annual cycle for many disease pathogens involves infecting the crop in the spring and summer, surviving the winter as spores in the soil or in plant litter/residues, and attacking the new crop the following year. Therefore, if the same crop is planted in the same field year after year, the pathogen populations (especially for some soil-borne pathogens) can continue to build up so that it becomes difficult to grow a crop. Thus, the cycle is broken by growing a non-host crop species after the susceptible plants in the same field. Over time, many pathogens will decline due to the lack of a suitable host. Rotating to non-host crops prevents the buildup of large populations of pathogens and makes it possible to plant the original crop in the same field again (McGrath, 2007).

Several factors that limit the effectiveness of crop rotations in suppressing diseases should be considered before rotating into another crop. The botanical classification should be figured out before deciding on the rotation crops. Soybean, peas, beans, lentils and alfalfa, all belonging to the legume family, can be rotated with maize, a species from the *Gramineae* family, to reduce the incidence of soil-borne diseases (McGrath, 2007).

Furthermore, how long the pathogen can survive in the crop residue/soil and which additional plant species (mainly weed species and cover crops) can be a host for the pathogens should be determined. The rotation length will be equal to the pathogen survival duration. Table 3 lists the sources of pathogen inoculum and recommended rotation periods for various diseases of maize, alfalfa and soybean because those crops are usually used in the maize-legume rotation system. The information in Table 3 should be regarded as general guidelines because the number of rotation years cannot be precisely determined for efficient control of many diseases due to other factors and a lack of extensive research (Munkwold, 2003; McGrath, 2007).

Table 3. Sources of inoculum for maize, alfalfa, and soybean diseases and corresponding suggested rotation year. Adapted from McGrath (2007)

Disease	**Pathogen name**	**Suggested rotation year** [a]	**Weed hosts** [b]
Alfalfa			
Anthracnose	*Colletotrichum trifolii*	Y (2-3)	no records located
Aphanomyces root rot	*Aphanomyces euteiches*	Y (2-3)	legumes
Bacterial wilt	*Clavibacter michiganensis* subsp. *insidiosus*	N	no records located
Brown root rot	*Phomasclerotioides*	Y (3); non-legumes	wide host range
Crown and root rot complex	*Fusarium* spp., *Phoma medicaginis*, *Pythium* spp., *Rhizoctonia solani*	Y (long)	wide host range
Fusarium wilt	*Fusarium oxysporum* f. sp. *Medicaginis*	Y (long)	no records located
Lepto leaf spot	*Leptosphaerulina trifolii*	N	medics, soybean
Lesion nematodes	*Pratylenchus* spp.	N	wide host range

Disease	Pathogen name	Suggested rotation year [a]	Weed hosts [b]
Phytophthora root rot	*Phytophthora medicaginis, P. megasperma* f. sp. *medicaginis*	Y (very long)	no records located
Sclerotinia crown and stem blight	*Sclerotinia trifoliorum, S. sclerotiorum*	Y (3-4)	wide host range
Spring black stem and leaf spot	*Phoma medicaginis*	Y (2-3)	legumes
Summer black stem and leaf spot	*Cercospora medicaginis*	Y (2-3)	legumes
Verticillium wilt	*Verticillium albo-atrum*	Y (3)	wide host range
Maize			
Anthracnose leaf blight and stalk rot	*Colletotrichum graminicola*	Y (2)	several grasses
Barley yellow dwarf virus	*Barley yellow dwarf luteovirus* (BYDV- PAV)	NA	witchgrass, Italian ryegrass, annual bluegrass
Common rust	*Puccinia sorghi*	N	yellow woodsorrel
Common smut	*Ustilago maydis*	Y (long)	no records located
Diplodia ear rot, stalk rot, seed rot and seedling blight	*Stenocarpella maydis* (*Diplodia maydis*)	Y (2-3); worst in no-till continuous maize	no records located
Diplodia leaf spot, Diplodia leaf streak	*Stenocarpella macrospora* (*Diplodia macrospora*)	Y (2-3)	no records located
Eyespot	*Aureobasidium zeae*	Y; worst in no-till continuous maize	no records located
Fusarium stalk and ear rot	*Fusarium moniliforme*	Y; worst in no-till continuous maize	grasses
Gibberella stalk and ear rot16	*Fusarium* spp.	Y (2-3); worst in no-till continuous maize	wide host range
Gray leaf spot, Cercospora leaf spot	*Cercospora sorghi*	Y (1-2); wirst in no-till continuous maize	a few grasses
Leaf blight	*Erwinia stewartii*	N	a few grasses
Lesion nematodes	*Pratylenchus* spp.	N	wide host range
Maize dwarf mosaic virus (MDMV)	Maize dwarf mosaic virus (MDMV)	N	witchgrass, johnsongrass, barnyardgrass, goosegrass
Northern maize leaf blight	*Exserohilum turcicum*	Y (2-3)	johnsongrass, green foxtail, wild-proso millet
Seed rots	*Fusarium* spp., *Diplodia* spp., etc	N	wide host range
Soybean			

Table 3. (Continued)

Disease	Pathogen name	Suggested rotation year [a]	Weed hosts [b]
Asian soybean rust	*Phakopsora pachyrhizi*	NA	kudzu, clover, lupin
Brown spot	*Septoria glycines*	Y	no records located
Brown stem rot	*Phialophora gregata* (*Cephalosporium gregatum*)	Y (2-3)	unknown; likely
Damping-off, stem rot	*Rhizoctonia* spp., *Fusarium* spp., *Phytophthora sojae*, *Pythium* spp.	Y (long)	wide host range
Diaporthe stem canker	*Diaporthe phaseolorum* (*Phomopsis phaseoli*)	Y	no records located
Green stem/soybean viruses	Bean pod mottle virus	NA	no records located
Phomoposis seed rot	*Phomopsis* spp.	Y	velvetleaf
Phytophthora root rot	*Phytophthora sojae*	Y (long)	unknown; likely
Sclerotinia white mold, sclerotinia stem rot	*Sclerotinia sclerotiorum*	Y (long)	wide host range
Soybean cyst nematode	*Heterodera glycines*	Y (3+)	many
Sudden death syndrome	*Fusarium solani* f. sp. *Glycines*	Y (3)	no records located

[a] "Y" means the disease can be managed by crop rotation. The number in parentheses is the suggested minimum rotation return time needed for successful management of the disease; "long" means more than 3 years; "N" means the disease cannot be managed by crop rotation because the pathogen can survive in the soil or the spores are dispersed widely by wind; "NA" means not applicable as the pathogen does not survive in the soil medium. [b] "wide host range" mean many common weeds can be the host of the disease, such as *Fusarium*, *Pythium*, *Rhizoctonia solani* species, etc.

More attention must also be paid to weeds that can host diseases harmful to the legumes or maize, and that will need to be controlled during the rotation period (Table 3). Suppression of weeds during the rotation can help to suppress diseases by preventing the reintroduction of the pathogen to a clear field when the rotated crop is planted again (Krupinsky et al., 2002).

Maize in rotation is often less damaged by various disease organisms than when grown as a monoculture crop, but effective avoidance of attack by disease often varies unpredictably (Munkwold, 2003; Reid et al., 2001). This is because rotation practices are not very effective in controlling some pathogens that have a wide host range such as *Rhizoctonia solani*, a common fungus in most soils worldwide, which can become a serious pathogen on susceptible maize cultivars. Crop rotations need to be carefully selected to reduce such pathogens (McGrath, 2007).

In addition, some diseases produce resting structures that are able to survive in the soil for long periods of time, such as *Fusarium* spp. in cereals; *Rhizoctonia solani, Phytophthora medicaginis* in alfalfa; *Ustilago maydis, Stenocarpella maydis* in maize; and *Rhizoctonia* spp.,

Phytophthora sojae and Sclerotinia sclerotiorum in soybean. Rotations of two to three years may have little effect on the population levels in the soil of certain diseases (Table 3).

Crop rotation also suppresses some insect pests, although the effectiveness depends on the life cycle of the target insect. To successfully use crop rotation for pest management, the pest must spend the period from the end of preceding crop to the beginning of the next in a stage with low mobility and must have a restricted range of host crops (Smith and McSorley, 2000; Stoner, 2007). Some key pests can be restricted through crop rotation, such as maize rootworms (*Diabrotica* spp.), wireworms (*Melanotus communis, Limonius* spp.) and white grubs (*Phyllophaga* spp.) (White, 1999). Adult beetles from western maize rootworm and northern maize rootworm are important worldwide pests and feed on maize silks in summer and lay their eggs on the surface soil of the maize fields.

Over winter the eggs and the newly hatched larvae feed on the maize roots in the following spring (Levine and Oloumi-Sadeghi, 1991; Ma et al., 2009). If some small legume grains or sorghum are rotated with maize (rather than continuously planting maize or some grassy weed species), the hatched larvae cannot survive. Thus, this pest has relied on widespread continuous maize for their survival and will experience interruption of their cycle in a rotation system (Stoner, 2007).

Past control strategies for most diseases and pests have centered on the use of fungicides. However, this option is not considered sustainable due to serious environmental issues and the potential risk of emergence of resistant populations (Brent and Hollomon, 1995). The use of crop rotation to suppress diseases and pests is one of the best, most widely practiced and cost effective methods (Munkwold, 2003).

Other options, such as soil amendments, crop residue management, tillage or crop management strategies, which influence the canopy structure, e.g., plant architecture and population density, can be used effectively (Reid et al., 2001). Canopy structure can determine a number of environmental factors within the crop field, such as humidity and temperature (Dwyer et al., 1996), which affect the spread of pest and disease development (Castilla et al., 1996). Plant resistance to diseases and pests is generally considered a polygenic trait, with wide variations in susceptibility levels among the different varieties (Munkwold, 2003).

Therefore, the development of cultivars resistant to pest and disease in a crop rotation system may also be a vital strategy for pest and disease suppression.

Crop rotation has also shown extensive efficiency in weed control (Liebman and Dyck, 1993; González-Díaz et al., 2012). Weed control as a benefit of crop rotation is gaining renewed interest due to the increasing need to develop sustainable control strategies with a smaller environmental impact. An effective crop rotation establishment can limit the demand for increasing herbicide applications for weed control, especially in intensified cropping systems (Liebman and Davis, 2000).

According to Liebman and Dyck (1993), "the success of rotation systems for weed suppression appears to be based on the use of crop sequences that employ varying patterns of resource competition, allelopathic interference, soil disturbance, and mechanical damage to provide an unstable and frequently inhospitable environment that prevents the proliferation of a particular weed species." Greater crop yield and less weed growth may be achieved in maize crops in rotation with legumes than in single maize crop monoculture because resources are usurped from weeds or the growth of weeds is suppressed through allelopathy. Generally, maize rotated with legumes and grown with the appropriate N amendment gave the highest leaf area index and light interception and hence the best weed control while also maintaining the highest N, P, K and other resource uptake and yield; continuous cropping of maize, however, produced plants possessing low leaf area index and left more space for weeds to grow (Schreiber, 1992; Shrestha et al., 2002; Subedi and Ma, 2009). González-Dìaz et al. (2012) established a landscape model for weed control and suggested that crop rotation implemented at the landscape level, similar to implementation at the field level, has the potential to control short- and long-term weed population densities.

3.4. Economic Benefit

A major challenge for maize production systems is to maximize economic returns while minimizing the environmental impact through an appropriate crop rotation system. Dual-purpose legumes that produce food and feed, such as soybean (*Glycine max*), cowpea (*Vignaunguiculata*), alfalfa (*Medicago sativa*) and clover (*Trifolium*), are particularly attractive to small-scale or subsistence farmers who practice integrated crop-livestock systems (Rao and Mathuva, 2000; Jama and Pizarro, 2008). Grain legumes usually command a higher price than the staple crop, such as maize, and are marketed easily due to an increasing demand worldwide. In addition, as maize is less sensitive to drought or heat than grain legumes (Subbarao et al., 1995), which is the case in most semiarid regions, diversified cropping systems involving grain legumes, are likely to be less risky and more attractive than continuous cropping of maize to small farmers (Lal et al., 1978).

An economic evaluation experiment at the research station of the International Centre for Research in Agroforestry (ICRAF) suggests that pigeon pea-based systems are 32-49% more profitable than continuous maize cropping (Rao and Mathuva, 2000). Generally, grain legume-maize rotations require less labor for land preparation, sowing and weeding, and are less likely to experience total crop failure in the event of drought stress, which indicates a lower risk even in harmful climates. Their study also illustrated that although green manure-maize rotation increased maize productivity significantly, this system was not more economical than the grain legume-maize rotation due to high labor costs for production and the application of prunings to the crop.

A similar study was conducted in a different agroecological zone with soil conditions of high, medium and low fertility, in western Kenya (Ojiem et al., 2014).

The researchers reported that maize crops rotated with soybean, groundnut and lablab generally led to the better returns for land (approximately $488 per ha, averaged from all experimental locations; Figure 8a) and labor ($1.6 per day, averaged from all experimental locations; Figure 8b) than continuous maize and even maize rotated with green manures, mainly due to the higher prices of the edible grains.

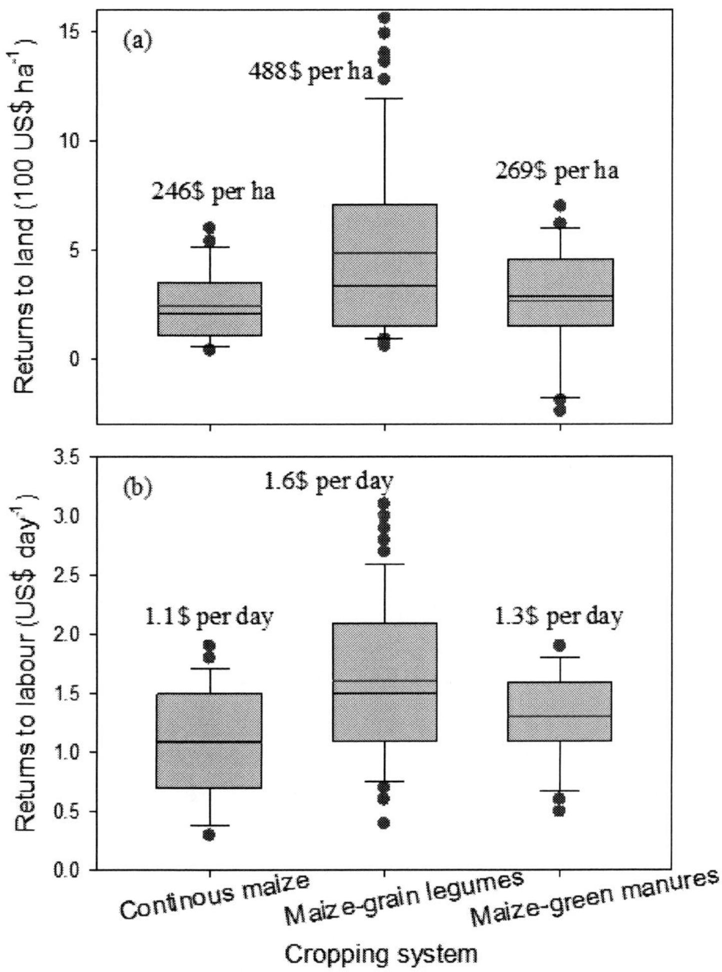

Figure 8. Comparisons of returns to land (100 US$ per ha) and labour (US$ per day) between continuous monoculture, maize-grain legumes and maize-green manure rotations. Data was adapted from Ojiem et al. (2014) and aggregated across from different kinds of grain legumes and green manures in various soil fertilities in different agro-ecological Zones in Kenya.

3.4. Reduction in GHG Emissions and Carbon Footprints

Global warming resulting from greenhouse gas (GHG) emissions of agricultural origin is considered an important current environmental impact issue (IPCC, 2014; Ma et al., 2012). The environmental impacts of agricultural activities are diverse, including the contribution of excess nutrients, sediments and pesticides to surface and ground waters, air pollution, and the production of GHGs, mainly CO_2, N_2O, and CH_4 (Hass et al., 2000). Crop producers are urged to adopt effective management practices to mitigate GHGs by lowering the C footprints of agricultural production on the farm (Gan et al., 2011a). The C footprint has been defined as the total GHGs per unit grain basis (kg CO_2eq kg^{-1} grain) to quantify the impact of GHGs on environmental sustainability. Consumers and the public want to know the C cost and wish to cut C footprints in their choice of food products. Legumes, introduced into maize-based cropping systems (Ma et al., 2003; 2012) or wheat-based cropping systems (Gan et al., 2011b) to diversify cropping systems, mitigate climate change effects by reducing fossil-fuel use, mineral fertilizer and pesticide production or by providing feedstock for the emerging bio-based economies, where fossil-fuel sources of energy and industrial raw materials are replaced in part by sustainable and renewable biomass resources. Thus, legumes are an important component of sustainable production systems for human prosperity. The contribution of various crop rotations to overall GHGs and the overall C footprints has been calculated recently (Adviento-Borbe et al., 2007; Ma et al., 2012; Gan et al., 2011a). A contextual view of the science of carbon footprints in agricultural crop production systems is presented in chapter 8 of this book.

3.6. Promotion of Biodiversity

Crop rotation is one way of introducing more biodiversity into agroecosystems (Davis et al., 2012). Higher biodiversity richness may be associated with improved nutrient cycling characteristics that often can regulate soil fertility (Russell, 2002), limit nutrient leaching losses and significantly reduce the negative impacts of pests and diseases (Gurr et al., 2003; Stoner, 2007). Rotations between alternating crops promote biodiversity by providing a habitat for a variety of insects (Thom et al., 2016) and soil organisms that would not otherwise be present in a single crop environment. Crop rotation is similar to a somewhat stable natural system in comparison with a continuous cropping system; rotating systems are typically diverse, and contain different types of plant species, arthropods, mammals and microorganisms. Thus, serious pest or disease outbreaks are rare because those controls can automatically bring populations back into balance (Altieri and Nicholls, 2004; Gurr et al., 2003). On-farm biodiversity can lead to agroecosystems that are capable of maintaining their soil fertility (Chan et al., 2013), regulating natural protection against pests and sustaining productivity (Ma et al., 2003; Davis et al., 2012; Ojiem et al., 2014). On-farm biodiversity with introducing specialty oilseeds into current crop rotations could also provide abundant floral resources for pollinating insects (Thom et al., 2016). From this point of view, crop rotations that improve the biodiversity of cropping systems can make crop ecosystems more stable and thereby reduce disease and pest incidence.

At the Iowa State University Marsden Farm, Davis et al. (2012) implemented traditional crop rotation patterns into more diverse cropping rotations with high biodiversity in a long-term field study. The typical 2-year maize/soybean rotation was compared with 3-year and 4-year rotations. Their data illustrated that cropping system diversification would promote ecosystem services that supplement and eventually displace the synthetic external inputs used to maintain crop productivity. Compared with the traditional 2-year maize-soybean rotation with conventional amounts of synthetic fertilizer and herbicide applications, the 3-year and 4-year diversified rotations added the production of a small grain (triticale and oat) along with the use of a legume and composted animal manure. Reductions in agrochemical inputs and increases in grain yields, harvested products and net profit in diversified systems were achieved compared with a conventional rotation system. Thus, a more diversified strategy of integrating crop and small-scale livestock production by the farm operator into a moderate-sized rotational cropping system will meet production needs and utilize opportunities beneficially.

CONCLUSION

This chapter explains the rationale and methodology of leguminous-cereal crop rotation systems and highlights the role of a legume-maize crop rotation as a system with high-yielding potential and as a more cost-effective and environmentally-friendly strategy, compared with continuous monoculture accompanied with the application of large amounts of inorganic fertilizer. Our meta-data analysis indicates that the maize yield advantage is approximately 9.6% in a maize-soybean rotation, and up to 40% under a maize-green manure (or legume forage) rotation, in comparison with continuous maize monoculture. Other benefits of this system include nutrient credits, economic profits, less environmental impact, resistance to diseases and pests, weed suppression, and the promotion of biodiversity.

Grain legumes, such as soybean, contribute less N than forage or green manure legumes to the subsequent maize in a rotation system because most of the N fixed biologically by grain legumes is translocated to the grain, which is removed from the field. Hence, the N requirement of maize is seldom met, and additional N based on appropriate soil amendments or other organic sources is required to exploit the potential of the maize crop in favorable seasons.

Conversely, green manure left behind more N in residues evidenced with high fertilizer replacement value (FRV; 149 kg N ha^{-1}) than 62 kg N ha^{-1} in grain legumes, resulting in greater maize yields in the subsequent season. The legume-maize rotation in combination with appropriate soil amendments represents the best source of the "ideal" fertilizer management and therefore commands great interest as a promising subject for future research.

ACKNOWLEDGMENTS

This work was financially supported, in part, by the Canola Council of Canada Canola/Flax Cluster program (CL08) of Agriculture and Agri-Food Canada (AAFC) Growing Forward II. We wish to thank Dr. Chang Liang of Environment Canada, who critically reviewed the first draft of this chapter and provided us with invaluable suggestions. AAFC-ECORC contribution No. 15-082.

REFERENCES

Adviento–Borbe, M.A.A., M.L. Haddix, D.L. Binder, D.T., Walters, and A. Dobermann. 2007. Soil greenhouse gas fluxes and global warming potential in four high–yielding maize systems. *Global Change Biol.* 13: 1972–1988.

Ainsworth, E.A., P.A. Davey, C.J. Bernacchi, O.C. Dermody, E.A. Heaton, D.J. Moore, B. Morgan, S.L. Naidu, H.Y. Ra, X. Zhu, P.S. Curtis, and S.P. Long. 2002. A meta–analysis of elevated [CO_2] effects on soybean (*Glycine max*) physiology, growth and yield. *Global Change Biol.* 8: 695–709.

Albrecht, S.L., J.M. Bennett, and K.J. Boote. 1994. Relationship of nitrogenase activity to plant water stress in field grown soybeans. *Field Crops Res.* 8: 61–71.

Altieri, M.A., and C.I. Nicholls (eds.). 2004. Biodiversity and Pest Management in Agroecosystems. Food Products Press, New York, USA, pp256.

Andrés, J.A., M. Rovera, L.B. Guiñazú, N.A. Pastor, and S.B. Rosas. 2012. Interactions between legumes and rhizobia under stress conditions. pp. 77-94. *In* D.K. Maheshwari (ed.) Bacteria in Agrobiology: Stress Management. Springer Berlin Heidelberg.

Anyanzwa, H., J.R. Okalebo, C.O. Othieno, A. Bationo, B.S. Waswa, and J. Kihara. 2010. Effects of conservation tillage, crop residue and cropping systems on changes in soil organic matter and maize–legume production: A case study in Teso District. *Nutr. Cycl. Agroecosyst.* 88: 39–47.

Aplin, J.D. 1991. Implantation, trophoblast differentiation and hemochorial placentation, mechanistic evidence in vivo and in vitro. *J. Cell. Sci.* 99: 681–692.

Aranjuelo, I., J. Aldasoro, C. Arrese–Igor, G. Erice, and A. Sanz–Sáez. 2015. How does high temperature affect legume nodule symbiotic activity? pp. 67-87. *In* S. Sulieman, L.-S.P. Tran (eds.) Legume nitrogen fixation in a changing environment: Achievements and Challenges. Springer International Publishing, Switzerland.

Balachandar, D., P. Raja, K. Kumar, and S.P. Sundaram. 2007. Non–rhizobial nodulation in legumes. *Biotechnol. Mol. Biol. Rev.* 2: 49–57.

Ben Salah, I., A. Albacete, A.C. Martinez, R. Haouala, N. Labidi, F. Zribi, V. Martinez, and A.C. Perez–Aflocea. 2009. Response of nitrogen fixation in relation to nodule carbohydrate metabolism in Medicago ciliaris lines subjected to salt stress. *J. Plant Physiol.* 166: 477–488.

Bethlenfalvay, G.J., and D.A. Phillips. 1977. Photosynthesis and symbiotic nitrogen fixation in *Phaseolus vulgaris* L. pp. 401–408. *In* A. Holaender et al. (eds.) Genetic Engineering for Nitrogen Fixation. Plenum Press, New York, USA.

Blevins, R.L., J.H. Herbek, and W.W. Frye. 1990. Legume cover crops as a nitrogen source for no–till corn and grain sorghum. *Agron. J.* 82: 769–772.

Blümmel, M., P. Ratnakumar, and V. Vadez. 2012. Opportunities for exploiting variations in haulm fodder traits of intermittent drought tolerant lines in a reference collection of groundnut (*Arachis hypogaea* L.). *Field Crops Res.* 126: 200–206.

Brent, K.J., and D.W. Hollomon. 2007. Fungicide resistance in crop pathogens: How can it be managed? FRAC Monograph 1, second edition, FRAC. 56 pp. Croplife International, Brussels, Belgium.

Brockwell, J., A. Pilka, and R.A. Holliday. 1991. Soil pH as a major determinant of the numbers of naturally–occurring Rhizobium meliloti in non–cultivated soil of New South Wales. *Aust. J. Exp. Agric.* 31: 211–219.

Bueckert, R., S. Wagenhoffer, G. Hnatowich, and T. Warkentin. 2015. Effect of heat and precipitation on pea yield and reproductive performance in the field. *Can. J. Plant Sci.* 95: 629–639.

Bundy, L.G., T.W. Andraski, and R.P. Wolkowski. 1993. Nitrogen credits in soybean–corn crop sequences on three soils. *Agron. J.* 85: 1061–1067.

Bünemann, E.K., P.C. Smithson, B. Jama, E. Frossard, and A. Oberson. 2004. Maize productivity and nutrient dynamics in maize–fallow rotations in western Kenya. *Plant Soil* 264: 195–208.

Castilla, N.P., R.M. LeaÑo, F.A. Elazegui, P.S. Teng, and S. Savary. 1996. Effects of plant contact, inoculation pattern, leaf wetness regime, and nitrogen supply on inoculum efficiency in rice sheath blight. *J. Phytopathol.* 144: 189–192.

Chan, Y.K., W.A. McCormick, and B.L. Ma. 2013. Fertilizer effects on soil archaeal abundance at two Ontario experimental farms during three consecutive cropping seasons. *Appl. Soil Ecol.* 68: 26-35.

Crookston, R.K., J.E. Kurle, P.J. Copeland, J.H. Ford, and W.E. Lueschen. 1991. Rotational cropping sequence affects yield of corn and soybean. *Agron. J.* 83: 108–113.

Curtis, P.S., and X. Wang. 1998. A meta–analysis of elevated CO_2 effects on woody plant mass, form, and physiology. *Oecologia* 113: 299–313.

Davidson, E.A., J.N. Galloway, N. Millar, and A.M. Leach. 2014. N–related greenhouse gases in North America: Innovations for sustainable future. *Curr. Opin. Env. Sust.* 9: 1–8.

Davis, A.S., J.D. Hill, C.A. Chase, A.M. Johanns, and M. Liebman. 2012. Increasing cropping system diversity balances productivity, profitability and environmental health. *PloS ONE* 7(10): e47149.

Ding, W., D.J. Hume, T.J. Vyn, and E.G. Beauchamp. 1998. N credit for soybean to a following maize crop in central Ontario. *Can. J. Plant Sci.* 78: 29–33.

Dommergues, Y.R. 1995. Nitrogen fixation by trees in relation to soil nitrogen economy. *Fert. Res.* 42: 215–230.

Dwivedi, S., K.L. Sahrawat, H.D. Upadhyaya, A. Mengoni, M. Galardini, M. Bazzicalupo, E.G. Biondi, M. Hungria, G. Kaschuk, M.W. Blair, and R. Ortiz. 2015. Advances in host plant and rhizobium genomics to enhance symbiotic nitrogen fixation in grain legumes. *Adv. Agron.* 129: 1–44.

Dwyer, L.M., B.L. Ma, D.W. Stewart, H.N. Hayhoe, D. Balchin, and J.L.B. Culley. 1996. Root mass distribution under conventional and conservation tillage. *Can. J. Soil Sci.* 76: 23-28.

Edwards, J.H., D.L. Thurlow, and J.T. Eason. 1988. Influence of tillage and crop rotation on yields of corn, soybean, and wheat. *Agron. J.* 80: 76–80.

Erickson, B. 2008. Crop/soybean rotation literature summary. http://www.agecon.purdue.edu/pdf/Crop_Rotation_Lit_Review.pdf.

Erman, M., E. Ari, Y. Togay, and F. Cig. 2009. Response of field pea (*Pisum sativum* sp. Arvense L.) to Rhizobium inoculation and nitrogen application in eastern Anatolia. *J. Anim. Vet. Adv.* 8: 612–616.

Feng, Z.Z., K. Kobayashi, and E.A. Ainsworth. 2008. Impact of elevated ozone concentration on growth, physiology, and yield of wheat (*Triticum aestivum* L.): A meta–analysis. *Global Change Biol.* 14: 2696–2708.

Fox, R.H., and W.P. Piekielek. 1988. Fertilizer N equivalence of alfalfa, birdsfoot trefoil, and red clover for succeeding corn crops. *J. Prod. Agric.* 1: 313–317.

Franke, A.C., G. Laberge, B.D. Oyewole, and S. Schulz. 2008. A comparison between legume technologies and fallow, and their effects on maize and soil traits, in two distinct environments of the West African savannah. *Nutr. Cycl. Agroecosyst.* 82: 117–135.

Gan, Y., C. Liang, C. Hamel, H. Cutforth, and H. Wang. 2011a. Strategies for reducing the carbon footprint of field crops for semiarid areas: A review. *Agron. Sust. Develop.* 31: 643–656.

Gan, Y., C. Liang, H. Wang, and B. McConkey. 2011b. Lowering carbon footprint of durum wheat by diversifying cropping systems. *Field Crops Res.* 122: 199–206.

Gentry, L.F., M.L. Ruffo, and F.E. Below. 2013. Identifying factors controlling the continuous corn yield penalty. *Agron. J.* 105: 295–303.

Giller, K.E., M.H. Beare, P. Lavelle, A.M. Izac, and M.J. Swift. 1997. Agricultural intensification, soil biodiversity and agroecosystem function. *Appl. soil ecol.* 6: 3–16.

Giller, K.E., and G. Cadisch. 1995. Future benefits from biological nitrogen fixation: An ecological approach to agriculture. *Plant Soil* 174: 255–277.

González–Díaz, L., F. Van Den Berg, F. Van Den Bosch, and J.L. González–Andújar. 2012. Controlling annual weeds in cereals by deploying crop rotation at the landscape scale: *Avena sterilis* as an example. *Ecol. Appl.* 22: 982–992.

Graham, P.H., and C.P. Vance. 2000. Nitrogen fixation in perspective: An overview of research and extension needs. *Field Crops Res.* 65: 93–106.

Griffith, D.R., E.J. Kladivko, J.V. Mannering, T.D. West, and S.D. Parsonsp. 1988. Long–term tillage and rotation effects on corn growth and yield on high and low organic matter, poorly drained soils. *Agron. J.* 80: 599–605.

Gurr, G.M., S.D. Wratten, and J.M. Luna. 2003. Multi–function agricultural biodiversity: pest management and other benefits. *Basic Appl. Ecol.* 4: 107–116.

Gyaneshwar, P., A.M. Hirsch, L. Moulin, W.M. Chen, G.F. Elliott, C. Bontemps, S.P. Estrad, and E. Gross. 2011. Legume–nodulating beta–proteobacteria: diversity, host range and future prospects. *Mol. Plant Microbe. Interact.* 24: 1276–1288.

Hass, G., F. Wetterich, and U. Geier. 2000. Life cycle assessment framework in agriculture on the farm level. *Int. J. Life Cycle Assess.* 7: 107–118.

Hedges, L.V., J. Gurevitch, and P.S. Curtis. 1999. The meta–analysis of response ratios in experimental ecology. *Ecol.* 80: 1150–1156.

Hesterman, O.B., M.P. Russelle, C.C. Sheaffer, and G.H. Heichel. 1987. Nitrogen utilization from fertilizer and legume residues in legume–corn rotations. *Agron. J.* 79: 726–731.

Howard, D.D., A.Y. Chambers, and G.M. Lessman. 1998. Rotation and fertilization effects on corn and soybean yields and soybean cyst nematode populations in a no–tillage system. *Agron. J.* 90: 518–522.

Hungria, M., J.C. Franchini, R.J. Campo, C.C. Crispino, J.Z. Moraes, R.N.R. Sibaldelli, I.C. Mendes, and J. Arihara. 2006. Nitrogen nutrition of soybean in Brazil: Contributions of biological N_2 fixation and N fertilizer to grain yield. *Can. J. Plant Sci.* 86: 927–939.

Hungria, M., and A.A. Franco. 1993. Effects of high temperature on nodulation and nitrogen fixation by Phaseolus vulgaris L. *Plant Soil* 149: 95–102.

IPCC - Intergovernmental Panel on Climate Change. 2014. Fifth assessment report: Climate change 2014: Mitigation of climate change. Working Group III and the 39th Session of the IPCC. Berlin, Germany.

Jama, B., and G. Pizarro. 2008. Agriculture in Africa: Strategies to improve and sustain smallholder production systems. Ann. New York Acad. Sci. 1136: 218–232.

Kang, B.T., L. Reynolds, and A.N. Atta–Krah. 1990. Alley farming. *Adv. Agron.* 43: 315–359.

Karmakar, K., A. Rana, A. Rajwar, M. Sahgal, and B.N. Johri. 2014. Legume–rhizobia symbiosis under stress. pp. 240–258. *In* N.K. Arora (ed.) Plant Microbes Symbiosis: Applied Facets. Springer, India.

Katsvairo, T.W., and W.J. Cox. 2000. Tillage rotation management interaction in corn. *Agron. J.* 92: 493–500.

Kishinevsky, B.D., D. Sen, and R.W. Weaver. 1992. Effect of high root temperature on Bradyrhizobium–peanut symbiosis. *Plant Soil* 143: 275–282.

Krupinsky, J.M., K.L. Bailey, M.P. McMullen, B.D. Gossen, and T.K. Turkington. 2002. Managing plant disease risk in diversified cropping systems. *Agron. J.* 94: 198–209.

Lal, R.B., R. De, and R.K. Singh. 1978. Legume contribution to fertilizer economy in legume–cereal rotations. *Indian J. Agric. Sci.* 48: 419-424.

Larrainzar, E., S. Weinkoop, C. Scherling, S. Kempa, R. Ladera, C. Arrese–Igor, W. Weckwerth, and E.M. Gonzalez. 2009. Carbon metabolism and bacteroid functioning are involved in the regulation of nitrogen fixation in Medicago truncatula under drought and recovery. *Mol. Plant–Microbe Interact.* 22: 1565–1576.

LaRue, T.A., and T.G. Patterson. 1981. How much nitrogen do legumes fix? *Adv. Agron.* 34: 15–38.

Lauer, J., P. Porter, and E. Oplinger. 1997. The corn and soybean rotation effect. Field Crops 27: 426. http://corn.agronomy.wisc.edu/AA/A014. aspx.

Lee, C., and G. John. 2003. University of Kentucky College of Agriculture Corn and Soybean Science Group Newsletter 3:1 May 2003. Lexington, KY. http://www.uky.edu/Ag/CornSoy/Newsletters/cornsoy%20vol3–1.pdf

Levine, E., and H. Oloumi–Sadeghi. 1991. Management of diabroticite rootworms in corn. *Ann. Rev. Entomol.* 36:229–255.

Levine, E., J.L. Spencer, S.A. Isard, D.W. Onstad, and M.E. Gray. 2002. Adaptation of the western corn rootworm to crop rotation: Evolution of a new strain in response to a management practice. American Entomologist–Lanham 48(2): 94–107.

Lichtfouse, E., M. Navarrete, P. Debaeke, V. Souchère, C. Alberola, and J. Menassieu. 2009. Agronomy for sustainable agriculture: A review. *Sustain. Agric. Dev.* 29: 1-6.

Liebman, M., and A.S. Davis. 2000. Integration of soil, crop and weed management in low–external–input farming systems. Weed Res. 40: 27–48.

Liebman, M., and E. Dyck. 1993. Crop rotation and intercropping strategies for weed management. *Ecol. Appl.* 3: 92–122.

Liu, A., B.L. Ma, and A.A. Bomke. 2005. Effects of cover crops on soil aggregate stability, total organic C and polysaccharides. *Soil Sci. Soc. Am. J.* 69: 2041-2048.

Lund, M.G., P.R. Carter, and E.S. Oplinger. 1993. Tillage and crop rotation affect corn, soybean, and winter wheat yields. *J. Prod. Agric.* 6: 207–213.

Ma, B.L., L.M. Dwyer, and E.G. Gregorich. 1999a. Soil nitrogen amendment effects on nitrogen uptake and grain yield of maize. *Agron. J.* 91: 650–656.

Ma, B.L., L.M. Dwyer, and E.G. Gregorich. 1999b. Soil nitrogen amendment effects on seasonal nitrogen mineralization and nitrogen cycling in maize production. *Agron. J.* 91: 1003–1009.

Ma, B.L., B.C. Liang, D.K. Biswas, M.J. Morrison, and N.B. McLaughlin. 2012. The carbon footprint of maize production as affected by nitrogen fertilizer and maize–legume rotations. *Nutr. Cycl. Agroecosyst.* 94: 15–31.

Ma, B.L., F. Meloche, and L. Wei. 2009. Agronomic assessment of Bt trait and seed or soil-applied insecticides on the control of corn rootworm and yield. *Field Crops Res.* 111: 189-196.

Ma, B.L., K.D. Subedi, and A. Liu. 2006. Variations in grain nitrogen removal associated with management practices in corn production. *Nutr. Cycl. Agroecosyst.* 76: 67-80.

Ma, B.L., and T.Y. Wu. 2008. Plant Available N in the soil: Relationships between preplant and presidedress nitrate tests for corn production. *J. Plant Nutr. Soil Sci.* 171: 458-465.

Ma, B.L., T.Y. Wu, N. Tremblay, W. Deen, N.B. McLaughlin, M.J. Morrison, E.G. Gregorich, and G. Stewart. 2010. Nitrous oxide fluxes from corn fields: On-farm assessment of the amount and timing of nitrogen fertilizer. *Global Change Biol.* 16: 156-170.

Ma, B.L., J. Ying, L.M. Dwyer, E.G. Gregorich, and M.J. Morrison. 2003. Crop rotation and soil nitrogen amendment effects on maize production in Eastern Canada. *Can. J. Soil Sci.* 83: 483-495.

Mallarino, A., E. Ortiz–Torres, and K. Pecinovsky. 2004. Effects of crop rotation and nitrogen fertilization on crop production. Iowa State University Publication: ISRF04–13. http://extension.agron.iastate.edu/ soilfertility/info/NERF_AnnRepo2004_Nrotation_Publ–2005.pdf

McGrath, M.T. 2007. Managing plant diseases with crop rotation. Available: http://www.sare.org/Learning–Center/Books/Crop–Rotation–on–Organic–Farms/Text–Version/Physical–and–Biological–Processes–In–Crop–Production/Managing–Plant–Diseases–With–Crop–Rotation#Table3.5

Meese, B.G., P.R. Carter, E.S. Oplinger, and J.W. Pendleton. 1991. Corn/soybean rotation effect as influenced by tillage, nitrogen, and hybrid/cultivar. *J. Prod. Agric.* 4: 74–80.

Michiels, J., C. Verreth, and J. Vanderleyden. 1994. Effects of temperature stress on bean nodulating Rhizobium strains. *Appl. Environ. Microbiol.* 60: 1206–1212.

Mikić, A., Ć. Branko, R.Diego, VM. ojislav, Š. Lina, F. Joëlle, A. Svetlana, D. Krstic, L. Bedoussac, L. Zoric, V. Dordevic, V. Peric, and M. Srebric. 2015. Models, developments, and perspectives of mutual legume intercropping. *Adv. Agron.* 130: 337–419.

Mora, L.E., and R.A. Moreno. 1984. Cropping pattern and soil management influence on plant diseases: I. Diplodia macrospora leaf spot of maize. *Turrialba* 34: 35–40.

Munevar, F., and A.G. Wollum. 1982. Response of soybean plants to high root temperature as affected by plant cultivar and Rhizobium strain. *Agron. J.* 74: 138–142.

Munkwold, G.P. 2003. Cultural and genetic approaches to managing mycotoxins in maize. *Ann. Rev. Phytopathol.* 41: 99–116.

Nadar, H.M., and W.A. Faught. 1984. Effects of legumes on the yield of associated and subsequent maize in intercropping and rotation systems without nitrogen fertilizer. *East African Agric. Forest. J.* 44: 127–136.

Ndakidemi, P.A., F.D. Dakora, E.M. Nkonya, D. Ringo, and H. Mansoor. 2006. Yield and economic benefits of common bean (*Phaseolus vulgaris*) and soybean (*Glycine max*) inoculation in northern Tanzania. *Aust. J. Exp. Agric.* 46: 571–577.

Ndiyae, M.A.F., M.M. Spencer, and M. Gueye. 2000. Genetic variability in dinitrogen fixation between cowpea [*Vigna unguiculate* (L.) Walp] cultivars determined using the nitrogen–15 isotope dilution technique. *Biol. Fert. Soils* 32: 318–320.

Ojiem, J.O., A.C. Franke, B. Vanlauwe, B., N. de Ridder, and K.E. Giller. 2014. Benefits of legume–maize rotations: Assessing the impact of diversity on the productivity of smallholders in Western Kenya. *Field Crops Res.* 168: 75–85.

Pedersen, P., and G.L. Joseph. 2002. Influence of rotation sequence on the optimum corn and soybean plant population. *Agron. J.* 94: 968–974.

Pedersen, P., and G.L. Joseph. 2003. Corn and soybean response to rotation sequence, row spacing, and tillage system. *Agron. J.* 95: 965–971.

Pena–Cabriales, J.J., and J.Z. Castellanos. 1993. Effect of water stress on N2 fixation and grain yield of *Phaseolus vulgaris* L. *Plant Soil* 152: 151–155.

Peoples, M. B., D.F. Herridge, and J.K. Ladha. 1995. Biological nitrogen fixation: An efficient source of nitrogen for sustainable agricultural production? *Plant Soil* 174: 3–28.

Peterson, T.A., and G.E. Varvel. 1989. Crop yield as affected by rotation and nitrogen rate. III. Corn. *Agron. J.* 81: 735–738.

Pikul Jr, J.L., L. Hammack, and W.E. Riedell. 2005. Corn yield, N use and corn rootworm infestation of rotations in the northern corn belt. *Agron. J.* 97: 854–863.

Pilbeam, C.J., M. Wood, and P.G. Mugane. 1995. Nitrogen use in maize–grain legume cropping systems in semiarid Kenya. *Biol. Fert. Soil* 20: 57–62.

Porter, P.M., J.G. Lauer, W.E. Lueschen, J.H. Ford, T.R. Hoverstad, E.S. Oplinger, and R.K. Crookston. 1997. Environment affects the corn and soybean rotation effects. *Agron. J.* 89: 441–448.

Rainbird, R.M., C.A. Akins, and J.J.S. Pate. 1983. Effect of temperature on nitrogenase functioning in cowpea nodules. *Plant Physiol.* 73: 392 394.

Rao, M.R., and M.N. Mathuva. 2000. Legumes for improving maize yields and income in semi–arid Kenya. *Agr. Ecosyst. Envir.* 78: 123–137.

Rao, M.R., P.K.R. Nair, and C.K. Ong. 1998. Biophysical interactions in tropical agroforestry systems. *Agrofor. Syst.* 338: 3–50.

Reid, L.M., X. Zhu, and B.L. Ma. 2001. Crop rotation and nitrogen effects on maize susceptibility to gibberella (*Fusarium graminearum*) ear rot. *Plant soil* 237: 1–14.

Rennie, R.J., and G.A. Kemp. 1983. N_2–fixation in field beans quantified by ^{15}N isotope dilution. II. Effect of cultivars of beans. *Agron. J.* 75: 645–649.

Riedell, W.E., T.E. Schumacher, S.A. Clay, M.M. Ellsbury, M. Pravecek, and P.D. Evenson. 1998. Corn and soil fertility responses to crop rotation with low, medium, and high inputs. *Crop Sci.* 38: 427–433.

Rodrigues, A.C., A. Bonifacio, J.E.L. Antunes, J.A.G. Silveira, and M.V.B. Figueiredo. 2013. Minimization of oxidative stress in cowpea nodules by the interrelationship between Bradyrhizobium sp. and plant growth–promoting bacteria. *Appl. Soil. Ecol.* 64: 245–251.

Rogel, M.A., E. Ormeno–Orillo, and E. Martinez Romero. 2011. Symbiovars in rhizobia reflect bacterial adaptation to legumes. *Syst. Appl. Microbiol.* 34: 96–104.

Roughley, R.J. 1970. The influence of root temperature, Rhizobium strain and host selection on the structure and nitrogen–fixing efficiency of the root nodules of Trifolium subterraneum. *Ann. Bot.* 34: 631–646.

Russell, A.E. 2002. Relationship between crop–species diversity and soil characteristics in southwest Indian agroecosystems. *Agr. Ecosyst. Environ.* 92: 235–249.

Sahrawat, K.L., M.K. Abekoe, and S. Diatta. 2001. Application of inorganic phosphorus fertilizer. pp. 225–246. *In*: Tian, G., F. Ishida, D. Keatinge (eds.), Sustaining fertility in West Africa. Soil Sci. Soc. Am. and Am. Soc. Agron, Madison, Wisconsin, USA, Soil Science Society of America Special Publication no. 58.

Schreiber, M.M. 1992. Influence of tillage, crop rotation, and weed management on giant foxtail (*Setaria faberi*) population dynamics and corn yield. *Weed Sci.* 40: 645–653.

Sellstedt, A., L. Staahl, M. Mattsson, K. Jonsson, and P. Hoegberg. 1993. Can the ^{15}N dilution technique be used to study N_2 fixation in tropical tree symbioses as affected by water deficit? *J. Exp. Bot.* 44: 1749–1755.

Shearer, G., and D.H. Kohl. 1986. N_2–fixation in field settings: estimations based on natural 15N abundance. *Aust. J. Plant Physiol.* 13: 699–756.

Shrestha, A., S.Z. Knezevic, R.C. Roy, B.R. Ball–Coelho, and C.J. Swanton. 2002. Effect of tillage, cover crop and crop rotation on the composition of weed flora in a sandy soil. *Weed Res.* 42: 76–87.

Sinclair, T.R., and V. Vadez. 2012. The future of grain legumes in cropping systems. *Crop Pasture Sci.* 63: 501–512.

Singer, J.W., and W.J. Cox. 1998. Corn growth and yield under different crop rotation, tillage, and management systems. *Crop Sci.* 38: 996–1003.

Singer, J.W., C.A. Chase, and D.L. Karlen. 2003. Profitability of various corn, soybean, wheat, and alfalfa cropping systems. Online. *Crop Management.* doi: 10.1094/CM–2003–0130–01–RS.

Singh, A., S. Eapen, and M.H. Fulekar. 2009. Potential of Medicago sativa for uptake of cadmium from contaminated environment. *Romanian Biotechnology Letters,* 14: 4164–4169.

Singh, U., and B. Singh. 1992. Tropical grain legumes as important human foods. *Economic Bot.* 46: 310–321.

Smith, H. A., and R. McSorley. 2000. Intercropping and pest management: A review of major concepts. *Am. Entomol.* 46:154–61.

Sogut, T. F. Oztturk, G. Temiz, O. Toncer, and B. Zaimoglu. 2013. The effects of rhizobial inoculation and nitrogen fertilizer application on nodulation, yield components of groundnut (*Arachis hypogaea* L.). *AWER procedia Adv. Appl. Sci.* 1: 156–165.

Stanger, T.F., J. Lauer, and C. Jean–Paul. 2008. Long–term cropping systems: The profitability and risk of cropping systems featuring different rotations and nitrogen rates. *Agron. J.* 100: 105–113.

Stoner, K.A. 2007. Management of insect pests with crop rotation and field layout. Available: http://www.sare.org/Learning–Center/Books/Crop–Rotation–on–Organic–Farms/Text–Version/Physical–and–Biological–Processes–In–Crop–Production/Management–of–Insect–Pests–with–Crop–Rotation–and–Field–Layout.

Subedi, K.D. and B.L. Ma. 2009. Assessment of the major yield limiting factors of maize production in a humid temperate environment. *Field Crops Res.* 110: 21-26.

Subbarao, G.V., C. Johansen, A.E. Slinkard, R.C. Nageswara, N.P. Saxena, Y.S. Chauhan, and R.J. Lawn. 1995. Strategies for improving drought resistance in grain legumes. *Crit. Rev. Plant Sci.* 14: 469–523.

Thom, M.D., C.A. Eberle, F. Forcella, R. Gesch, S. Weyers, and J.G. Lundgren. 2016. Nectar production in oilseeds: Food for pollinators in an agricultural landscape. *Crop Sci.* 56: (*in press*).

Tilman, D., C. Balzer, J. Hill, and B.L. Befort. 2011. Global food demand and the sustainable intensification of agriculture. *Proc. Nat. Acad. Sci.* 108: 20260–20264.

University of Illinois Extension. 2002. 23rd Edition Illinois Agronomy Handbook. http://www.ag.uiuc.edu/iah/pdf/Agronomy_HB/06chapter.pdf

van Kessel, C. and C. Hartley. 2000. Agricultural management of grain legumes: has it led to an increase in nitrogen fixation? *Field Crop Res.* 65: 165–181.

Vanotti, M.B., and L.G. Bundy. 1995. Soybean effects on soil nitrogen availability in crop rotations. *Agron. J.* 87: 676–680.

Varvel, G.E., and W.W. Wilhelm. 2003. Soybean nitrogen contribution to corn and sorghum in western corn belt rotations. *Agron. J.* 95: 1220–1225.

Vyn, T. 2006. Meeting the ethanol demand: Consequences and compromises associated with more corn on corn in Indiana. Purdue extension bulletin, ID–336. http://www.ces.purdue.edu/extmedia/ID/ID–336.pdf

Walters, D.T., A. Dobermann, K.G. Cassman, R. Drijber, J. Lindquist, J. Specht, and H. Yang. 2005. Proceedings Indiana CCA Conference, December 14–15, Indianapolis, IN.

Wang, C., B.L. Ma, J. Han, Y. Wang, Y. Gao, X. Hu, and C. Zhang. 2008. Photoperiod effect on phytochrome and abscisic acid in alfalfa varieties differing in fall dormancy. *J. Plant Nutr.* 31: 1257 - 1269.

Wang, C., B.L. Ma, X. Yan, J. Han, Y. Guo, Y. Wang, and P. Li. 2009. Yields of alfalfa varieties with different fall-dormancy levels in a temperate environment. *Agron. J.* 101: 1146-1152.

Weisany, W., Y. Rael, and K.H. Allahverdipoor. 2013. Role of some mineral nutrients in biological nitrogen fixation. *Bull. Env. Pharmacol. Life Sci.* 2: 77 84.

White, D. G. (ed.) 1999. Compendium of Corn Diseases, 3rd ed. St. Paul, MN: APS Press.

Wilhelm, W.W., and C.S. Wortmann. 2004. Tillage and rotation interactions for corn and soybean grain yield as affected by precipitation and air temperature. *Agron. J.* 96 : 425–432.

Wu, T., B.L. Ma, and B.C. Liang. 2008. Quantification of seasonal soil nitrogen mineralization for corn production in eastern Canada. *Nutr. Cycl. Agroecosyst.* 81: 279-290.

Wu, W., and B.L. Ma. 2015. Integrated nutrient management (INM) for sustaining crop productivity and reducing environmental impact: A review. *Science of the Total Environment* 512:415-427.

Yusuf, A.A., E.N.O. Iwuafor, R.C. Abaidoo, O.O. Olufajo, and N. Sanginga. 2009. Grain legume rotation benefits to maize in the northern guinea savanna of Nigeria: fixed–nitrogen versus other rotation effects. *Nutr. Cycl. Agroecosyst.* 84: 129–139.

Zahran, H.H. 1999. Rhizobium–legume symbiosis and nitrogen fixation under severe conditions and in arid climate.Microbial. *Mol. Biol. Rev.* 63: 968–989.

Zahran, H.H., and J.I. Sprent. 1986. Effects of sodium chloride and polyethylene glycol on root hair infection and nodulation of Vicia faba L. plant by Rhizobium leguminosarum. *Planta* 167: 303–309.

In: Crop Rotations
Editor: Bao-Luo Ma

ISBN: 978-1-63484-496-3

Chapter 2

CROP ROTATION TRENDS: PAST, PRESENT AND FUTURE BENEFITS AND DRIVERS

William Deen[1], Ralph C. Martin[1], Dave Hooker[1] and Amélie Gaudin[2]

[1]University of Guelph, Department of Plant Agriculture, ON, Canada
[2]University of California Davis, Department of Plant Sciences, CA, US

ABSTRACT

In temperate production regions, crop rotation complexity is declining. Before 1950, complex crop rotations provided such benefits as weed, pest and insect management, nutrient supply and labour distribution. More recently however, technological advancements in nitrogen fertilizers, pesticides, plant genetics and equipment have reduced the apparent need for crop rotation complexity in favour of more "simple" rotations. Simple rotations, consisting of only two crops or continuous planting of one crop, are perceived to be "most profitable" when intensively managed. Long-term rotation trials however, demonstrate that simple rotations are associated with reduced yields and resiliency of a system, along with negative environmental impacts such as reduced soil organic matter, reduced nutrient use efficiency and increased nutrient loss to air and water. The costs of these negative impacts are often not borne by the producer but by other segments of society. Future effects of a changing climate, emerging biomass industries, and intensification of production systems could increase the overall costs associated with simple rotations, thereby compromising long-term profitability and leading to the need for development of more rotation diversity with associated environmental benefits.

INTRODUCTION

Crop rotation is the practice of growing a series of dissimilar/different types of crops in the same field in sequential seasons. It is in contrast to continuous monoculture, which is the practice of growing a single species repeatedly on the same land. It is also in contrast to intercropping, which is the practice of growing two or more crops simultaneously on the same land. The term "crop rotation" is often used interchangeably with "cropping system diversity." The latter term, which can be applied to landscape as well as a field scale, represents a broader concept which includes crop rotation, intercropping, and cover crop usage.

Complexity of crop rotation can vary from a simple two-species rotation to rotations that involve numerous species. Despite numerous benefits associated with greater crop rotation complexity, monocultures or simple two-species rotations are becoming increasingly prevalent in many key agricultural production regions of the world. In the Northern Corn Belt, including Ontario, there has been a dramatic shift in the past century to a crop rotation system dominated by corn and soybean (Gaudin et al., 2015a). While this shift to simple systems may appear to be a contradiction given the many potential benefits of rotation, the choice and sequence of crops that a farmer uses in a rotation is the result of a decision making process that is informed by various drivers and constraints.

The purpose of this chapter is to: 1) discuss trends in rotation diversity in the province of Ontario and compare these trends to the broader Northern Corn Belt, 2) demonstrate benefits farmers associate with crop rotation in response to changing economic, social and environmental drivers, and 3) present evidence that present and future drivers of crop rotation diversity may lead to greater benefit.

TRENDS IN ROTATION DIVERSITY

In general, the trend towards more simple rotations is a growing concern. This concern is not unique to Ontario, but to the Northern Corn Belt states and many of the major field crop production regions of the world. In the United States, harvested areas of corn and soybean increased by 500% between 1950 and 2003 (Karlen, 2004) despite the known benefits of crop rotations. During that same period, area under oat production declined by 90% and areas devoted to forage decreased more than 15%. The displacement of small grains, hay, and pasture and the resulting dominance of corn and soybean has been repeatedly identified (Brown and Schulte, 2011; Hatfield et al., 2009; Johnston, 2013). In Iowa, corn and soybean now occupy 63% of the state's total land area and 82% of its cropland (National Agricultural Statistics Service, 2014). In 2013, corn and soybean represent approximately 92.1, 74.2, 60.3% of the total harvested area grown in Iowa, Minnesota and Michigan, respectively (Gaudin et al., 2015a). In comparison, corn, soybean and hay represented approximately 99.5, 86.3, and 74.8% of the total harvested area grown in Iowa, Minnesota and Michigan, respectively. Harvested hay acreage is relatively low.

The trends towards simplification, observed in the Midwest United States, are similar to those observed in the province of Ontario (Figure 1). In Ontario, corn, soybean and winter wheat acreage has increased by approximately 25%, 900%, and 135%, respectively, from 1970 to 2014. During the same period, forage and spring cereal acreage declined by approximately 35 and 80% respectively.

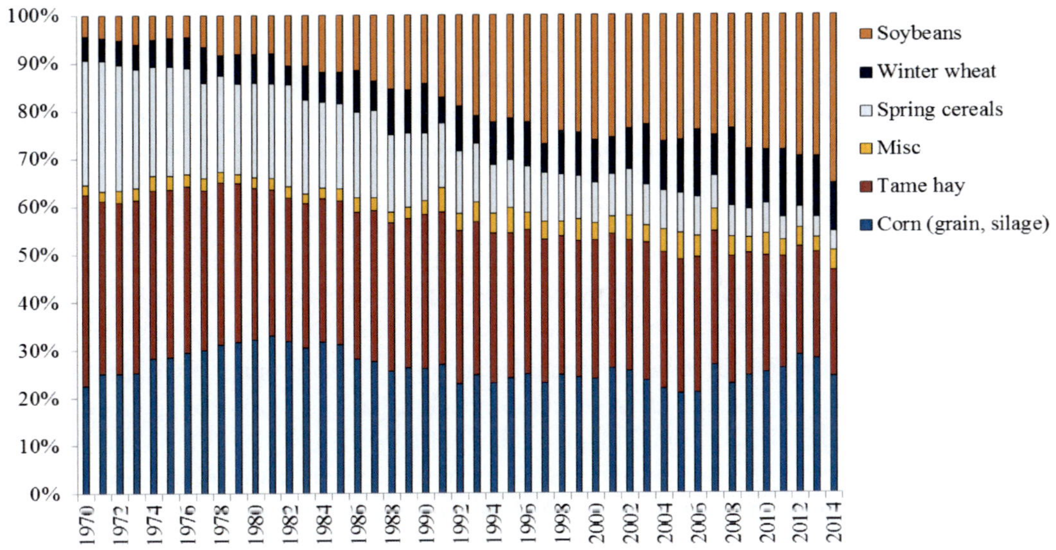

Figure 1. Harvested areas (hectares) of major field crops shown as % of total harvested area from 1970 to 2014 for Ontario. Source: Statistics Canada. Table 001-0010 - Estimated areas, yield, production and average farm price of principal field crops, in metric units, annually (accessed: November 08, 2014).

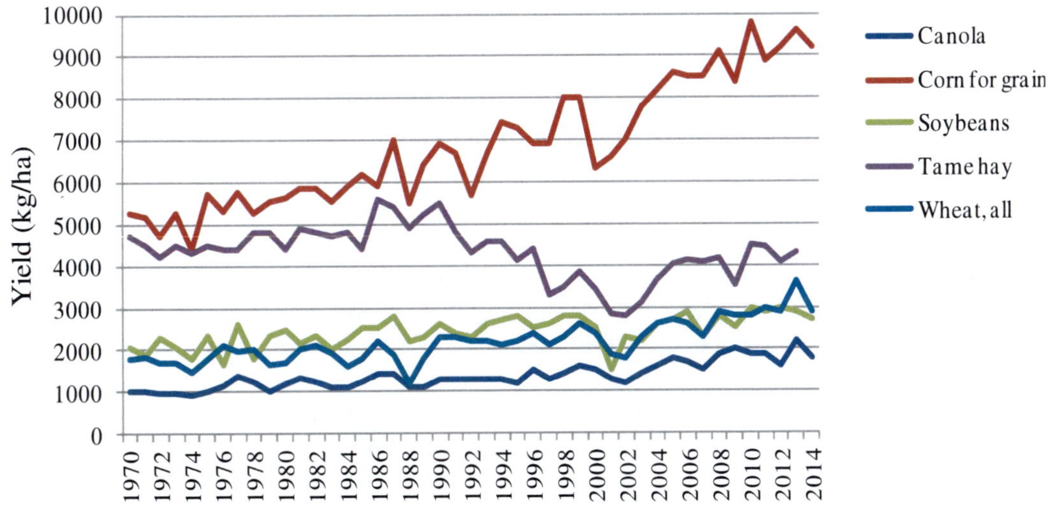

Figure 2. Average yield of major field crops from 1970 to 2014 for Canada. Source: Statistics Canada. Table 001-0010 - Estimated areas, yield, production and average farm price of principal field crops, in metric units, annual (http://www5.statcan.gc.ca/ cansim/a26?lang=eng&id=10010 accessed: November 08, 2014).

A confounding, and perhaps contributing factor related to the trend of forage removal is that average forage yields have declined during the period from approximately 1985 to 2000. Between 1990 and 2014, across Canada, there is no evidence of yield increases in forages compared to other major crops (Figure 2). One explanation for stagnant forage yield during this period is that forages may have been displaced by corn and soybean primarily from Class

1-2 land[1], and have been relegated to being produced on Class3-6 lands. Secondly, breeding and management advancements in annual row crops may have been disproportionately less for forages (Brummer and Casler, 2014). Consequently, the concentration of corn and soybean on Class 1 and 2 lands may not be accurately represented by Figure 1. If forages are removed, similar to Iowa, Minnesota and Michigan, corn and soybean represent approximately 80% of the annual row crops grown in Ontario.

BENEFITS INFLUENCING CROP ROTATION COMPLEXITY: PAST

Increasing crop rotation complexity provides a range of economic, environmental and social benefits (Table 1). The list of crop rotation benefit has not really changed over time, but what has changed is the perceived magnitude of influence each benefit has on a farmer's decision regarding crop rotations. In the past, a farmer's decisions regarding crop rotation were motivated by a differing set of benefits than in the present or the future.

The benefits of crop rotation have been recognized for centuries even though the mechanisms underlying the benefits of this practice may not have been fully understood. For example, crop rotation was reportedly used during the Han dynasty of China more than 2000 years ago (MacRae and Mehuys, 1985). Some of the earliest descriptions of crop rotations seem to suggest that soil productivity was the main impetus underlying crop rotation. The British Agricultural Revolution (1750 -1880), was, in a large part, associated with the development of the Norfolk four-crop rotation (Overton, 1996). This four-crop rotation replaced a two-field crop rotation system where one field was left fallow or turned into pasture; a common practice during the Middle Ages. The four-crop rotation was observed to greatly increase crop and livestock yields by improving soil fertility and reducing fallow. George Washington Carver promoted the use of crop rotation over a century ago, teaching Southern United States farmers to rotate soil-depleting crops like cotton with soil-enriching crops like peanuts and peas (McMurry, 1982). In 1927, it was reported that "rotation is 75% as efficient as fertilizers in maintaining and increasing crop yields" (Weir, 1927). In the following statements, the consequences of poor rotation in Minnesota at the turn of the past century are summarized (Pond et al., 1931):

> "The one-crop system, followed so persistently during the first thirty years of farming in the Valley, developed the usual hazards that are inevitable with single-crop farming-weed pests, plant diseases, insects, and poor physical condition and lowered fertility of the soil."

---

[1] In Canada, land capability determined using the Canada Land Inventory (CLI) for agriculture. CLI is an interpretative system for assessing the effects of climate and soil characteristics on the limitations of land for growing common field crops. Class 1 soils have no significant limitations in use for crop production. The soils are deep, are well to imperfectly drained, hold moisture well, and in the virgin state were well supplied with plant nutrients. They can be managed and cropped without difficulty. Under good management they are moderately high to high in productivity for a wide range of field crops. Class 2 soils have moderate limitations, and moderate conservation practices required to achieve moderately high to high in productivity. Class 3 soils have moderately severe limitations and the range of crops is restricted or special conservation practices required. Class 4 soils have severe limitations. Class 5 soils are restricted to forage crops and improvement practices are feasible. Class 6 soils are restricted to forage crops and improvement practices not feasible. (http://sis.agr.gc.ca/cansis/nsdb/cli/class.html).

"Wheat became a crop of uncertain yield, except as it was grown in a crop rotation."

"Most of these problems of crop production ordinarily can be met most effectively through a crop rotation program extending over a series of years ...."

"It is the general opinion of farmers that the continuous growing of spring grains has gradually lowered the yields of these crops and that the reduced yields are partly due, in addition to the effects of weeds and diseases, to the gradual depletion of the physical condition and to some extent the fertility of the soil."

Yield reductions from 3 to 57% for major crops grown in short rotation sequences and monocultures relative to yields in more complex rotations were documented in a recent review (Bennett et al., 2012). Other documented benefits of crop rotation complexity include reduced prevalence and reduced damage from insect pests and weeds, beneficial interactions with soil microbes and nematodes, reduced soil compaction, reduced nutrient depletion and increased soil water availability (Karlen et al., 1994), and reduced risk (Helmers et al., 2001). Crop rotation complexity when integrated with livestock production, enabled inclusion of perennial forage crops and the application of manure on crop fields (Russelle, 2007). Inclusion of forages in a crop rotation increases nitrogen supply (Cavigelli et al., 2008; Magdoff and van Es, 2000; Stanger and Lauer, 2008), improved soil structure and carbon (Riedell, 2009), and reduced weed pressure (Liebman, 2008).

In summary, decisions regarding crop rotation in the past were driven by improvements related to soil fertility, control of pests, meeting livestock feed requirements, increasing yields and distribution of labour; however, over time, these have been displaced with a more narrow set of benefits, which has ultimately resulted in simpler rotations that are currently practised.

Drivers of Crop Rotation Complexity: Past to Present

Over the past century, there have been a number of "significant" drivers" causing change in farmer decisions regarding crop rotation complexity and ultimately leading to the simplification of rotations observed currently.

Prior to the discovery of the Haber–Bosch process in 1909, legumes and livestock represented the primary means to introduce nitrogen into a crop production system. The Haber–Bosch process enabled artificial nitrogen fixation and production of ammonia. In 2010, the fixation of nitrogen through Haber–Bosch (120 Tg N yr^{-1}) was double the natural terrestrial sources (63 Tg N yr^{-1}) (Fowler et al., 2013). Large-scale industrial fixation of nitrogen has displaced the main value of legumes from the rotation and enabled crop production without livestock. In the latter part of the 20th century, nitrogen and phosphorus use increased by 800 and 300%, respectively (Tilman et al., 2001).

Table 1. Crop rotation complexity and underlying benefits and drivers

Crop Rotation Benefits	Influence of Benefits - "Past"	Drivers of Crop Rotation Complexity: Past to Present →	Influence of Benefits - "Present"	Drivers of Crop Rotation Complexity: Present to Future →	Influence of Benefits - "Future:
yield improvement	++++	synthetic fertilizers, herbicides, insecticides, fungicides, improvements in crop genetics, specializationof farms (eg. separation of crop and livestock production), increase in equipment scale and performance, government programs, market demand, large %age rented land	+	high input costs, climate change, increasing drought sensitivity, emerging biomass markets,pest resistance, declining water quality, increasing poulation concentration	++++
system resilience	na		++		+++
improve fertilizer use efficiency	na		+		+++
nitrogen supply	+++++		++		+++
weed, disease and insect management	+++++		+		+++
integration with livestock	+++++		++		++
labour distribution	++++		++		++
scales of economy	na		+++++		++
risk mitigation	++		+++		+++
soil quality enhancement	+		++		+++
GHGs reduction	na		na		+++
water quality improvement	na		na		+++
habitat biodiversity	na		na		+++
increased success of no-till/reduced till	na		+		+++
cover crop niches	na		+		++
sustainable biomass removal	na		na		+++

na - not applicable, + low influence ------ +++++ high influence

The role of crop rotation for weed suppression has largely been eliminated by the introduction of synthetic herbicides and genetically modified crop species that are resistant to herbicide modes of action. Prior to the introduction of synthetic herbicides, such as 2,4-D, which was first introduced in 1945 by the American Chemical Paint Company, weed control was achieved by a combination of methods including crop rotation, tillage, hand weeding and other cultural methods. Herbicides such as 2,4-D revolutionized weed control. It could selectively control broadleaf plants, while not injuring most monocots. It reduced the need for complex rotations. Where labour was scarce, it replaced the hoe and facilitated larger scale production. Numerous other synthetic herbicides were introduced in subsequent years including the triazine family of herbicides, which includes atrazine, in the 1950s and glyphosate in 1974. Reliance on crop rotation for weed management was further reduced by the introduction of genetically modified herbicide tolerant crops whose growth and development are not significantly affected by herbicides thus facilitating herbicide based weed management. Whereas, complex crop rotations in the past were a necessary tool to control weeds, it is no longer a primary weed control strategy. Herbicides and genetically modifies herbicide tolerant crops have also facilitated larger scale farms which, in turn, has encouraged further simplification of rotations.

Similarly, in the past, crop rotation was an essential tool for managing insects and diseases. Introduction of various pesticides and pest resistant crops developed either through traditional or genetic modification, have effectively enabled simplification of crop rotations.

Livestock production and field crop production are increasingly separated at both the farm and regional scale (Beaulieu et al., 2001; Boschma et al., 2015). A movement away from integrated livestock and crop production systems has been enabled, in part, by the increased use of synthetic fertilizers that replace the requirement for nutrients from livestock manure. Concentration of livestock production has resulted in the removal of forages from rotations, as well as increased production of corn, soybean and corn silage (Rankin, 2015).

Advancements in mechanization further promoted the adoption of short rotations (Sprugeon and Grisson, 1965). A reduced number of crops in rotations enable farmers to own fewer pieces of equipment and reduce capital investment in equipment (Colvin et al., 1990), to specialize in the production of fewer crops (Actams et al., 1970), and to take advantage of economies of scale for production (Johnston, 2013; McGranahan, 2014).

Other drivers for reducing crop rotation diversity include government policies, and favorable economics driven, in part by growing demand for a narrow spectrum of crops, commodity programs that emphasized short-term profit, increased prevalence of rented land, public and private research and development efforts devoted to genetic improvement of corn and soybean, and increased food and industrial uses for both corn and soybean oils and various by-products (Francis and Clegg, 1990; Karlen, 2004; Porter et al., 2003).

Benefits Influencing Crop Rotation Complexity: Present

While crop rotation has not been entirely abandoned in Ontario and the rest of the Northern Corn Belt, the present situation is that simple rotations dominate the landscape. Simple two-crop rotations of corn and soybean are prevalent, particularly on Class 1 and 2 soils. Forages have been displaced and, to a large extent, are relegated to lower classes of land. With the dominance of corn and soybean in rotation, opportunities to effectively deploy cover crops in a rotation and to realize their benefits have been reduced. This is particularly true in shorter season regions where cover crop biomass potential is low due to the use of full-season crop varieties aimed at maximizing corn and soybean yield potential (Vanhie et al., 2015). In previous studies, conducted at two locations in Southwestern Ontario, annual rye grass, cereal rye and oats were planted into a corn-soybean rotation either at pre-soybean leaf drop, post soybean harvest or post corn grain harvest. Over the three years, cover crop biomass did not exceed 0.5 t ha^{-1} when measured either in the fall or spring prior to application of a herbicide to control the cover crop (Deen and Hooker, data unpublished). Due to relatively low cover crop biomass, rotational benefits due to inclusion of cover crops were not observed.

While annual corn and soybean rotations consistently produce high yields of commodities for marketing and livestock feed, it relies extensively on external inputs of synthetic pesticides and fertilizers (Liebman et al., 2008). Past drivers of crop rotation (weed, disease and insect control, nutrient supply, integration with livestock and labour distribution) presently do not have a significant impact on farmer decision making regarding crop rotation complexity. This does not mean that these drivers have been entirely negated and that there is no benefit of rotation related to them. Synthetic fertilizers and pesticides do not entirely replace crop rotation (Bullock, 1992 and references therein). Crop rotation benefits are observed even when external inputs appear to be non-limiting. For example, crop rotation can improve weed control even with synthetic herbicide use (Forcella and Lindstrom, 1988). Corn-soybean rotations compared to continuous monocultures of these crops do realize a degree of benefit associated with weed, disease and insect control, nutrient supply, integration with livestock and labour distribution; however, these benefits contribute little to farmer decision making regarding rotation. Presently, a farmer's decision to produce a monoculture versus a two-crop rotation versus a more complex rotation appears to be driven largely by other factors.

Farmers in general recognize that a simple rotation may result in a yield reduction relative to more complex rotations. They also recognize that input costs associated with simple rotations may be somewhat higher than for complex rotations. Their decision to adopt simple rotations reflects a reality or perception that simple rotations provide greater economic return and stability in the short term and perhaps in the longer term. This reality or perception of greater economic returns associated with simple rotations may arise from a number of drivers: government policies, favorable economics, commodity programs as well as public and private research and development efforts devoted to genetic improvement of corn and soybean, and increased food and industrial uses for both corn and soybean oils and various by-products (Karlen, 2004; Porter et al, 2003; Francis, and Clegg, 1990) may have resulted in greater net return per acre and economic advantage to simple corn-soybean rotations versus more complex rotations. Net return per acre may be lower for a simple rotation than for a complex rotation, but farmers may still have an economic advantage to adopting simple rotations. This is due to shorter rotations enabling simplification of management strategies and increased scale of production (Johnston, 2013; McGranahan, 2014), especially on leased land that may only be under the farmer's control for a year or two. With a simple rotation, although returns per acre may be lower, the ability to increase farm size results in higher economy of scale returns, over the entire farm.

Crop rotation is generally thought to reduce risk compared with monoculture cropping (Helmers et al., 2012). Diversification with crop rotation complexity reduces risk by offsetting low returns in one year for one crop with relatively high returns from a different crop. Crop rotation also reduces risk by reducing yield variability (Gaudin et al., 2015and references therein) and consequently variability in returns (Helmers et al, 2012). However, it was demonstrated that a corn- soybean rotation was the most profitable rotation even when risk is taken into consideration (Stanger and Lauer, 2008).

DRIVERS OF CROP ROTATION COMPLEXITY: PRESENT TO FUTURE

While the adoption of simple rotations has provided economic advantages, there are a number of drivers emerging that could ultimately lead farmers increase crop rotation complexity (Table 1).

Long term rotation trials have quantified yield benefits of crop rotation complexity. Two long term rotation trials in Ontario show yield benefits with more complex rotations [Gaudin et al., 2015a; Gaudin et al., 2015b), which corroborates work summarized elsewhere (Bennett et al., 2012). There is evidence, however, that the yield penalty associated with simple rotations increase with the duration of the simple rotation (Drury and Tan, 1995).

There are a number of potential reasons for why the rotation effect accentuates over time. First, the longer the time period under a simple rotation, there is a greater probability that yield reductions associated with weed, insects or diseases will result. As mentioned previously, while external inputs such as pesticides and improved genetics reduce the impact of these pests, they are not eliminated. Secondly, simple crop rotations are associated with reduced soil organic matter (West and Post, 2002; Karlen et al., 2006) and reductions in overall soil quality (Congreves et al., 2015; Munkholm et al., 2013). As crop productivity is associated strongly with soil organic matter, these effects increase in magnitude the longer the simple rotation is practised. Lower soil organic matter tends to reduce other measures of soil quality. For example, a reduction in soil organic matter reduces aggregate stability, which is positively correlated with water infiltration and compaction resistance (Congreves et al., 2015; Munkholm et al., 2013). Reduced soil organic matter results in reduced soil biological activity. Under more complex rotations, the quantity, quality and chemical diversity of residues increased and more diverse soil biological communities were sustained, with positive effects on soil organic matter and soil fertility (Tiemann et al., 2015). Reductions in soil organic matter and aggregate stability with simple rotations increased soil erosion (Stonehouse et al., 1998). Simple rotations also require greater amounts of tillage to maintain yield (Pittelkow et al., 2014), thereby resulting in tillage exacerbating the effects of simple rotations on soil quality. Third, yield of crops under simple rotations appear to be more vulnerable to abnormal moisture conditions, particularly drought (Gaudin et al., 2015b). Yield stability significantly increased when corn and soybean were integrated into more diverse rotations consisting of small grains, red clover and/or alfalfa. In hot and dry years, diversification of corn-soybean rotations with additional crop species and reduced tillage increased yield by 7 and 22% for corn and soybean, respectively. Yield reduction and variability of simple rotations associated with weather abnormalities, particularly drought, may be increasing due to: 1) a trend towards warmer, drier summers and more variable precipitation patterns in the mid latitude regions (IPPC, 2014), and 2) increases in yield potential of corn and soybean that increase crop water requirements (Richards, 2000). Soils associated with more complex crop rotations tend to have greater plant water availability.

Widespread water quality concerns persist for watersheds impacted by agricultural production in Ontario and the Northern Corn Belt. Nitrogen deposition from agricultural lands into the Gulf of Mexico has led to a large coastal hypoxic zone (Alexander et al., 2008; Broussard and Turner, 2009; Zhou et al., 2014) and in the Great Lakes. Simple rotation impacts on soil quality and soil erosion contribute to this water quality problem. Crops like corn and soybean are much less effective in reducing soil erosion, thus reducing water infiltration and nutrient retention, in contrast to perennial species (Asbjornsen et al., 2014). Integration of diverse, deep-rooted communities of perennial plants into landscapes and watersheds dominated by corn and soybean fields resulted in a 95% reduction in sediment export, a 90% reduction in total phosphorus export, and an 85% reduction in total nitrogen export (Helmers et al., 2012). Furthermore, diversification of simple corn-soybean cropping systems with small grain crops and perennial forages reduces nutrient requirements (Davis et al., 2012; Gaudin et al., 2015) thereby reducing risk of loss.

Over the past century the global nitrogen cycle has been profoundly altered by human activity. The amount of 'reactive' nitrogen produced by humans is now greater than the amount created through natural processes (Fowler et al., 2013). Currently N use efficiency for world cereal grain production systems is estimated between 20-50% (Mosier et al., 2004; Raun and Johnson, 1999). Low nitrogen use efficiency, in addition to being an economic concern, is a significant environmental concern. Agricultural sources of nitrogen can contaminate water sources through N runoff or nitrate leaching, and contribute to greenhouse gas levels through carbon dioxide emissions associated with nitrogen fertilizer production and soil emissions. There is also growing evidence of the 'nitrogen cascade' effect which occurs as a result of the same nitrogen atom contributing to multiple negative effects in the air, on land, in freshwater and marine systems, and on human health.

As a result of these nitrogen related concerns, the benefit of crop rotation from reducing nitrogen fertilizer input and increasing nitrogen fertilizer use efficiency may increase. Simple corn - soybean rotations require greater nitrogen inputs than more complex rotations that include forages. In a review of red clover it was determined that red clover can provide an average nitrogen fertilizer equivalent of 41-64 kg N ha^{-1} (Gaudin et al., 2013). Other studies have similarly demonstrated the benefits of including forage legumes in rotation with grain crops (Baldock et al., 1981; Cavigelli et al., 2008). The benefit of rotation complexity in terms of nitrogen benefit is not just associated with legume inclusion in the rotation: nitrogen requirements for corn were reduced in a corn-soybean-winter wheat rotation compared to a corn-soybean rotation (Gaudin et al., 2015a). The authors attribute reduced N requirement associated with winter wheat inclusion to improvements in corn nitrogen use efficiency.

Global warming is an issue that has emerged in the past 1-2 decades and could significantly impact discussions related to crop rotation complexity. Field crop production contributes to greenhouse gas emissions and global warming primarily through processes related to fossil fuel consumption for nitrogen fertilizer production and crop production (tillage, grain drying), nitrous oxide emissions, and soil organic matter dynamics. Simple crop rotations are associated with greater tillage requirements (Pittelkow et al., 2014), reduced diesel use efficiency due to lower yield potential, increased nitrogen fertilizer requirements and lower soil organic matter. Adoption of complex rotations represents a mitigation strategy to reduce GHG emissions and increase carbon sequestration through elevation of soil organic matter (Meyer-Aurich et al., 2006). Complex rotations also represent an adaptation strategy to the weather extremes anticipated with global warming. As already mentioned above, complex rotations are associated with increased resilience to weather extremes, including drought.

Weed management may re-emerge as a driver of crop rotation. The trend to simple rotations has been facilitated by herbicides and herbicide resistant crops that are effective across a range of weed species, soil types, timings and crops which enable a farmer to increase farm size. Simple rotations are, however, associated with larger weed seedbanks, increased weed pressure and increased rapidity of herbicide resistance. Herbicide resistance is now widespread for many commonly used herbicides (Shaner, 2014) and represents an emerging challenge to simple crop rotations in Ontario and the Northern Corn Belt (Mortensen et al., 2014). The ability to rely on herbicides as the sole weed management strategy could be compromised. Complex crop rotation represents a means to practice more integrated weed management strategies with less reliance of herbicides for controlling weeds. Complex rotations consisting of diverse crops and management practices significantly increase the options for integrated weed control (Swanton et al., 2008). In summary, complex rotations: 1) increase crop yield and thus improve crop competition with weeds, 2) increase herbicide options and modes of actions, 3) enable greater diversity in herbicide timing, 4) diversify both the timing of competition of crop and weeds and the nature of competition, 5) reduce weed seed banks, 6) increase tillage options, and 7) increase cover crop options for added weed suppression.

Benefits Influencing Crop Rotation Complexity: Future

In the future, a broader set of benefits will impact farmer decisions regarding crop rotation (Table 1). This is due to the drivers outlined in the previous section. Yield benefits of rotation will increase due to increasing soil moisture constraints resulting from increasing transpiration demand from higher yields and increased variation of precipitation resulting from a changing climate. Greater emphasis will be placed on system resiliency, again due to changing climate effects. Crop rotation complexity will be encouraged in an effort to obtain resiliency from improved soil quality, cover crops and no-till production. Increasing occurrence of pest resistance will necessitate the reintroduction of greater crop rotation complexity as a pest management tool. Since soil organic matter is more easily maintained with more complex rotations (Kludze et al., 2013), greater economic and environmental benefits will be associated with complex rotations, especially as commercial biomass markets develop. Environmental benefits of complex rotations (increased nutrient use efficiency, reduced nitrogen use, GHG reduction, improved water quality, and habitat diversity) represent externalities that should be incorporated into farmer's crop rotation decisions.

Conclusion

Although for the past number of decades there has been an obvious trend towards simplification of rotations, different drivers are emerging that may change the trajectory of this trend back toward more complex rotations. While simple monocrop or two-crop systems of corn and soybean currently dominate the landscape in Ontario and the Northern Corn Belt, there is growing evidence that crop rotation complexity could provide a wider set of benefits than present, or in the past.

A radical increase in crop rotation complexity from the current simple corn and soybean rotation is not necessarily required. Similar to the relationship between biological diversity and ecosystem function which resembles an asymptotic hyperbola (Cardinale et al., 2012), the addition of species complexity to a rotation follows the law of diminishing returns. The incremental benefit of a two-species rotation over continuous monoculture is typically greater than the incremental benefit of a three-species rotation. Above some intermediate and undetermined level of rotation complexity, incremental benefits are small. Given the asymptotic hyperbola relationship between crop rotation complexity and benefits, introduction of a single additional crop species to the dominant corn-soybean system could have significant benefit. Inclusion of winter wheat, a forage species, or a 3-5 year dedicated perennial biomass crop could provide significant benefits in Ontario, and other parts of the Northern Corn Belt.

Concerns that are emerging regarding simple rotations have economic, environmental and social implications that impact the farmer as well as other groups within society. Climate change, biomass markets based on crop residue removal, and increasing intensity of production systems may exacerbate these concerns. Decisions regarding crop rotation complexity are made by the farmer; however, that decision is informed by various incentives, economic and otherwise. Given the emerging interests of other stakeholders in the crop rotation complexity discussion, and the growing awareness of potential benefits associated with more complex rotations, the various stakeholders need to investigate means of encouraging farmers to increase crop rotation complexity.

REFERENCES

Actams, W. E., H. D. Morris, and R.N. Dawson. 1970. Effect of cropping systems and nitrogen levels on corn (Zea mays) yields in the southern Piedmont Region. *Agron. J.* 62: 655.

Alexander, R.B., R.A Smith, G.E. Schwarz, E.W. Boyer, and J.V. Nolan. 2008. Differences in phosphorus and nitrogen delivery to the Gulf of Mexico from the Mississippi River Basin. *Environ. Sci. Technol.* 42: 822–830.

Asbjornsen, H., V. Hernández-Santana, M. Liebman, J. Bayala, and J. Chen. 2014. Targeting perennial vegetation in agricultural landscapes for enhancing ecosystem services. Renewable Agriculture and Food Systems 29: 101–125. doi: 10.1017/S17421705 12000385.

Baldock, J.O., R.L. Higgs, W.H. Paulson, J.A. Jakobs, and W.D. Shrader. 1981. Legume and mineral N effects on crop yields in several crop sequences in the upper Mississippi Valley. *Agron. J.* 73:885–890.

Beaulieu, M.S., F. Bédard, and P. Lanciault. 2001. Distribution and Concentration of Canadian Livestock. Agriculture and Rural Working Paper Series Working Paper No. 47.

Statistics Canada. Ottawa, ON. http://ageconsearch.umn.edu/bitstream/28016/1/wp010047.pdf.

Bennett, A.J., G. D. Bending, D. Chandler, S. Hilton and P. Mills. 2012. Meeting the demand for crop production: the challenge of yield decline in crops grown in short rotations. *Biol. Rev.* 87: 52–71. doi: 10.1111/j.1469-185X.2011.00184.x.

Boschma, M., A. Joaris, C. Vidal. Concentration of livestock production. http://ec.europa.eu/agriculture/envir/report/en/live_en/report.htm.

Broussard, W., and R.E. Turner. 2009. A century of changing land-use and water-quality relationships in the continental US. Frontiers in Ecology and the Environment **7**: 302–307.

Brown P.W., and L.A. Schulte. 2011. Agricultural landscape change (1937–2002) in three townships in Iowa, USA. Landscape Urban Plan 100: 202–212.

Brummer, E.C. and M.D. Casler. 2014. Cool-season forages. In S. Smith, B. Diers, J. Specht, and B. Carver (eds.). Genetic gain in major U.S. field crops. CSSA Spec. Publ. 33. ASA, CSSA, and SSSA. Madison, WI. doi:0.2135/cssaspecpub33.c3.

Bullock, D. G. 1992. Crop Rotation. Critical Reviews in Plant Sciences, 11(4):309-326.

Cardinale, B.J., J.E. Duffy, A. Gonzalez, D.U Hooper, C. Perrings. 2012. Biodiversity loss and its impact on humanity. Nature doi: 10.1038/nature11148.

Cavigelli, M.A., J.R. Teasdale, and A.E. Conklin. 2008. Long-term agronomic performance of organic and conventional field crops in the mid-Atlantic region. Agron. J. 100:785–794.

Colvin, T. S., D. C. Erbach, and W. D. Kemper. Socioeconomic aspect of machinery requirements for rotational agriculture. In Sustainable Agricultural Systems, Edwards, C. A., Ed., Soil and Water Conservation Society, Ankeny, IO, 1990, 532.

Congreves, K.A., A. Hayes, E.A. Verhallen, and L.L Van Eerd. 2015. Long-term impact of tillage and crop rotation on soil health at four temperate agroecosystems. *Soil* and *Tillage Res.* 152: 17 – 28.

Davis, A.S., J.D. Hill, C.A. Chase, A.M. Johanns, and M. Liebman. 2012. Increasing cropping system diversity balances productivity, profitability and environmental health. *PLoS ONE*. doi: 10.1371/journal.pone.0047149.

Drury, C. F., and C.S. Tan. 1995. Long-term (35 years) effects of fertilization, rotation and weather on corn yields. *Can. J. Plant Sci.* 75: 355-362.

Forcella, F. and M. J. Lindstrom. 1988. Weed seed populations in ridge and conventional tillage, *Weed Sci.* 36, 500.

Fowler, D., M. Coyle, U. Skiba, M.A. Sutton, J. N. Cape, S. Reis, L.J. Sheppard, A. Jenkins, B. Grizzetti, J.N. Galloway, P. Vitousek, A.Leach, A.F. Bouwman, K. Butterbach-Bahl, F. Dentener, D. Stevenson, M. Amann, and M. Voss. 2013 The global nitrogen cycle in the twenty-first century. Phil Trans R Soc B 368: 20130164. http://dx.doi.org/10.1098/rstb.2013.0164.

Francis, C. A. and M. D. Clegg. 1990. Crop rotations in sustainable production systems, in Sustainable Agricultural Systems, Edwards, C. A., Ed., Soil and Water Conservation Society, Ankeny, IO, 107.

Gaudin A.C.M., S. Westra, C.E.S. Loucks, K. Janovicek, R.C. Martin, and W. Deen. 2013. Improving Resilience of Northern Field Crop Systems Using Inter-Seeded Red Clover: A Review. *Agronomy.* 3:148-180.

Gaudin, A.C.M., K. Janovicek, B. Deen, and D. C. Hooker. 2015a. Wheat improves nitrogen use efficiency of maize and soybean-based cropping systems. *Agriculture, Ecosystems and Environment.* 210:1–10.

Gaudin A.C.M., T.N. Tolhurst, A.P. Ker, K. Janovicek, C. Tortora, R.C. Martin, and W. Deen. 2015b. Increasing Crop Diversity Mitigates Weather Variations and Improves Yield Stability. *PLoS ONE* 10. (2): e0113261. doi:10.1371/journal.pone.0113261.

Hatfield J.L., L.D. McMullen, and C.S. Jones. 2009. Nitrate-nitrogen patterns in the Raccoon River Basin related to agricultural practices. *J. Soil Water Conserv.* 64(3): 190–199.

Helmers M.J., X. Zhou, H. Asbjornsen, R. Kolka, and M.D. Tomer. 2012. Sediment removal by prairie filter strips in row-cropped ephemeral watersheds. *J. Environ. Qual.* 41: 1531–1539. doi: 10.2134/jeq2011.0473.

Helmers, G.A., C. F. Yamoah, and G. E. Varvel. 2001. Separating the Impacts of Crop Diversification and Rotations on Risk. *Agron. J.* 93: 6: 1337-1340. doi:10.2134/agronj2001.1337.

IPPC - Intergovernmental Panel on Climate Change. 2013. The physical science basis. Contribution of Working Group I to the Fifth Assessment Report of the Intergovernmental Panel on Climate Change. Cambridge, United Kingdom and New York: Cambridge University Press. 1535 p.

Johnston, C.A. 2013. Agricultural expansion: land use shell game in the U.S. Northern Plains. *Landscape Ecol* 29: 81–95.

Karlen, D.L. 2004. Cropping systems: Rain-fed maize-soybean rotations of North America. p. 358–362. In R.M. Goodman (ed.) Encyclopedia of plant and crop science. Marcel Dekker, New York.

Karlen, D.L., G.E. Varvel, D.G. Bullock, and R. M. Cruse. 1994. Crop rotations for the 21st Century. *Adv. Agron.* 53, 1–45.

Karlen, D.L., E.G. Hurley, S.S. Andrews, C.A. Cambardella, D.W. Meek, and M.D. Duffy. 2006. Crop rotation effects on soil quality at three northern corn/soybean belt locations. *Agron. J.* 98:484–495. doi:10.2134/agronj 2005.0098.

Kludze, H., B. Deen, A. Weersink, R. van Acker, K. Janovicek, A. De Laporte, and I. McDonald. 2013. Estimating sustainable crop residue removal rates and costs based on soil organic matter dynamics and rotational complexity. *Biomass and Bioenergy* 56: 607-618.

Liebman, M., L.R. Gibson, D.N. Sundberg, A.H. Heggenstaller, P.R. Westerman, C.A. Chase, R.G. Hartzler, F.D. Menalled, A.S. Davis, and P.M. Dixon. 2008. Agronomic and economic performance characteristics of conventional and low-external input cropping systems in the central Corn Belt. *Agron. J.* 100:600–610.

MacRae, R. J. and G. R Mehuys. 1985. The effect of green manuring on the physical properties of temperate-area soils. *Adv. Soil Sci.* 3: 71.

Magdoff, F., and H. van Es. 2000. Crop rotations. p. 99–108. In Building soils for better crops. 2nd ed. Sustainable *Agric. Publ.,* Univ. of Vermont, Burlington.

McGranahan DA. 2014. Ecologies of scale: multifunctionality connects conservation and agriculture across fields, farms, and landscapes. Land 3: 739–769. doi:10.3390/land3030739.

McMurry, L. 1982. George Washington Carver: Scientist and Symbol, pp. 9–10, New York: Oxford University Press (ISBN 0-19-503205-5).

Meyer-Aurich, A., A. Weersink, K. Janovicek, W. Deen. 2006. Cost efficient rotation and tillage options to sequester carbon and mitigate GHG emissions from agriculture in Eastern Canada. Agric. Ecosys. and Env. 117:119-127.

Mortensen D.A., J.F. Egan, B.D. Maxwell, M.R. Ryan, and R.G. Smith. 2012. Navigating a critical juncture for sustainable weed management. *BioScience* 62: 75–84.

Mosier, A., J. K. Syers, and J.R. Freney. 2004. Agriculture and the nitrogen cycle: Assessing the impacts of fertilizer use on food production and the environment. Island Press. Washington, D.C. 344pp.

Munkholm, L.J., R.J. Heck, and B.Deen. 2013. Long-term rotation and tillage effects on soil structure and crop yield. Soil and Tillage Res. 127:85 – 91.

National Agricultural Statistics Service (NASS). 2014. Iowa 2013 Annual Statistical Bulletin. NASS, U.S. Department of Agriculture, Washington, DC. http://www.nass.usda.gov/Statistics_by_State/Iowa/Publications/Annual_Statisti-cal_Bulletin/2013/index.asp.

Overton, M. 1996. Agricultural Revolution in England: The transformation of the agrarian economy 1500-1850. Cambridge University Press. p. 206. ISBN 978-0-521-56859-3.

Pittelkow, C.M., X. Liang, B.A. Linquist, K. Jan van Groenigen, J. Lee, M.E. Lundy, N. van Gestel, J. Six, R.T. Venterea, and C. van Kessel. 2014. Productivity limits and potentials of the principles of conservation agriculture. *Nature* 517: 365–368. doi:10.1038/nature13809.

Pond, G.A., G.A Sallee, C.W. Crickman. 1931. Planning Systems of Farming for the Red River Valley of Minnesota. Minnesota Agricultural Experiment Station. Retrieved from the University of Minnesota Digital Conservancy, http://hdl.handle.net/11299/163728.

Porter, P.M., D.R. Huggins, C.A. Perillo, S.R. Quiring, and R.K. Crookston. 2003. Organic and other management strategies with two- and four-year crop rotations in Minnesota. *Agron. J.* 95:233–244.

Rankin, M. Choosing between Alfalfa and Corn Silage or when do you trade Medicago for Maize? Forage Resources and UW Extension Forage Resource. http://cdp.wisc.edu/jenny/crop/choosing.pdf.

Raun, W.R., and G.V. Johnson. 1999. Improving nitrogen use efficiency for cereal production. *Agron. J.* 91:357–363.

Richards, R. 2000. Selectable traits to increase crop photosynthesis and yield of grain crops. *J. Exp. Bot.* 51: 447–458. PMID: 10938853.

Riedell, W.E., J.L. Pikul, Jr., A.A. Jaradat, and T.E. Shumacher. 2009. Crop rotation and nitrogen input effects on soil fertility, maize mineral nutrition, yield, and seed composition. *Agron. J.* 101:870–879.

Russelle, M.P., M.H. Entz, and A.J. Franzluebbers. 2007. Reconsidering integrated crop–livestock systems in North America. *Agron. J.* 99:325–334.

Shaner, D. L. 2014. Lessons learned from the history of herbicide resistance. *Weed Sci.* 62:427–431.

Sprugeon, W. I. and P. H. Grisson. 1965. Influence of cropping systems on soil properties and crop production, *Miss. Agr. Stat. Bull.* 710.

Stanger, T.F. and J. G. Lauer. 2008. Corn Grain Yield Response to Crop Rotation and Nitrogen over 35 Years. *Agron. J.* 100: 643-650. doi:10.2134/agronj2007.0280.

Stonehouse, D.B., B.D. Kay, J.K. Baffoe, and D.L. Johnston-Drury. 1998. Economic choices of crop sequences on cash cropping farms with alternative crop yield trends. *J. Soil Water Conserv.* 43:266–270.

Swanton, C.J., K.J. Mahoney, K. Chandler, and R.H. Gulden. 2008. Integrated weed management: Knowledge-based weed management systems. *Weed Sci.* 56:168–172. DOI: 10.1614/WS-07-126.1.

Tiemann L. K., A. S. Grandy, E. E. Atkinson, E. Marin-Spiotta, and M. D. McDaniel. 2015. Crop rotational diversity enhances belowground communities and functions in an agroecosystem. Ecology Letters doi: 10.1111/ele.12453.

Tilman, D., J. Fargione, B. Wolff, C. D'Antonio, A. Dobson, R. Howarth, D. Schindler, W. H. Schlesinger, D. Simberloff, and D. Swackhamer. 2001. Forecasting agriculturally driven global environmental change. Science 292(5515):281-284. http://dx.doi.org/10.1126/science.1057544.

Vanhie, V., W. Deen, J. D. Lauzon, and D. C. Hooker. 2015. Effect of increasing levels of maize (Zea mays L.) residue on no-till soybean (Glycine max Merr.) in Northern production regions: A review. *Soil and Tillage Research* 150: 201–210.

Weir, W. W. 1927. Crop Rotation. Agron. J. 19: 6: 517. doi:10.2134/agronj1927.00021962001900060006x.

West, T.O., and W.M. Post. 2002. Soil organic carbon sequestration rates by tillage and crop rotation: A global data analysis. *Soil Sci. Soc. Am. J.* 66:1930–1946. doi:10.2136/sssaj2002.1930.

Zhou, X., M.J. Helmers, H. Asbjornsen, R. Kolka, M.D. Tomer. 2014. Nutrient removal by prairie filter strips in agricultural landscapes. *J. Soil Water Conserv.* 69: 54–64.

In: Crop Rotations
Editor: Bao-Luo Ma

ISBN: 978-1-63484-496-3

Chapter 3

LEGUME-CEREAL CROP ROTATION SYSTEMS IN CHINA

Zhao-Hai Zeng[1,*], Zhan-Yuan Lu[2], Ying Jiang[1], Kai Zhang[1], Ya-Dong Yang[1] and Pei-Yi Zhao[2]

[1]College of Agriculture and Biotechnology,
China Agricultural University, Beijing, China
[2]Inner Mongolia Academy of Agricultural and Animal
Husbandry Sciences, Hohhot, Inner Mongolia, China

ABSTRACT

Crop rotation, a universal management practice with yield benefits that have been recognized and exploited for centuries, is the practice of growing a series of different crop types in the same area in sequential seasons. China has a long agricultural history, and rotation cropping systems have been practiced for millennia. Because of the large differences in climate and soil types, cropping systems from different regions in China are highly diverse. The research on the effects of legume-cereal crop rotation on crop yields has been conducted for many years. The current literatures indicate that the soybean-corn crop rotation increases grain yields and is a kind of cropping system for sustainable agriculture. Understanding changes in soil physical, chemical and biological properties is important in explaining the mechanisms involved in the direct and indirect effect of crop rotation on grain yields as well as the benefit to the environment. This chapter collates and synthesizes the available literature on different legume-cereal crop rotation studies conducted in diverse agroecological regions of China. Sustainable production and effective adoption of legume-cereal crop rotations on a site-specific basis involve many factors, including crop structure, variety selection, conservation tillage and integrated pest and weed management practices, and thus require a multi-disciplinary and multi-sectoral cooperation. Therefore, the introduction and adoption of any cropping systems and new technologies must be tested under site-specific conditions and gradually extended from demonstration plots to other regions.

[*] Corresponding author: zengzhaohai@cau.edu.cn.

1. INTRODUCTION

Crop rotation is the practice of growing a sequence of different crop species on the same land (Yates, 1954; Shen and Liu, 1983). This is in contrast to intercropping, which is the practice of growing two or more crops simultaneously on the same land (Stinner and Blair, 1990), or continuous monoculture, which is the practice of growing a single species repeatedly on the same land (Power, 1990). Crop rotations have been repeatedly shown to be effective methods of minimizing soil erosion, improving water use efficiency, and maintaining high yields (Zhu et al., 1994; Li et al., 2000). In this chapter, emphasis is placed on main types of crop rotation in China, especially for legume-cereal crop rotations. Crop rotation may include two or more crops in sequence over several growing seasons. A two-crop rotation, such as corn (*Zea mays*) and soybean (*Glycine max*) or corn and alfalfa (*Medicago sativa*) in alternate years, uses the legume to provide complementary inorganic nitrogen in the soil for the succeeding crops. For a long time, legumes have been known as the "soil building crops" because the biological, physical, and chemical properties of the soil are significantly improved when legumes are grown on it. It therefore makes good sense agriculturally to alternate them with cereals and other crops that require large amounts of N. Figure 1 illustrates some typical crop rotation practices including legumes in China.

Figure 1. (Continued).

Figure 1. Examples of crop rotation in China: (A) a corn-soybean-wheat rotation in Northeastern China; (B) a legume-cereal rotation in North China Plain; (C) a green manure with recycled straws for paddy rice in Southern China.

China has a long agricultural history; the practice of crop rotation can be dated back to antiquity, as early as the West Han Dynasty, more than 2000 years ago (206 B.C. to year 24 A.D.) when implementing fallow rotation (Han, 2000). The alternate crop-fallow cultivation method was reflected in Jia Sixie's *Qi Min Yao Shu* narration (Cao and Xian, 1985). The book also described a method of growing green manure in a crop production system as the way of raising soil fertility. It was observed that the effect of green manure can be the same as the silkworm feces and overrotten dung. Green manure crops such as mung bean (*Vigna radiata*) and red bean (*Vigna angularis*) have been widely used to rotate with millet for grain and vegetable production. On the basis of this thought, a dry farming system in northern China was formed and considered a relatively mature green manure crop rotation in Wei (220-265 A.D.), Jin (265-420 A.D.) and Northern and Southern Dynasties (420-589 A.D.). Green manure-rice rotation cropping was one of the more ancient grass field rotation patterns in China. Green manure has long been planted on major production areas of paddy rice (Zhu et al., 2012).

Crop rotation, an important cropping practice, has been considered as an effective approach to increase yield and profit and allow for sustainable production (Mitchell et al., 1991). Positive rotation effects on crop yields have been reported by many scientists for years (Bullock, 1992; Carsky et al., 1997; Fan et al., 2012; Kelley et al., 2003; Yusuf et al., 2009). This beneficial effect, in the legume-cereal rotation system, has been attributed to the availability of extra nitrogen (N) through biological nitrogen fixation and other rotation effects (Sanginga et al., 2002). Legume-cereal rotation improves the nutrients in soil. Legumes, for instance, have nodules on their roots which contain nitrogen-fixing bacteria called rhizobia (Danga et al., 2009). Grain legumes grown in rotation with annual cereal crops contribute to the total pool of nitrogen in the soil and improve the yields of cereals.

Crop rotation systems make full use of soil resources such as nutrients, water and biodiversity through inter-annual rotation of different crops based on their heterogeneous compensation and thus promote crop production and land use efficiency (Bullock, 1992). For example, farmers can use some forms of crop rotation to keep their fields under continuous production, instead of letting them fallow and reducing the need for chemical fertilizers. In addition, a general effect of crop rotation is that there is a geographic mixing of crops, which can slow down the spread of pests and diseases during growing season (Kutcher et al., 2011; Larkin, 2008; Satti, 2012). The different crops can also reduce the effects of adverse weather for the individual farmers, and a proper allocation of resources for planting and harvesting at different times allows more land to be farmed with the same amount of machinery and labor. There has been no obvious scientific basis for the sometimes 10%-25% yield increase in a crop grown in rotation, compared to monoculture of the same crop, as grain legumes such as soybean would need more nitrogen for grain filling than the amounts of nitrogen the plants actually fix through symbiotic N fixation during the growing season. The factors related to the increase are simply described as alleviation of the negative stressors in a monoculture cropping system. Explanations, due to improved nutrition; pest, pathogen, and weed stress reduction; and improved soil structure, have been found in some case studies, but direct cause-effect relationship has not been determined for the majority of cropping systems. Other benefits of rotation cropping systems include production cost advantages. Overall financial risks are more widely distributed over more diverse production of crops and/or livestock. Less reliance is placed on purchased inputs and over time crops can maintain production goals with fewer inputs. This in tandem with greater short and long term yields make rotation a powerful tool for improving agricultural systems.

The application of synthetic nitrogen fertilizers and pesticides, changes of crop rotation in China occurred rapidly over the past decades. Monoculture became popular when it appeared that chemical fertilizers and pesticides could be used as a substitute for rotation after the mid- to late-1980s (Liu et al., 2013). However, more and more input of fertilizers for crop yield improvement also brought a series of environmental problems, including air pollution, degraded water quality and increased greenhouse gas emissions (Duan et al., 2011; Liu et al., 2011). As a result, crop rotation becomes a new favorable practice in the farming systems again, especially, legume-cereal rotation system in China. Large differences in climate, geography, soil types as well as farming systems result in a wide variety of cropping systems including single, double and triple crops in one year (Figure 2). There are also diverse crop rotation patterns among various regions of China (Figure 3), such as the northeast plain of China, the North China Plain and the Loess Plateau of northern China (Zhang et al., 2014).

The northeast black soil region of China is an important grain production area. Single cropping system (one crop per year) is practiced only in the Northeast and some parts of the Northwest China. In the northeast plain of China, soybean and corn are the main grain crops. About one half of China's corn production and a third of soybean production occur on the highly productive "Black Soil" (*Mollisol*) zone in the northeastern provinces of Heilongjiang, Jilin, and Liaoning provinces and the Inner Mongolian autonomous region (Liu et al., 2013). In recent years, diverse crop rotations with conservation tillage have gained renewed interest in the semiarid farming systems, especially in the Mongolian autonomous region, likely due to an increased net income (Table 1).

As one of the most primary agricultural regions in China, the North China Plain plays an important role in securing the supply of grain. Crop production in this region accounted for 35.3% and 69.2% of China's total corn and winter wheat production, respectively, from 1996 to 2007 (Duan et al., 2011). Multiple cropping systems, such as double cropping (two crops per year) are dominant systems in this area of China. A wide range of crop rotation systems, such as soybean-corn, soybean-wheat, peanut-wheat, has been practiced over the past years (Yang et al., 2015).

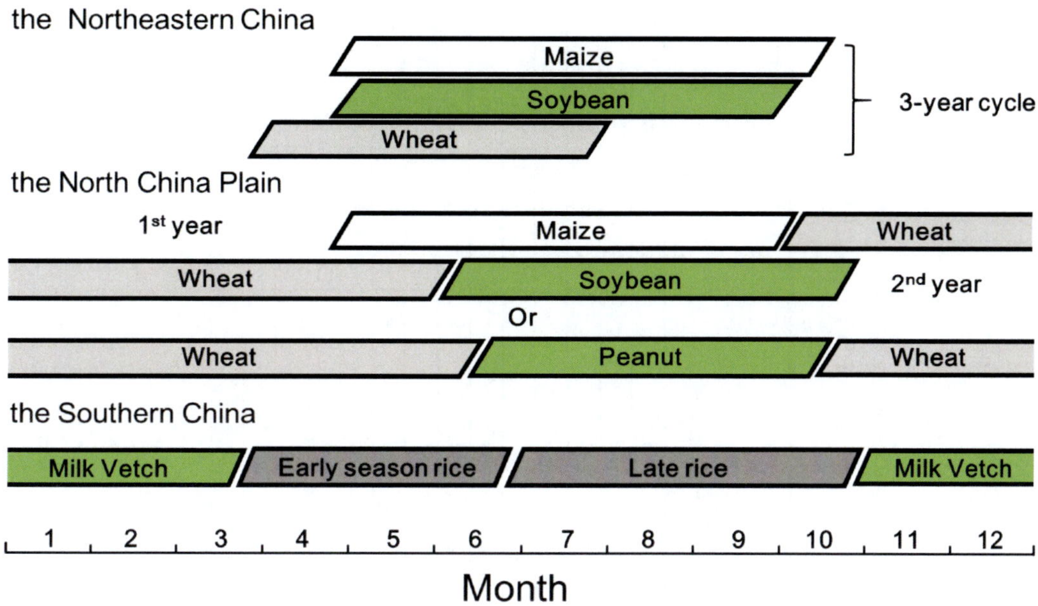

Figure 2. Main patterns of legume-cereal rotation cropping in three regions of China: (1) a 3-year corn-soybean-wheat rotation in Northeastern China; (2) a corn-wheat-soybean rotation on a 2-year cycle or peanut-wheat double cropping system in North China Plain; (3) milk vetch-rice-rice in the one-year crop rotation system in Southern China. The forward slash of the figure represented crop stubbles.

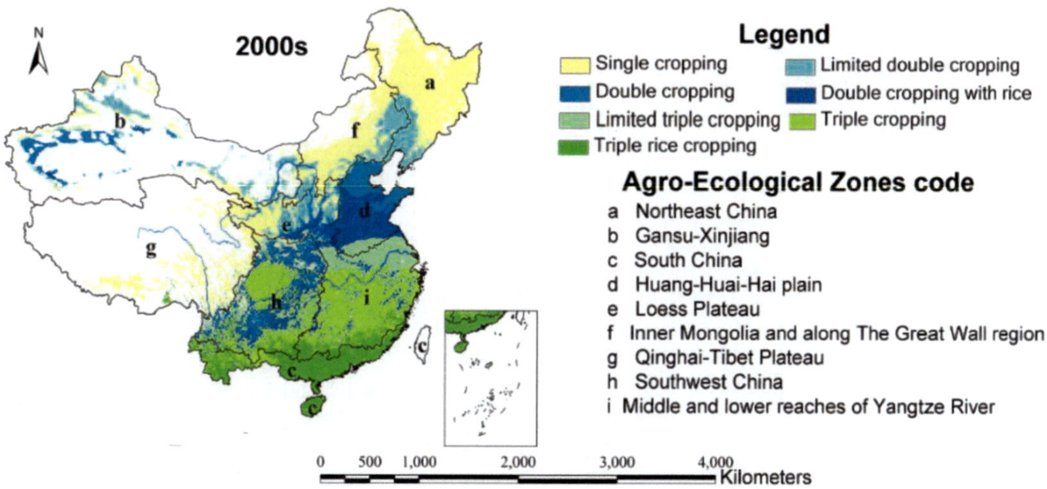

Figure 3. Distribution of multiple cropping systems and agro-ecological systems in China. Adapted from Liu et al. (2013).

Table 1. Comparison of yield, output values and net returns among different rotation patterns from a long-term field study (2004-2014) conducted in Wuchuan, Inner Mongolia, China. Results are the average of 2013 and 2014

Cropping System	Crop	Yield kg/ha	Yield increase %	Price $/kg	Output value $/ha	Input cost（$/ha）				Net income $/ha
						Seeds	Fertilizer	Tillage/ harrowing	Labour management	
Oilseed rape - Potato	Oilseed rape	1638	20.8	0.81	1317	7.2	192.6	264.8	216.7	768
	Potato	16752	20.4	0.21	3622	987.0	192.6	300.9	818.5	1540
Foxtail millet - Potato	Foxtail millet	12959	34.8	0.14	1653	46.9	192.6	264.8	216.7	1065
	Potato	16874	21.7	0.21	3627	987.0	192.6	300.9	818.5	1545
Continuous maize	Maize (silage)	18288	-	0.10	1961	110.7	192.6	264.8	433.3	1092
Common Vetch	Common Vetch	2417	-	0.79	1782	144.4	192.6	264.8	216.7	1096
Oat—Common Vetch	Oat	2757	16.4	0.39	1004	120.4	192.6	264.8	216.7	400
	Common Vetch	2596	-	0.79	2041	144.4	192.6	264.8	216.7	1355
Sunflower	Sunflower	2733	-	1.31	3522	313.0	192.6	288.9	361.1	2523
-Potato	Potato	16731	22.3	0.21	3521	987.0	192.6	300.9	818.5	1439
Continuous cropping pattern	Foxtail millet	9496	-	0.14	1253	46.9	192.6	264.8	216.7	665
	Oat	2357	-	0.39	908	120.4	192.6	264.8	216.7	246
	Oilseed rape	1357	-	0.81	1091	7.2	192.6	264.8	216.7	542
	Potato	13883	-	0.21	2990	987.0	192.6	300.9	818.5	908

- Not applicable.

The Loess Plateau is the semiarid region, one of the most important agricultural regions of China. It is situated between 34 and 40°N, and 102 and 112°E at an altitude ranging from 700 to 2,200 m. It comprises parts of Gansu, Qinghai, Ningxia, Shanxi, and Shaanxi provinces and the autonomous Inner Mongolia with a total area of 620,000 km^2, approximately 6% of China's territory (Liu et al., 2010). In this area, the dry land is used mainly for growing cereals or cash crops in rotation with legume forage or green manure for a long time, such as corn–soybean–wheat rotation in a 3-year-cycle (Qiao et al., 2014). Alfalfa, as a primary rotation legume, is widely grown for animal feed for the fast growing livestock industry (Wang et al., 2009) in the Loess Plateau of northwestern China, while also reducing soil erosion and improving soil fertility and quality (Fan et al., 2014).

For a further understanding of the legume-cereal rotation cropping systems in China, there is a strong need to review and synthesize the most recent findings from the literature. This chapter describes the current understanding and advances on legume-cereal crop rotation systems in China.

2. Legume-Cereal Crop Rotation in China

Crop rotation has been used for thousands of years because of the yield-benefits (Crookston et al., 1991). Farmers in ancient cultures as diverse as those of China, Greece, and Rome shared a common understanding on crop rotations. They learned from experience that growing the same crop year after year on the same piece of land leads to the decrease of yields, and that they could dramatically increase productivity on the land by cultivating a sequence of different crops over several seasons. They came to understand how crop rotations, combined with such practices as cover crops and green manures, enhanced soil organic matter, fertility, and tilth (Bullock,1992).

Legume-cereal crop rotation refers to a practice of growing a legume crop in one or several years (perennial legume, such as alfalfa) and growing cereal crops in the same area in sequential seasons. In China, rotating legume crops with cereal crops is a universal management practice with yield benefits that have been recognized and exploited for centuries. No one disputes the fact that rotations are beneficial. There is a large diverse in legume-cereal crop rotation systems due to the different climate, farming systems and soil properties throughout China. Describing all systems is beyond the scope of this chapter. Only brief account of the main legume-cereal crop rotation systems will be discussed.

2.1. Soybean-Corn Rotation Cropping System

Soybean-corn rotation has been recommended as a good cropping practice for soil quality and crop productivity improvement (Smith et al., 2007). Accordingly, this rotation system has been focused and further examined under China's agriculture and the environmental conditions. The following sections are focused on the changes in crop yields, soil properties, and greenhouse gas emissions as affected by this cropping system with various studies conducted in China.

2.1.1. Crop Yields

The yield increases of cereals following legumes in rotation have been reported by many studies in the past years (Hairiah et al., 2000; Shah et al., 2003; Yusuf et al., 2009). The rate of yield improvement of corn and soybean from 1978 to 2012 is showed in Figure 4. Over the past 35 years, grain yield in China has been increasing at a rate of 77.8 kg ha^{-1} $year^{-1}$ for corn and 20.1 kg ha^{-1} $year^{-1}$ for soybean, respectively. In general, the improvement in yield can be attributed to plant breeding and better agronomic practices (Kou et al., 2012). Twenty-five percent of yield enhancement can be attributed to improved agronomic practices (Tollenaar and Lee, 2006), including crop rotation such as legume-cereal crop rotation. Legumes, described as "nitrogen fixing" plants have the ability to synthesize atmospheric nitrogen in the root systems in the form of nodules, and add nitrogen to the soil after the plant is harvested. Soybean, as a typical legume, is a good preceding crop to alternate with heavier feeding plants such as corn. Compared to those in a rotation system, Xu et al. (1996) observed that soybean yield was 35.5% lower in a 3-year continuous soybean cropping and 18.6% in a 2-year monoculture, averaged across five field experiments. Similarly, Fan et al. (2012) found the soybean-corn rotation under no-till produced better yield and profitability, particularly in drier years, than the corn-corn-corn and corn-corn-soybean rotations. They also observed that the average corn yield under no-till condition was significantly greater under soybean-corn than under corn-corn-soybean rotation (9.7%) or under continuous corn (9.8%) from an 8-year study (Table 2). Therefore, grain legumes grown in rotation with cereal crops contribute to improve the yields of cereal crops in China.

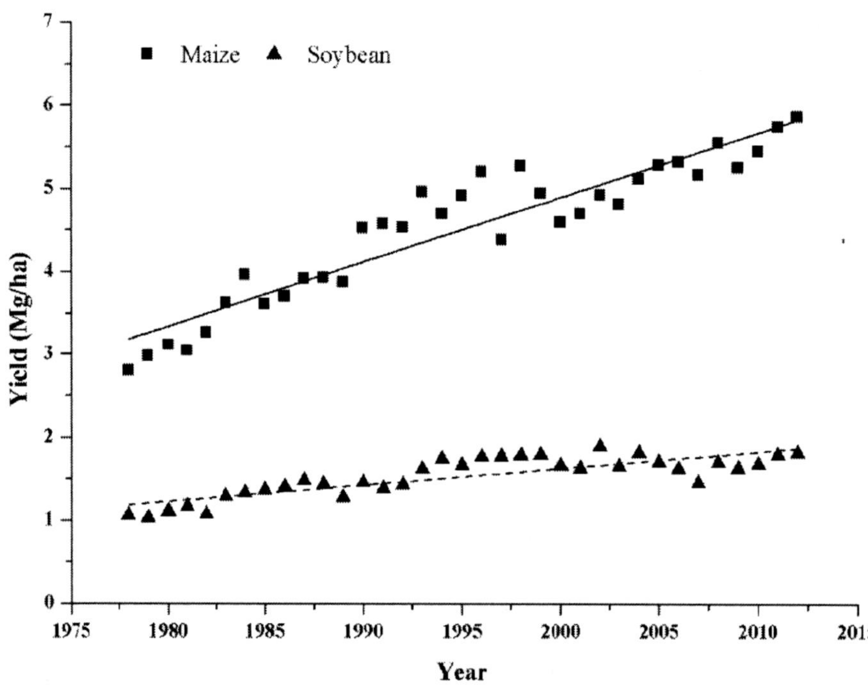

Figure 4. Trends in the average corn and soybean yields in China between 1978 and 2012, Solid line, fitted linear equation of corn yield ($y = 0.0778x - 150.67$, $R^2 = 0.882$); dashed line, fitted linear equation of soybean yield ($y = 0.0201x - 35.52$, $R^2 = 0.6898$). Source: National Bureau of Statistics of China (2015).

2.1.2. Soil Properties and Microbial Community

The benefits associated with the inclusion of a legume in a crop rotation can be partitioned into the N effect and the non-N effect (Bagayonko et al., 1992; Stevenson et al., 1996). Several studies have shown that increase in the yield of cereals following the legumes mainly due to the N contribution (Herridge et al., 1995; Lo′pez-Bellido et al., 2004; McGuire et al., 1998; Turpin et al., 2002), associated with the symbiotic N_2 fixation in the legume (the N effect). In addition, crop rotation has been considered as an effective approach to sequestrate C in soil and to enhance soil fertility (Liu, 1999; Hassink, 1995; Jarecki and Lal, 2003; Madari et al., 2005). Drinkwater et al. (1998) demonstrated that legume-based cropping systems have reduced carbon and nitrogen losses. Sun (2008) observed that organic matter and nutrient content in the rhizosphere of continuous soybean were lower than those of rotational soybean soil. However, the effect of legume-cereal crop rotation on soil carbon sequestration is also controversial. By comparing results from different studies, Kou et al. (2012) suggest that soybean–corn rotation may not be the best cropping practice for soil carbon sequestration in the rain-fed farmland Mollisol (Cumulic Hapludoll) in Northeast China, but this system could possibly sustain soil fertility for a long time, and in turn, supporting continuously high yields under intensive cropping practices with manure application.

Table 2. Average crop yields under all tillage and rotation treatments from 2002 to 2009 in Dehui County, Jilin Province, China (adapted from Fan et al., 2012)

Crop	Treatment	NT	MP	RT
		Yield (kg/ha)		
Corn	C-S	10143A	9804A	10239
	C-C-S	9248B	9377A	NA[z]
	C-C-C	9241B	9296A	NA
Soybean	C-S	2330A	2277A	2275
	C-C-S	2143B	2166A	NA

MP, moldboard plow; NT, no-tillage; RT, ridge tillage; NA, not applicable.

Yield means in the same column within the same crop followed by the same letter are not significantly different at $P = 0.05$.

Crop rotation influences the soil microbial activities. Soil microorganism populations in the rhizosphere contribute to plant health by mediating nutrient acquisition. Sun (2008) observed a significant change of rhizosphere soil microbial community in continuous cropping system. In Qingdao, China, Chen et al. (2014) illustrated the succession dynamics of soil microbial communities in a continuous cropping system. Li et al. (2006) reported that corn in rotation with soybean in alternate years (corn-soybean-corn-soybean) can alleviate the effect of nitrogen fertilizer on rhizosphere soil microbial diversity and richness, but the effect of nitrogen application significantly changed its bacterial community structure (Chan et al., 2013). Corn-corn-soybean was less affected by nitrogen fertilizer and showed relatively high stability (Zhou et al., 2013). This indicates that microbial diversity in the soil community has improved through crop rotation. In addition, it was also observed that the population of beneficial bacteria increased in gramineae and leguminous rotation system (Liu et al., 2007; Liu et al., 2009), which alleviated the barriers of continuous cropping on soil nutrient cycling, leading to the improved crop yields and soil quality. Thus, improvement in soil physical properties would allow the soil to sustain life and health conditions. Moreover, the prevention of erosion with deep roots and rich soil in the rotation cropping system could hold water and nutrients and improve soil nutrient cycling conditions and water usage (Duan et al., 2011).

In Inner Mongolia, corn, potato (*Solanum tuberosum*), soybean and wheat are the main field crops. New technologies have been developed for conservation tillage under semiarid dryland farming conditions in the past decades. Scientists and innovative farmers have developed four kinds of technological patterns, including high stubble with dibble seeding, low stubble with drill, roots with tubers and grassland improvement. These technologies have been applied in the main crop rotations, such as corn-soybean, soybean-wheat-potato, legume-corn-wheat. The improved production efficiency under crop rotation systems has been clearly shown (Table 1).

2.1.3. Greenhouse Gas Emissions

Current global climate change is evidenced by ongoing increases in ambient temperature, rising sea level, and decreasing snow cover worldwide (IPCC, 2007). These accelerated processes have been mostly attributed to progressively increasing concentration of greenhouse gases in the atmosphere, such as CH_4, CO_2 and N_2O (Hernandez-Ramirez et al., 2011). Major science efforts are now focusing on identifying and developing suitable land-management strategies to effectively mitigate or decelerate these detrimental global environmental effects (Suyker et al., 2004; Johnson et al., 2005; Russell et al., 2005; Hutchinson et al., 2007). Gan et al. (2011, 2014) contended that there are opportunities to reduce agricultural emissions by developing numerous cleaner production technologies including improved farming practices, applying reasonable amounts of fertilizers, herbicides and pesticides, and more importantly, adopting diversified cropping systems.

Much of the research on the effects of various crop types on N_2O emissions has focused on the differences between N-fixing legumes and non-leguminous crops (Pelster et al., 2011). Even using the same rotation patterns in different studies however, results have varied. For example, in corn-soybean rotations, the reported greenhouse gas emissions ranged from higher cumulative N_2O emissions from corn (Bavin et al., 2009; MacKenzie et al., 1997), to no difference in emissions to higher emissions from soybeans (Parkin et al., 2006; Sey et al., 2008). In general however, it seems that the presence of legumes increased N_2O emissions, likely because of greater N release from root exudates during the growing season, as well as additional N inputs via decomposition of root nodules and crop residue in the autumn (Rochette et al., 2004; Rochette and Janzen, 2005). In China, the effects of crop rotation on the greenhouse gas emissions have been given extensive studies in recent years. It was observed that N_2O fluxes started to increase in May and lasted until October in the corn and soybean phases with significantly lower emissions in the wheat phase of the corn- soybean-wheat rotation system. Qiao et al. (2014) reported that with the rainfed dominant corn–soybean–wheat rotation system in Northeastern China, there is a potential for improved soil N management to reduce N_2O emissions with addition of P fertilizer. Crop type would affect the C: N ratio of the residue, with typically lower C: N ratio in legumes than the non-legume crops, and thus C: N ratio of the crop residue tended to be negatively correlated with N_2O emissions (Baggs et al., 2000).

2.2. Soybean – Wheat/Rice/Oat Rotation Cropping System

Wheat (*Triticum aestivum* L.) is cultivated widely in the Loess Plateau of northwest China, North China Plain and some parts of the Northeast Plain. Winter wheat and spring wheat are grown in the regions according to the different climatic conditions and soil properties. Thus, various rotation systems with wheat are usually carried out in the different growing areas. For example, in the North China Plain, summer corn is the main crop in rotation with winter wheat, while in the Loess Plateau of Northwest China, legume-wheat rotation is more popular. Spring wheat production is generally located in the north of the Great Wall and the northeast China. In these areas, spring wheat is often grown in rotation with soybean, or other grain legumes. Several studies conducted in these regions suggest that enhanced UV-B radiation may lead to a decrease in soil respiration and in N_2O emissions, while straw incorporation may increase soil respiration, and the combined treatment may have no significant influence on soil respiration and N_2O emissions from soybean-winter wheat rotation systems (Zhao et al., 2015). Yang et al. (2014) observed that the benefits of crop rotation with soybean on wheat grain yields became more evident with time, in the second and third years, the grain yields of wheat amended with 108 kg N/ha fertilizer, following preceding soybean crop, reached 4871 and 5089 kg/ha. These yields were 21% and 12% higher than the highest yields of wheat under a fallow-winter wheat rotation.

Soybean and some legume green manures are also considered as the rotation crop or catch crop planted in the double-rice crop production regions. Considering the long-term benefits that legumes can have on N fertilizer savings, improved C and N cycling and soil fertility (Drinkwater et al., 1998), Zhao et al. (2015) concluded that substituting an N-fixing legume for winter wheat is a feasible method for mitigating N pollution from heavily-fertilized rice/wheat cropping system in the Taihu Lake Plain of southeast China.

Soybean is also considered as an appropriate rotation crop for oat (*Avena sativa*) cultivated in the cool and mountain regions in China. Zang (2014) found that under field conditions, oat plant N content increased by 20% in rotation with soybean, compared to that under continuous oat. For example, the N transfer from soybean to oat was about 1.1-1.9 times more than that from the oat to oat field. In continuous oat system, about 5.6% of the total N in oat was derived from rhizodeposition (NdfR) of preceding oat, 3.3% from returning straws, and root residues have little contribution (Zang, 2014). In comparison, about 10% of the total N in oat originated from the decomposition of roots and straw residues of the preceding soybean.

2.3. The Other Grain Legume-Cereal Rotation Cropping System

There are other grain legume-cereal crop rotation systems, such as peanut (*Arachis hypogaea* L.), mungbean (*Vigna radiate* L.) and faba bean (*Vicia faba* L.), in rotation with cereal crops. These minor legume crops have a long production history in China. They are extensive in distribution, plentiful in germplasm resources and varied in cultivation pattern (Lang et al., 1993). Consequently, these have offered a suitable opportunity to practice the beneficial legume- cereal crop rotations in the region where peanut, mungbean and faba bean are cultivated.

As shown in Figure 5, peanut is grown on nearly 11.8 million ha in Asia with the total production of 35 850 tonnes (7 863 tonnes from China) and an average yield of 2556 kg/ha in 2013 (FAO, 2014). China, India, Nigeria, USA and Myanmar are the major peanut growing countries worldwide. In Asia, peanut accounted for 50% of global production area and 64% of global production. Peanut-corn is the recommended crop rotation system in the North China Plain region. Few studies have been conducted to examine the effects of peanut rotation with different cereal crops on crop yields in China. Jeranyama et al. (2007) showed that corn grain yields increased by 0.7 t/ha when it was following peanut compared with continuous corn, when no fertilizer was applied to both cropping systems. Corn yield was more responsive to fertilizer-N after peanut as preceding crop than continuous corn. Fertilizer requirement by corn were also reduced by up to 64 kg N/ha when corn followed peanut (Jeranyama et al., 2007). These results indicated that a suitable legume-cereal crop rotation pattern should be chosen according to different soil properties and environmental conditions in China. Many studies have been conducted to show the impacts of peanut-cereal rotation on the soil properties, potential productivity, water requirement, as well as soil microbial community (Aulakh and Pasricha, 1991; Chauhan, 2010; Gil et al., 2008; Ibañez et al., 2014; Siri-Prieto et al., 2007). Jin et al. (2007) observed that the suppressing effect of peanut proceeding crop on winter wheat yield was mainly attributable to the unfavorable soil moisture conditions after growing peanut in summer. This is inconsistent with the consensus that peanut-cereal rotations improve cereal crop yields. Therefore, more research is needed to determine whether current peanut-cereal crop rotations are sustainable for peanut production over a long-term period.

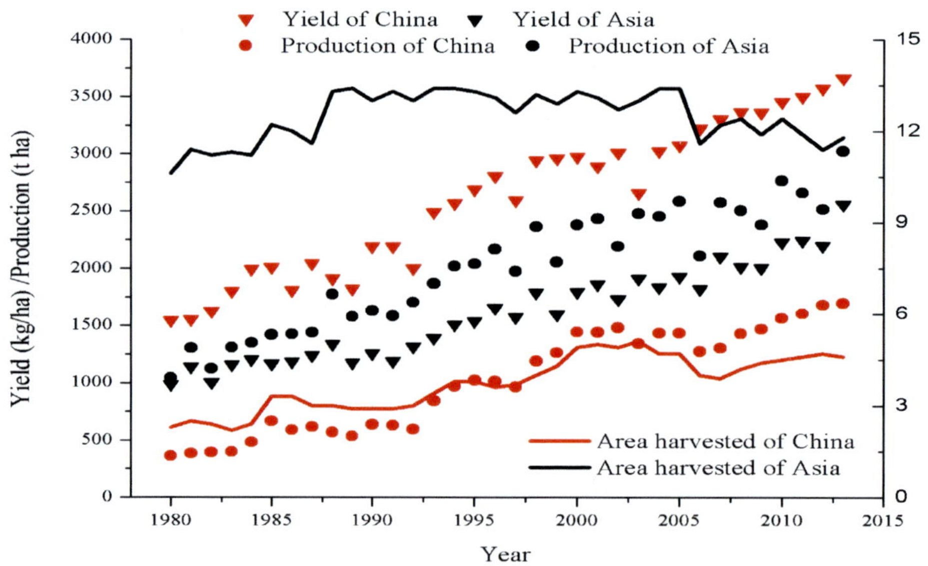

Figure 5. Changes in the harvested area, production and yield (with shell) of groundnut in Asia and China (FAO, 2015).

China is also by far the world's largest producer of faba bean. Cultivation of the crop can be traced back to the ancient times. The crop is now widely distributed throughout the country — especially along the Yangtze River region where close to 90% of faba bean production is concentrated. Faba bean is an important winter and spring food legume used in a wide range of traditional dishes, or for feed and for green manure (Lang et al., 1988). In the 1990s, the planting patterns of faba bean in different rotation patterns have been examined with different cropping frequencies, in triple, double and single-crop cropping areas in order to avoid the disadvantage of continuous monoculture cropping (Lang et al., 1993). Excluding faba bean grown for green manure, the current planted area is over one million ha, and production is close to 2 million metric tonnes with average yields of about 1700 kg ha^{-1}.

Mungbean is also an important food and cash crop throughout China. This crop grows over a wide range of agro-climatic zones in the country (Zang et al., 2015). In 2012, it is grown on nearly 694,000 ha with a total production of 867,000 tonnes and an average yield of 1248 kg/ha (NBSC, 2013). The previous research on the mungbean-cereal crop rotation in China has not clearly demonstrated the beneficial effect of mungbean on the yield of succeeding cereal crop and the mechanism of yield improvement by mungbean in the rotation. However, in the southeast countries of Asia, such as India and Vietnam (Xuan et al., 2012; Ranjan et al., 2014), some studies have shown a positive effect of mungbean cereal crop rotation on the cereal crop yield. Therefore, more research is needed to illustrate its role in sustainable production development.

2.4. Forage Legume/Green Manures -Cereal Crop Rotation System

Forage legume - cereal crop rotation system options with different characteristics and possible niches are available to farmers (Schulz et al., 2001). Green manure and fodder legumes, such as *Mucuna pruriens*, *Crotalaria* spp. or *Stylosanthes* spp., have primarily been selected for their ability to contribute large quantities of residual biomass and N to the soil and/or to livestock as feed (Ojiem et al., 2014). The green manure and forage legume technologies have been adopted by farmers in China for years. The common legumes used as green manure include alfalfa (*Medicago sativa* L.), Chinese milk vetch (*Astragalus membranaceus*) and common vetch. Alfalfa has long been recognized as a source of N for the subsequent crops in a rotation (Kelner et al., 1997).

The beneficial effect of preceding alfalfa on the yield of a non-legume crop has been shown in many studies (Basso et al., 2005; Jia et al., 2009; Wang et al., 2008). In general, soil water in dry soil layers can be quickly replenished, and crop yields in the alfalfa-crop systems are equal to those of the conventional system. Wang et al. (2008) pointed out that the N replacement value of alfalfa for succeeding corn crop is still underestimated or neglected by many farmers, resulting in enrichment of nitrate in the groundwater (Peterson et al., 1991). Zhang et al. (2009) also suggested that proper management of alfalfa fields can maintain or even improve chemical and physical quality of converted reed meadows soils. As a winter cover crop, Chinese milk vetch is common crop in paddy soils for rice production in southern China, as a cover crop followed by rice, rice grain yield and N yield are increased without chemical N applied during the double crop rice seasons. Zhu et al. (2012) observed that soil microbial biomass N was highest for milk vetch as cover crop (CMV,15.4 mg/kg), followed by ryegrass (RG, 11.3 mg/kg) and fallow without weed (CK, 6.1 mg/kg), and grain yield and total N yield of early rice were 0.6 Mg/ha and 11 kg/ha higher for CMV, respectively, and 1.0 Mg/ha and 20 kg/ha lower for RG, as compared with the continuous rice monoculture, averaged across years.

Conclusion and Future Perspectives

Crop rotation is a critical feature of all sustainable cropping systems because it provides the principal mechanism for building healthy soils, a major way to control pests along with a variety of other benefits. As being mentioned in the book called *Qi Min Yao Shu*, "the field growing cereal should be changed crop in the next growth period." Crop rotation in conjunction with conservation tillage in the semiarid dryland farming conditions of China has made significant contributions to agricultural and pasturing production, increasing rural residents' income, and improving the ecological environment.

Considering the long-term benefits that legumes-cereals can have on crop yield, improvement of soil properties, and environmental quality, an international emphasis for the sustainable agriculture should be continued to focus on the potential role of introducing legumes into cereal cropping systems in the future agriculture of the whole world. Improving the ecological and efficient cultivation of maize-legume crops on a site-specific basis involves many factors, including crop structure, variety selection, cultivation techniques, fertilization, plant protection, irrigation and conservation tillage, and requires a multi-disciplinary, multi-sectoral cooperation. In any specific regions, we should adhere to the principle of testing, demonstration and promotion as the three steps give a balanced consideration of economic, ecological and social benefits. Over-emphasis on one type of crop rotation such as corn-soybean rotation could lead to a reduction in the potential benefit in the long run, wasting government investment and loss of farmers' confidence on the potential benefits of crop rotation, and thus reducing large-scale adoption of crop rotation systems as a whole. Therefore, the introduction and adoption of any cropping systems and new technologies must be tested under site-specific conditions and gradually extended from demonstration plots to other regions. In view of the characteristics and demand of China's agriculture, research priorities include the following: (1) optimize and improve legume-cereal systems, taking into consideration of diverse soils, climate, crops, and cropping systems; (2) establish legume-cereal research networks and links involving multidisciplinary teams; (3) identify suitable rotational patterns for the small land holders and diverse farming areas; and (4) link food security with environmental protection, sustainable soil management, and climate change.

References

Aulakh, M.S., and N.S. Pasricha. 1991. Transformations of residual fertilizer P in a semi-arid tropical soil under eight-year peanut-wheat rotation. Fertilizer Res. 29: 145-152.

Bagayonko, M., S.C. Mason, and R.J. Sabata. 1992. Effect of previous cropping systems on soil nitrogen and grain sorghum yield. Agron. J. 84: 863-867.

Baggs, E.M., R.M. Rees, K.A. Smith, and A.J.A. Vinten. 2000. Nitrous oxide emission from soils after incorporating crop residues. Soil Use Manag. 16: 82-87.

Basso, B., and J.T. Ritchie. 2005. Impact of compost, manure and inorganic fertilizer on nitrate leaching and yield for a 6-year corn–alfalfa rotation in Michigan. Agric. Ecosys. Environ. 108: 329-341.

Bavin, T.K., T.J. Griffis, J.M. Baker, and R.T. Venterea. 2009. Impact of reduced tillage and cover cropping on the greenhouse gas budget of a corn/soybean rotation ecosystem. Agric. Ecosys. Environ. 134: 234-242.

Cao, L.G., and J.S. Xian. 1985. Combined on the dry land use and improve soil fertility of historical experience. Agric. History China 4: 63-77.

Chan, Y.K., W.A. McCormick, and B.L. Ma. 2013. Fertilizer effects on soil archaeal abundance at two Ontario experimental farms during three consecutive cropping seasons. Appl. Soil Ecol. 68: 26-35.

Chauhan, Y.S. 2010. Potential productivity and water requirement s of maize-peanut rotations in Australian semi-arid tropical environments - A crop simulation study. Agric. Water Manag. 97: 457-464.

Chen, M., L. X. Li, Q. Yang, X. Chi, L. Pan, N. Chen, Z. Yang, T. Wang, M. Wang, and S. Yu. 2014. Dynamic succession of soil bacterial community during continuous cropping of peanut (*Arachis hypogaea* L.). PLoS One 9(7): e101355.

Crookston, R.K., J.E. Kurle, P.J. Copeland, J.H. Ford, and W.E. Lueschen. 1991. Rotational cropping sequence affects yield of corn and soybean. Agron. J. 83: 108-113.

Danga, B.O., J.P. Ouma, I.I.C. Wakindiki, and B.T. Asher. 2009. Legume–wheat rotation effects on residual soil moisture, nitrogen and wheat yield in tropical regions. Adv. Agron. 101: 315-349.

Drinkwater, L.E., P. Wagoner, and M. Sarrantonio. 1998. Legume-based cropping systems have reduced carbon and nitrogen losses. Nature 396: 262-265.

Duan, X., Y. Xie, T. Ou, and H. Lu. 2011. Effects of soil erosion on long-term soil productivity in the black soil region of northeastern China. Catena 87: 268-275.

Fan, R., X. Zhang, A. Liang, X. Shi, X. Chen, K. Bao, X. Yang, and S. Jia. 2012. Tillage and rotation effects on crop yield and profitability on a Black soil in northeast China. Can. J. Soil Sci. 92: 463-470.

Fan, J., Y. Gao, Q. Wang, S.S. Malhi, and Y. Li. 2014. Mulching effects on water storage in soil and its depletion by alfalfa in the Loess Plateau of northwestern China. Agric. Water Manage. 138: 10-16.

Gan, Y., B.C. Liang, X. Wang, and B. Mcconkey. 2011. Lowering carbon footprint of durum wheat by diversifying cropping systems. Field Crops Res. 122: 199-206.

Gan, Y.T., B.C. Liang, Q. Chai, R.L. Lemke, C.A. Campbell, and R.P. Zentner. 2014. Improving farming practices reduces the carbon footprint of spring wheat production. Nature Commun. 11(5): 1-13.

Gil, S.V., J.M. Meriles, R. Haro, C. Casini, and G.J. March. 2008. Crop rotation and tillage systems as a proactive strategy in the control of peanut fungal soilborne diseases. BioControl 53: 685-698.

Hairiah, K., M. Van Noordwijk, and G. Cadisch. 2000. Crop yield, C and N balance of three types of cropping systems on an Ultisol in Northern Lampung. Neth. J. Agric. Sci. 48: 3-17.

Han, M.L. 2000. A brief view about the ancient system of crops in China. Agric. History China 3: 91-64.

Hassink, J. 1995. Decomposition rate constants of size and density fractions of soil organic matter. Soil Sci. Soc. Am. J. 59: 1631-1635.

Hernandez-Ramirez, G., J.L. Hatfield, T.B. Parkin, T.J. Sauer, and J.H. Prueger. 2011. Carbon dioxide fluxes in corn-soybean rotation in the midwestern US: Inter- and intra-annual variations, and biophysical controls. Agric. For. Meteorol. 151: 1831–1842.

Herridge, D.F., H. Marcellos, W.L. Felton, G.L. Turner, and M.B. Peoples. 1995. Chickpea increase soil-N fertility in cereal systems through nitrate sparing and nitrogen fixation. Soil Biol. Biochem. 27: 545-551.

Hutchinson, J.J., C.A. Campbell, and R.L. Desjardins. 2007. Some perspectives on carbon sequestration in agriculture. Agric. For. Meteorol. 142: 288-302.

Ibañez, F., M.E. Arroyo, J. Angelini, M.L. Tonelli, V. Muñoz, L. Ludueña, L. Valetti, and A. Fabra. 2014. Non-rhizobial peanut nodule bacteria promote maize (*Zea mays* L.) and peanut (*Arachis hypogaea* L.) growth in a simulated crop rotation system. Appl. Soil Ecol. 84: 208-212.

IPCC. 2007. Climate Change: The physical science basis. In: Solomon, S., D. Qin, M. Manning, Z. Chen, M. Marquis, K.B. Averyt, M. Tignor, and H.L. Miller (eds.). Contribution of working group I to the fourth assessment report of the intergovernmental panel on climate change. Cambridge University Press, Cambridge, U.K.

Jarecki, M.K., and R. Lal. 2003. Crop management for soil carbon sequestration. Crit. Rev. Plant Sci. 22: 471-502.

Jeranyama, P., S.R. Waddington, O.B. Hesterman, and R.R. Harwood. 2007. Nitrogen effects on corn yield following groundnut in rotation on smallholder farms in sub-humid Zimbabwe. African J. Biotechnol. 6: 1503-1508.

Jia, Y., F. Li, Z. Zhang, X. Wang, R. Guo, and K.H.M. Siddique. 2009. Productivity and water use of alfalfa and subsequent crops in the semiarid Loess Plateau with different stand ages of alfalfa and crop sequences. *Field Crops Res.* 114: 58-65.

Jin, K., W.M. Cornelis, W. Schiettecatte, J. Lu, Y. Yao, H. Wu, D. Gabriels, S. de Neve, D. Cai, J. Jin, and R. Hartmann. 2007. Effects of different management practices on the soil-water balance and crop yield for improved dryland farming in the Chinese Loess Plateau. Soil Till. Res. 96: 131-144.

Johnson, J.M.F., D.C. Reicosky, R.R. Allmaras, T.J. Sauer, R.T. Venterea, and C.J. Dell. 2005. Greenhouse gas contributions and mitigation potential of agriculture in the central USA. Soil Till. Res. 83: 73-94.

Kelley, K.W., J.H. Long Jr., and T.C. Todd. 2003. Long-term crop rotations affect soybean yield, seed weight, and soil chemical properties. Field Crops Res. 83: 41-50.

Kelner, D.J., J.K. Vessey, and M.H. Entz. 1997. The nitrogen dynamics of 1-, 2- and 3-year stands of alfalfa in a cropping system. Agric. Ecosyst. Environ. 64: 1-10.

Kou, T., P. Zhu, S. Huang, X. Peng, Z. Song, A. Den, H. Gao, C. Peng, and W. Zhang. 2012. Effects of long-term cropping regimes on soil carbon sequestration and aggregate composition in rainfed farmland of Northeast China. Soil Till. Res. 118: 132-138.

Kutcher, H.R., A.M. Johnston, K.L. Bailey, and S.S. Malhi. 2011. Managing crop losses from plant diseases with foliar fungicides, rotation and tillage on a Black Chernozem in Saskatchewan, Canada. Field Crops Res. 124: 205-212.

Lang, L., and Z. Zheng. 1988. Production of faba bean and pea in China. World crops: Cool season food legumes. Current Plant Sci. Biotechnol. Agric. 5:1135-1152.

Lang, L., Z. Yu, Z. Zheng, M. Xu, and H. Ying. 1993. Faba Bean in China: State-of-the-art review. International Center for Agricultural Research in the Dry Areas (CARDA) press, Aleppo, Syria, pp: vii, 84-88.

Larkin, R.P. 2008. Relative effects of biological amendments and crop rotations on soil microbial communities and soilborne diseases of potato. Soil Biol. Biochem. 40: 1341-1351.

Li, F., S. Zhao, and G.T. Geballe. 2000. Water use patterns and agronomic performance for some cropping systems with and without fallow crops in a semi-arid environment of northwest China. Agric. Ecosyst. Environ. 79: 129-142.

Li, C., X. Li, and J. Wang. 2006. Effect of soybean continuous cropping on bulk and rhizosphere soil microbial community function. Acta Ecologica Sinica 26: 1144-1150.

Liu, X. J., Y. L. Xu, C. J. Li, and Q. Meng. 2007. Effect of soybean rotation system on the bacterial physiological groups. Soybean Sci. 26: 723-727.

Liu, J., Y. Xu, G. Lv, C. Li, Z. Zhao, and W. Wei. 2009. Black soil region fusarium population structure and quantity in soybean rhizosphere of different rotation system. Soybean Sci. 28: 97-102.

Liu, L., X. Xu, D. Zhuang, X. Chen, and S. Li. 2013. Changes in the potential multiple cropping system in response to climate change in China from 1960-2010. *PLoS One* 8: 1-12.

Liu, S., X. Mo, Z. Lin, Y. Xu, J. Ji, G. Wen, and J. Richey. 2010. Crop yield responses to climate change in the Huang-Huai-Hai Plain of China. Agric. Water Manage. 97: 1195-1209.

Lo´pez-Bellido, R.J., L. Lo´pez-Bellido, J. E. Castillo, and F. J. Lo´pez-Bellido. 2004. Chickpea response to tillage and soil residual nitrogen in a continuous rotation with wheat II. Soil nitrate, N uptake and influence on wheat yield. Field Crops Res. 88: 201-210.

MacKenzie, A.F., M.X. Fan, and F. Cadrin. 1997. Nitrous oxide emission as affected by tillage, corn-soybean-alfalfa rotations and nitrogen fertilization. Can. J. Soil Sci. 77: 145-152.

McGuire, A.M., D.C. Bryant, and R.F. Denison. 1998. Wheat yields, nitrogen uptake, and soil moisture following winter legume cover crop vs Fallow. Agron. J. 90: 404-410.

Mitchell, C.C., R.L. Westerman, J.R. Brown, and T.R. Peck. 1991. Overview of long-term agronomic research. Agron. J. 83: 24-29.

Ojiem, J.O., A.C. Franke, B. Vanlauwe, N. de Ridder, and K.E. Giller. 2014. Benefits of legume–maize rotations: Assessing the impact of diversity on the productivity of smallholders in Western Kenya. Field Crops Res. 168: 75-85.

Parkin, T.B., and T.C. Kaspar. 2006. Nitrous oxide emissions from corn-soybean systems in the midwest. J. Environ. Qual. 35: 1496-1506.

Pelster, D.E., F. Larouche, P. Rochette, M.H. Chantigny, S. Allaire, and D.A. Angers. 2011. Nitrogen fertilization but not soil tillage affects nitrous oxide emissions from a clay loam soil under a corn-soybean rotation. Soil Till. Res. 115-116: 16-26.

Peterson, T.A., and M.P. Russelle. 1991. Alfalfa and the nitrogen cycle in the Corn Belt. J. Soil Water Conserv. 44: 240-243.

Qiao, Y., S. Miao, X. Han, M. You, X. Zhu, and W.R. Horwath. 2014. The effect of fertilizer practices on N balance and global warming potential of corn-soybean-wheat rotations in Northeastern China. Field Crops Res. 161: 98-106.

Ranjan, L., S. Sharma, M. Idris, A.K. Singh, S.S. Singh, B.P. Bhatt, Y. Saharawat, E. Humphreys, and J.K. Ladha. 2014. Integration of conservation agriculture with best management practices for improving system performance of the rice-wheat rotation in the Eastern Indo-Gangetic Plains of India. Agric. Ecosyst. Environ. 195: 68-82.

Rochette, P., D.A. Angers, G. Belanger, M.H. Chantigny, D. Prevost, and G. Levesque. 2004. Emissions of N_2O from alfalfa and soybean crops in eastern Canada. Soil Sci. Soc. Am. J. 68: 493-506.

Rochette, P., and H.H. Janzen. 2005. Towards a revised coefficient for estimating N_2O emissions from legumes. Nutr. Cycl. Agroecosyst. 73: 171-179.

Russell, C., and I. Fillery.1996. In situ ^{15}N labelling of lupin below-ground biomass. Aust. J. Agric. Res. 47: 1035-1046.

Sanginga, N., K. Dashiell, J. Diels, B. Vanlauwe, O. Lyasse, R.J. Carsky, S. Tarawali, B. Asafo-Adjei, A. Menkir, S. Schulz, B.B. Singh, D. Chikoye, D. Keatinge, and O. Rodomiro. 2003. Sustainable resource management coupled to resilient germplasm to provide new intensive cereal-grain-legume-livestock systems in the dry savanna. Agric. Ecosyst. Environ. 100: 305-314.

Satti, A. 2012. Combating agricultural pests and diseases through cultural means. The Experiment 5: 304-314.

Schulz, S., R.J. Carsky, and S. Tarawali. 2001. Herbaceous legumes: The panacea for west african soil fertility problems? Sustaining soil fertility in West Africa, vol. 58. SSSA Special Publication, pp. 179-196.

Sey, B.K., J.K. Whalen, E.G. Gregorich, P. Rochette, and R.I. Cu. 2008. Carbon dioxide and nitrous oxide content in soils under corn and soybean. Soil Sci. Soc. Am. J. 72: 931-938.

Shaha, Z., S.H. Shahb, M.B. Peoplesc, G.D. Schwenked, and D.F. Herridge. 2003. Crop residue and fertiliser N effects on nitrogen fixation and yields of legume-cereal rotations and soil organic fertility. Field Crops Res. 83: 1-11.

Shen, X., and X. Liu. 1983. Multiple cropping. Beijing: China agriculture press, 2-3.

Siri-Prieto, G., D.W. Reeves, and R.L. Raper. 2007. Tillage systems for a cotton–peanut rotation with winter-annual grazing: Impacts on soil carbon, nitrogen and physical properties. Soil Till. Res. 96: 260-268.

Smith, P., D. Martino, Z. Cai, D. Gwary, H. Janzen, P. Kumar, B. McCarl, S. Ogle, F. O'Mara, C. Rice, B. Scholes, and O. Sirotenko. 2007. Agriculture. In climate change 2007: mitigation. In: Metz, B., Davidson, O.R., Bosch, P.R., Dave, R., Meyer, L.A. Contribution of Working Group III to the Fourth Assessment Report of the Intergovernmental Panel on Climate Change. Cambridge University Press, Cambridge, United Kingdom and New York, NY, USA, pp. 506-507.

Stevenson, F.C., and C. van Kessel. 1996. A landscape-scale assessment of the nitrogen and non-nitrogen rotation benefits of pea. Soil Sci. Soc. Am. J. 60: 1797-1805.

Stinner, B.R., and J.M. Blair. 1990. Ecological and agronomic characteristics of innovative cropping systems. pp123. *In* Edwards, C.A. (ed.) Sustainable Agricultural Systems. *Soil and Water Conservation Society,* Ankeny, IO.

Suyker, A.E., S.B. Verma, G.G. Burba, T.J. Arkebauer, D.T. Walters, and K.G. Hubbard. 2004. Growing season carbon dioxide exchange in irrigated and rainfed corn. Agric. For. Meteorol. 24: 1-13.

Sun, L. 2008. Effect of soybean continuous cropping on the rhizosphere soil nutrition. Chinese Agric. Sci. Bull. 24: 266-269.

Tollenaar, M., and E. Lee. 2006. Dissection of physiological processes underlying grain yield in corn by examining genetic improvement and heterosis. Maydica 51: 399-408.

Turpin, J.E., D.F. Herridge, and M.J. Robertson. 2002. Nitrogen fixation and soil nitrate interactions in field-grown chickpea (*Cicer arietinum*) and faba bean (*Vicia faba*). Aust. J. Agric. Res. 53: 599-608.

Wang, C., B.L. Ma, X., Yan, J. Han, Y. Guo, Y. Wang, and P. Li. 2009. Yields of alfalfa varieties with different fall-dormancy levels in a temperate environment. Agron. J. 101: 1146-1152.

Wang, X., G. Sun, Y. Jia, F. Li, and J. Xu. 2008. Crop yield and soil water restoration on 9-year-old alfalfa pasture in the semiarid Loess Plateau of China. Agric. Water Manage. 95: 190-198.

Xu, Y., A. Liu, X. Han, and G. Wang. 1996. Study of the influence on yield and quality of soybean by different rotation system in chernozemic soil area of North East of China. Soybean Sci. 15: 48-55.

Xuan, D.T., V.T. Guong, A. Rosling, S. Alström, B. Chai, and N. Högberg. 2012. Different crop rotation systems as drivers of change in soil bacterial community structure and yield of rice, *Oryza sativa*. Biol. Fertil. Soils 48: 217-225.

Yang, N., Z. Wang,Y. Gao, H. Zhao, K. Li, F. Li, and S.S. Malhi. 2014. Effects of planting soybean in summer fallow on wheat grain yield, total N and Zn in grain and available N and Zn in soil on the Loess Plateau of China. Eur. J. Agron. 58: 63-72.

Yang, X., Y. Chen, S. Pacenka, W. Gao, L. Ma, G. Wang, P. Yan, P. Sui, and T.S. Steenhuis. 2015. Effect of diversified crop rotations on groundwater levels and crop water productivity in the North China Plain. J. Hydrol. 522: 428-438.

Yates, F. 1954. The analysis of experiments containing different crops rotations. Biometrics 10: 324-346.

Yusuf, A.A., R.C. Abaidoo, E.N.O. Iwuafor, O.O. Olufajo, and N. Sanginga. 2009. Rotation effects of grain legumes and fallow on maize yield, microbial biomass and chemical properties of an Alfisol in the Nigerian savanna. Agric. Ecosyst. Environ. 129: 325-331.

Zang, H., X. Yang, X. Feng, X. Qian, Y. Hu, C. Ren, and Z. Zeng. 2015. Rhizodeposition of nitrogen and carbon by mungbean (*Vigna radiata* L.) and its contribution to intercropped oats (*Avena nuda* L.). PLoS One 10(3): e0121132. doi:10.1371/ journal. pone.0121132

Zang, H. 2014. Nitrogen rhizodeposition and transfer in mungbean-oat inter-cropping system and soybean-oat rotation system. China Agriculture University. MS thesis, pp 51.

Zhang, T., Y. Wang, X. Wang, Q. Wang, and J. Han. 2009. Organic carbon and nitrogen stocks in reed meadow soils converted to alfalfa fields. Soil Till. Res. 105: 143-148.

Zhang, H., R. Lal, X. Zhao, J. Xue, and F. Chen. 2014. Opportunities and challenges of soil carbon sequestration by conservation agriculture in China. Adv. Agron. 124: 1-36.

Zhao, X., S. Wang, and G. Xing. 2015. Maintaining rice yield and reducing N pollution by substituting winter legume for wheat in a heavily-fertilized rice-based cropping system of southeast China. Agric. Ecosyst. Environ. 202: 79-89.

Zhu, B., L. Yi, L. Guo, G. Chen, Y. Hu, H. Tang, C. Xiao, X. Xiao, G. Yang S.N. Acharya, and Z. Zeng. 2012. Performance of two winter cover crops and their impacts on soil properties and two subsequent rice crops in Dongting Lake Plain, Hunan, China. Soil Till. Res. 124: 95-101.

In: Crop Rotations
Editor: Bao-Luo Ma

ISBN: 978-1-63484-496-3

Chapter 4

ROTATION OF PEANUT AND COTTON WITH BAHIAGRASS TO IMPROVE SOIL QUALITY AND CROP PRODUCTIVITY

Duli Zhao[1,*], David Wright[2], Jim Marois[2] and Diane Rowland[3]

[1]USDA-ARS Sugarcane Field Station, Canal Point, FL, US
[2]North Florida Research and Education Center, University of Florida, Quincy, FL, US
[3]Agronomy Department, University of Florida, Gainesville, FL, US

ABSTRACT

Perennial grasses are important for the U.S. livestock industry and have been widely recognized as a key in conserving soils and improving agricultural sustainability. Most perennial grasses have a wide range of tolerance to soil fertility, moisture, pH and other environmental conditions making them a good choice for various uses and marginal soils. Research has shown the benefits of perennial grasses to following agronomic crops. However, perennial grasses are seldom reintroduced into the rotations once fields are taken out for row crop production. Bahiagrass (*Paspalum notatum*), a perennial grass, has been used to rotate with other row crops in the southeastern USA. Research has focused on a short-term rotation system that keeps bahiagrass in the rotation with row crops and has been found to be economically and environmentally advantageous. This system has been found to increase soil organic matter content and water infiltration along with improving growth, yields, and profits of peanut (*Arachis hypogea* L.) and cotton (*Gossypium hirsutum* L.). One of the main factors contributing to the improved profit potential of the sod-based rotation is reduction in input costs compared to the conventional rotation. The system incorporates a short term bahiagrass-bahiagrass-peanut-cotton rotation (sod-based rotation) system as compared with the conventional peanut-cotton-cotton rotation in the region. In both the conventional and the sod-based rotations, reduced tillage techniques have been utilized as an added benefit to water and soil conservation. In this chapter, we review and update recent research and provide

[*] Correspondence to: duli.zhao@ars.usda.gov.

information about rotation of row crops (peanut and cotton) with bahiagrass to improve soil quality, crop physiology, growth, and productivity.

INTRODUCTION

Crop rotation is an important practice in conserving agricultural resources and sustaining development of crop production in agricultural systems worldwide. Integrating row crops with perennial grasses result in reduced irrigation, fertilizer and pesticide use, increased profitability, favorable environmental impacts, and risk aversion through diversified farming as compared with standard rotations using conservation farming technology (Wright et al., 2013). Early agricultural trends in the United States of America followed the European ancestral example of an integrated livestock/row crop system until modern mechanization began in the 20th century. Rapid advancements in plant genetics, machinery, fertilizer, pesticide and other new technologies, along with inexpensive energy and transportation have greatly improved productivity in the last half of the 20th century resulting in a shift from diversified farming operations to specialized production. This specialization resulted in regionalized infrastructure set up to handle and ship large quantities of agricultural commodities around the world. However, as fuel and input prices have increased and scientific evidence indicates impaired environmental quality as a result of current farming practices, scientists and governments are looking for ways to reduce environmental risk and to mitigate the disruptions caused by these specialized farming systems.

There is a growing awareness of the inequity in food distribution for many areas of the world and the costs and shortcomings of a global food system that has fallen short of expectations. Many areas of the world have gone from growing food locally for nearby markets to large commercial operations that grow one or two commodities that are shipped hundreds and even thousands of miles to other areas of the country for central processing to be shipped back as processed food as well as to world markets. This kind of compartmentalization and integration of food and feed production has disrupted nutrient cycling on farms by removing the more sustainable agroecosystem that was in place for generations with diversified farming (Gates, 2003; Franzluebbers, 2007).

Crop rotations have been recommended as useful farming components to improve agricultural sustainability and yields (Bruns, 2012). Mitchell et al. (2008) reported that in Alabama of the southeastern USA, rotation of cotton (*Gossypium hirsutum* L.) with winter legumes were as effective as application of fertilizer N in producing high cotton lint yields and increasing soil organic carbon levels. Crop rotations with proper reduced tillage methods, which have increased dramatically across the USA, are associated with improvement in soil and water quality as well as other favorable environmental impacts. Growers have accepted these systems due to the possibility of increased profits as well as improved natural resource stewardship. Reduced tillage has also increased in all regions of the USA over the past two decades because of the advent of herbicide tolerant crops and better equipment. Improved pesticides have allowed growers to control weeds with over-the-top applications of herbicides; while technological advancements in spray equipment have allowed growers to cover many more hectares in a short time period. However, herbicide resistant weeds have resulted in a significant increase in use of residual herbicides.

A sod-based rotation has been defined as a farming system that incorporates two or more consecutive seasons of a perennial grass into a conventional row-crop rotation with reduced tillage (Wright et al., 2012). The sod-based rotation system described here takes reduced tillage technology to a new level with benefits to the environment, crop yield and quality, risk management, and reduction in pesticide and water utilization. Several long term studies of more than 100 years have demonstrated the depletion of soils after long term cultivation and the negative impact of intensive tillage on soil organic carbon and nitrogen (Ranney, 1969; Boman et al., 1996; Raun, 2006; Girma et al., 2007). These studies are the basis for the sod-based rotation impacts on current farming systems and the benefits of including perennial grasses, such as bahiagrass, into cropping systems in the southeastern USA.

Peanut (*Arachis hypogea* L.), cotton and corn (*Zea mays* L.) in the southeastern USA are major summer agronomic crops requiring a long growing season. However, farmers face great challenges in maintaining production sustainability and profitability using the traditional crop rotation system of peanut-cotton-cotton (Katsvairo et al., 2006; Wright et al., 2013). The major challenges associated with this less diverse cropping system include multiple pests, infertile, deep sandy soils with low soil organic matter (OM), and low soil water holding capacity. Integration of perennial grasses, such as bahiagrass, into the traditional rotation system of peanut and cotton has been proposed and proved by many studies in the region (Katsvairo et al., 2006; 2007a; 2007b; Wright et al., 2008). For example, including perennial grasses in the typical agronomic rotation adds significantly to the soil organic carbon and long-term organic nitrogen pools (Franzluebbers et al., 2000; Tsigbey et al., 2009) as well as helps to diminish nematodes and other pests normally found with annual row crops (Tsigbey et al., 2009; Marois et al., 2009). Growers may know these benefits but once lands are put into row crop production, there is limited information showing that growers can benefit economically by keeping the perennial grass in a rotation scheme. It takes many years to establish these types of rotation systems. Research has sought to clearly quantify the outcomes of incorporation of a sod-based rotation to the soil, water, crops, livestock, and whole farm economics (Marois et al., 2002; Katsvairo et al., 2006; 2007b; Zhao et al., 2008a; 2008b; 2008c; Wright et al., 2008; Wright et al., 2012). In this chapter, we mainly report the impacts of rotating peanut and cotton with bahiagrass in the southeastern USA to improve soil quality and crop productivity from aspects of soil physical and biochemical properties, crop physiology, growth, yield, and sustainability.

THE SOD-BASED ROTATION

Numerous studies have shown that crop rotations are beneficial to farm systems (Hesterman et al., 1986; Katsvairo and Cox, 2000; Stanger et al., 2008). There are many types of crop rotation formats, depending on geographic locations, crops, specific demand requirement, and profits. In a 15-year study using 7 different rotations of continuous corn, continuous alfalfa (*Medicago sativa* L.), corn-soybean [*Glycine max* (L.) Merr.], corn-alfalfa, corn-corn-corn-alfalfa-alfalfa, corn-corn-oat (*Avena sativa* L.) with alfalfa seeding-alfalfa-alfalfa, and corn-soybean-corn-oat with alfalfa seeding-alfalfa in Wisconsin, Stanger et al. (2008) found that the corn-soybean rotation was the most profitable based on annual market prices and production costs. Rotation has been practiced for centuries with the more complex rotations with livestock showing the most environmental and economic benefit (Gardner and Faulkner, 1991; Loison et al., 2012; Wright et al., 2013). Growers tend to be reluctant to incorporate livestock into a traditional agronomic system because fences may be required with livestock management and new equipment may be needed when moving to a rotation system that incorporates crops not previously included in the existing cropping system. Consequently, the same crops are often grown in short rotations (1-2 years) or without rotation (Katsvairo et al., 2006). These short rotations or monocultures result in soil erosion, stagnant or decreasing yields, soil and water degradation, and increased pest pressure (Crookston, 1995; Zentner et al., 2001; Tanaka et al., 2002). Cotton and peanut have become the primary crops grown in the USA southeast region although corn and soybean acreages have increased since 2005 due to high prices for these commodities. A well-chosen diversified farming system with bahiagrass (livestock) - peanut - cotton can alleviate some of these issues through efficient utilization of resources and introduction of buffers against extreme climate or environmental events and price fluctuations (Tanaka et al., 2002; Zentner et al., 2002). This approach leads to risk management from not only economic markets but also weather extremes (Wright et al., 2013).

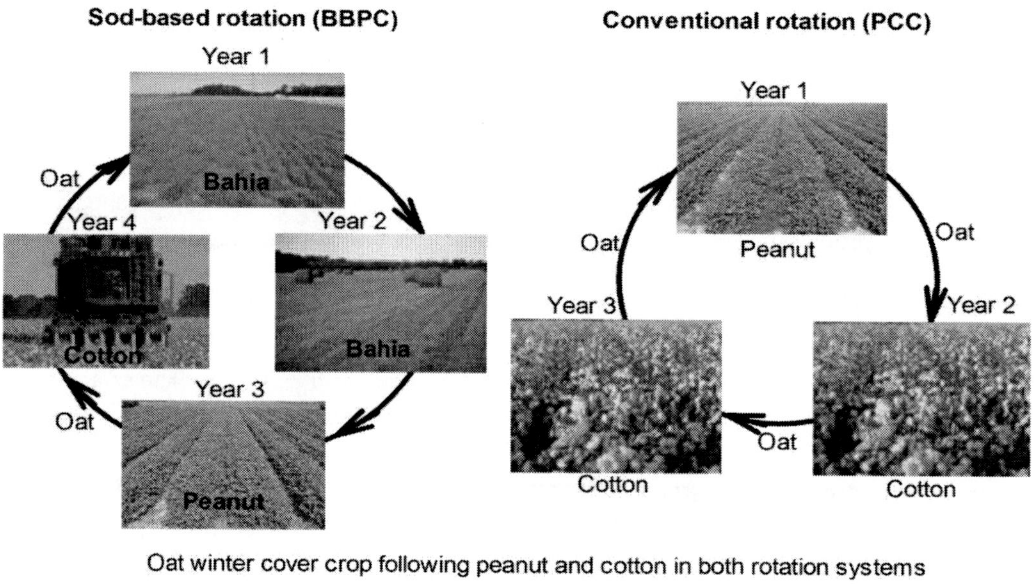

Figure 1. A long-term field experiment established in 2000 at the University of Florida, North Florida Research and Education Center, Quincy, FL to investigate the sod-based rotation effects on soil quality, crop growth and yields.

Conventional tillage degrades soil and causes loss of organic matter (Reeves, 1997; Reddy et al., 2004). Moreover, sandy soils of the Southeast Coastal Plain are inherently prone to erosion (both wind and water), which leads to loss of nutrients and environmental pollution, especially lowering water quality, along with decreasing productivity. Conservation tillage technology currently being used, such as strip tillage, can alleviate some of the water and air pollution but may not reduce pest and disease buildup or experience increases in yield as effectively as sod-based reduced tillage systems (Wright et al., 2013). However, short term crop rotations with bahiagrass (i.e., sod-based rotation, Figure 1) result in higher yields of subsequent row crops, which is probably due to pest reduction and the positive impacts on both chemical and physical properties of the soil (Elkins et al., 1977; Hagan et al., 2003; Wright et al., 2004; Katsvairo et al., 2007b). All of these factors contribute to the economic and natural resource sustainability of the farm. Growing bahiagrass for two years prior to row crops in a reduced tillage farming system has shown numerous economic and environmental advantages over the standard conservation farming practices with annual cover crops (Wright et al., 2013).

Background Information About a Sod-Based Rotation Study

A sod-based peanut-cotton rotation study was initiated in 2000 at the University of Florida's North Florida Research and Education Center in Quincy, FL (84°33' W, 30°36' N) (Katsvairo et al., 2006). The soil type at the experimental location is Dothan sandy loam (fine-loamy, kaolinitic, thermic Plinthic Kandiudult). Treatments included two rotation systems (sod-based and conventional peanut-cotton rotations), two levels of irrigation (irrigated and non-irrigated), and two levels of N rates (0 and 95 kg N ha^{-1}) for cotton. The sod-based system was a 4-yr rotation with bahiagrass-bahiagrass-peanut-cotton (BBPC) and the conventional system was a 3-yr rotation with peanut-cotton-cotton (PCC) using cover crops and conservation tillage methods in both systems (Figure 1). This study was a split-split plot field arrangement with three replications. Irrigation was the main plot, rotation crops were subplots, and N rate was sub-subplot. A total of 84 specific plots were included with all phases of rotation (Katsvairo et al., 2006; Zhao et al., 2010). The irrigation plots were irrigated to meet crop demand using a lateral move irrigation system based on the Florida cotton production guidelines (Rhoads, 2002) or crop leaf water potential (2007-2008) (Zhao et al., 2008a; 2008b). Irrigation was applied to cotton when the lowest leaf water potential (LWP) of the uppermost fully expanded main-stem leaves was about -1.5 MPa during squaring and fruiting (Zhao et al., 1989).

Bahiagrass was planted in spring in designated plots of the sod-based rotation system (Katsvairo et al., 2006) and cut one (first year) or three (second year) times during the growing season and hay yield of forage was estimated and added to the overall system economic analysis. The second year bahiagrass of the sod-based rotation was killed in late October of each year with Roundup Weather Max for the subsequent peanut crop. In late March of each year, about 3 weeks prior to cotton planting, the oat cover crop was killed with Roundup and plot rows were strip-tilled using a Brown Ro-till implement (Brown Manufacturing Co., Ozark, AL).

Cotton was planted between late April and early May depending on weather conditions in each year with a Monosem pneumatic planter (ATI Inc., Lenexa, KS). Rows were orientated west to east with a row spacing of 0.91 m and about 18 seeds per meter row. Starting in 2006, two levels of 0 and 95 kg N ha^{-1} (0 and 95 N) were applied each year for cotton in each rotation because excessive cotton growth was an issue in the first few years in the study. For the 95 N treatment, N (28 kg N ha^{-1}), P (56 kg P ha^{-1}), and K (84 kg K ha^{-1}) from a compound fertilizer (5-10-15) were band applied adjacent to each row at planting and an addition of 67 kg N ha^{-1} was sidedressed with ammonium nitrate at first square stage. For the 0-N treatment, the equal amount of P and K was band adjacent to each row at planting time by using commercial triple super phosphate (0-46-0) and Muriate of potash (0-0-60), but no N was used during growing season. Plant growth regulator PIX (mepiquate chloride) was split applied at first square (FS) and first flower (FF) stages if needed based on cotton plant growth.

Peanut was seeded with a two-row, twin-row pattern vacuum planter (ATI Inc., Lenexa, KS) in May of each year with a row spacing of 91 cm. No fertilizer was applied to the peanut crop, which reflected the grower practices in the region. An oat cover crop was planted using a no-till drill in late November to early December of each year after cotton and peanut were harvested (Zhao et al., 2010). Before oat planting, cotton stalks were shredded with a rotary mower. Oat cover crops were not fertilized or cultivated, and were killed with glyphosate N-(Phosphonomethyl) prior to reaching maturity in April of each year and before planting cotton or peanut. Details of bahiagrass, peanut, cotton, and oat cover crop management practices, including disease and insect control, herbicide application, and chemical defoliation, were employed according to standard University of Florida crop production recommendations (Ferrell et al., 2006) and in the published articles (Katsvairo et al., 2006; Zhao et al., 2010).

During the growing seasons of certain years, soil penetration resistance, bulk density, moisture, earthworms, organic matter, and nutrients were measured to determine effects of the sod-based rotation on soil physical, chemical, and biological properties. Crop growth, yield, and product quality data were collected for each crop in all seasons. Additionally, several physiological traits, such as leaf chlorophyll level, leaf area index, leaf photosynthesis characteristics, leaf water potential, dry matter accumulation and partitioning, were measured in 2007 and 2008 to estimate crop physiological responses to the sod-based rotation and to better understand physiological mechanisms of yield improvement by the sod-based rotation. Most of these growth and physiological data from this rotation system study are first reported in this chapter.

Soil Quality

Soil quality has been defined as the capacity of a soil to function within ecosystem boundaries to sustain biological productivity, maintain environmental quality, and promote plant and animal health (Doran and Parkin, 1994). Adoption of conservation farming methods in the southeastern USA has helped reduce erosion of sandy soils in Coastal Plain area (Belvins et al., 1994; Boquet et al., 2004). Annual cover crops have played a big role in reducing soil erosion. However, the root mass of perennial grasses, such as bahiagrass (19,000 kg ha^{-1}), is much higher with deeper roots than those of winter annual cover crop (3,000 - 4,000 kg ha^{-1}), which substantially and positively impacts subsequent row crops (Wright et al., 2013).

Soil Physical Properties

Bahiagrass has been used in sod-based rotations in the southeastern USA because it is widely grown on all types of soils and is drought tolerant (Field and Taylor, 2002). The biggest benefits of including bahiagrass in the traditional peanut-cotton rotation in this region could be derived from improved soil quality (Reeves, 1997; Katsvairo et al., 2007a). Bahiagrass can be grown under environmental conditions that are less than ideal for many agronomic crops and thus can be used to conserve soils under unfavorable conditions (such as drought or excess moisture) and to improve soil conditions, such as soil penetration and soil water content. Bahiagrass is adapted to the environmental conditions in the southeastern USA (Field and Taylor, 2002) and has a deep rooting system which can improve soil conditions. Much of the farmland in this region suffers from a natural compaction layer starting at 15- to 20-cm depth and continuing to 30 cm (Kashirad et al., 1967; Campbell et al., 1974). The soil compaction limits crop root growth and development and use of deep soil moisture and nutrients (USDA-NRCS, 2003). Thus, crop plants are sensitive to occasional water and nutrient stresses and yields have been negatively affected by the soil compaction layer. Bahiagrass develops a deep root system that can penetrate through the compaction layer (Elkins et al., 1977). When bahiagrass roots die, they decay and leave root channels which impart many positive attributes to soil structure and health (Elkins et al., 1977; Long and Elkins, 1983; Wright et al., 2003; Wright et al., 2004). In a comparison study of the sod-based and conventional rotations, soil penetration resistance in the sod-based rotation was substantially lower than that in the conventional rotation system (Figure 2).

Bahiagrass can improve plant tolerance to water deficit stress and reduce the need for irrigation in the following crop (Zhao et al., 2008c) because crop rooting depth is directly and positively correlated to the number of days without water stress following rainfall (Elkins et al., 1977; Katsvairo et al., 2007a). For instance, a crop with a rooting depth of 30 cm on the average Coastal Plain soil will experience a 60-day drought period from May through August in 5 out of 10 years. However, if rooting depth were 152 cm deep, the crop would experience only 11 days of drought (Elkins et al., 1977). Row crops following bahiagrass have increased rooting depth resulting in access to more soil moisture and thereby decreasing irrigation needs. Results of soil bulk density and soil moisture measured in oat cover crop plots in early spring in a long-term sod-based rotation study revealed that soil water content in the previous peanut plots did not differ in most soil depths between the two rotations, while soil water content in the previous cotton plots for the sod-based rotation was considerably higher as compared with the conventional rotation from soil surface through 0.8-m depth (Figure 3). Soil bulk density was different between the two rotation systems.

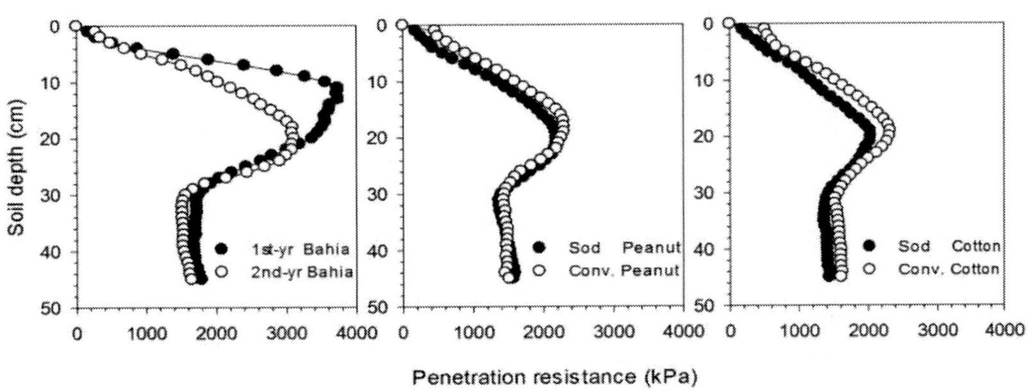

Figure 2. Soil penetration resistance profiles for different plot treatments in the long-term study of the sod-based (bahiagrass-bahiagrass-peanut-cotton) and conventional (peanut-cotton-cotton) rotation systems established in Quincy, FL in 2000. Measurements were taken in all plots in February 2008.

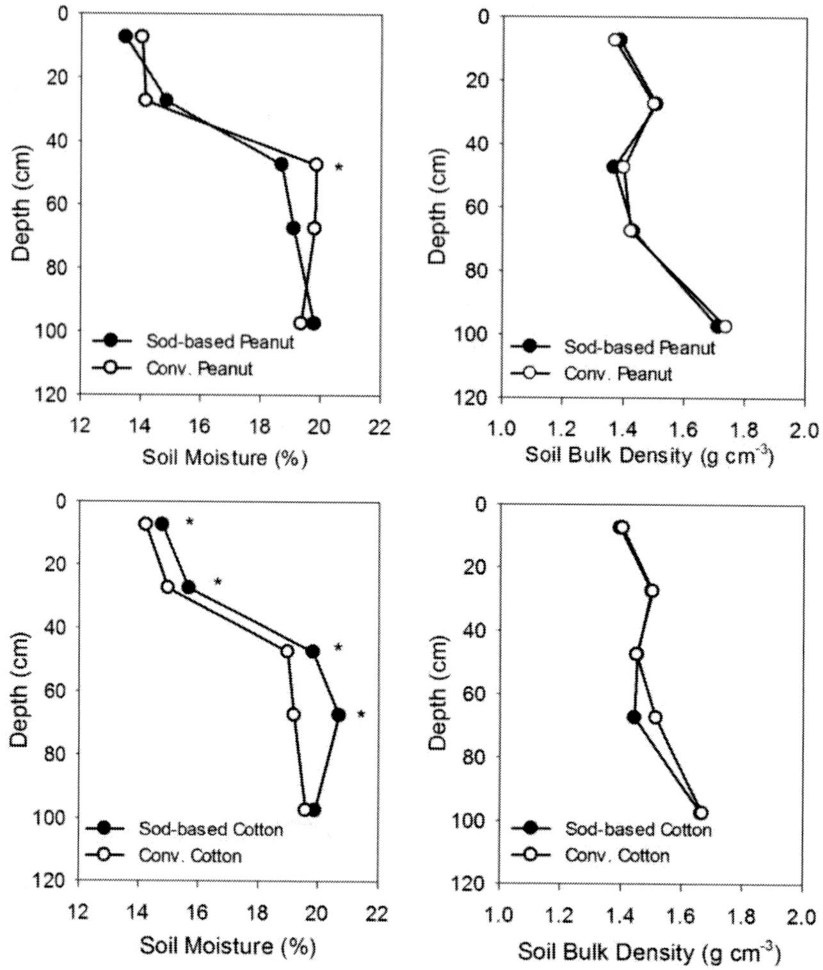

Note: * indicates difference between the two rotations at $P = 0.05$ level.

Figure 3. Soil moisture and bulk density profiles for plots where peanut and cotton previously planted in the long-term study of the sod-based (bahiagrass-bahiagrass-peanut-cotton) and conventional (peanut-cotton-cotton) rotations established in 2000 in Quincy, FL. Measurements were taken in February 2008.

Soil Organic Matter and Nutrients

Establishment of perennial forage on previously eroded cropland can lead to significant improvement in soil organic matter (SOM) and nutrient cycling (Franzluebbers, 2007). The potential increase in SOM is much higher for perennial bahiagrass than for annual cover crops. Roots play a dominant role in the soil carbon cycle (Wedin and Tilman, 1990; Gale et al., 2000; Puget and Drinkwater, 2001) and may have a greater influence on soil organic carbon level than the aboveground plant biomass (Milchumas et al., 1985). Historically, agricultural practices that were used in conventional tillage depleted SOM, leading to loss in soil productivity. The loss of SOM and productivity has resulted in the decline in farmers and farmlands (Wright et al., 2013). Bahiagrass in the cotton-peanut rotation increased the SOM content from 1.3 to 2.1% over a five year period (2003-2007) averaged from samples collected from the soil surface down to 90 cm deep when cover crops and conservation tillage were utilized together (Wright et al., 2003; Wright et al., 2013). This considerable increase in SOM is probably associated with not only bahiagrass, but also with winter cover crop and other conservation management practices (Gamble et al., 2014) because winter oat cover crop and strip tillage are important components of the sod-based rotation. Gamble et al. (2014) demonstrated that bahiagrass contributed up to 30% of soil organic carbon (SOC), but decreased to approximately 20% after the last cotton phase of the sod-based rotation in the top soil layers. The majority of SOC inputs from bahiagrass were reduced in the subsequent peanut crop, indicating bahiagrass–derived SOC is labile and does not further improve SOC storage (Gamble et al., 2014). They found that winter cover crop and other conservation practices are contributors of SOC accumulation in sod-based rotation system.

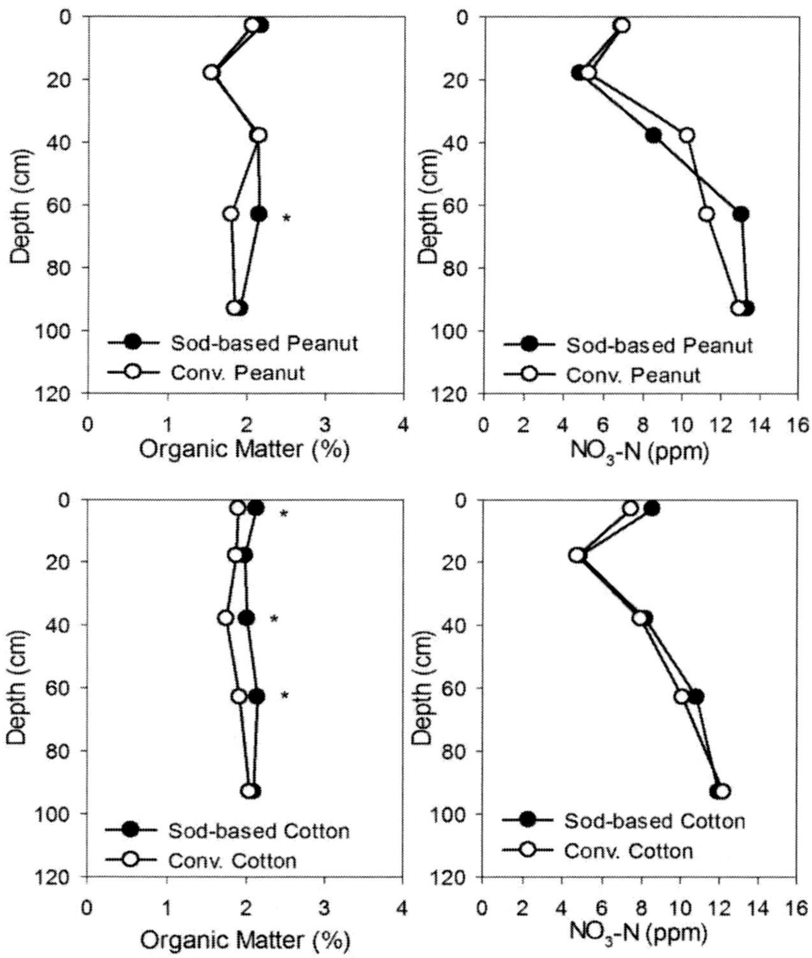

Note: * indicates difference between the two rotations at $P = 0.05$ level.

Figure 4. Soil organic matter and NO_3-N contents in 0 – 100 cm profile for plots where peanut and cotton previously planted in the long-term study of the sod-based (bahiagrass-bahiagrass-peanut-cotton) and conventional (peanut-cotton-cotton) rotations established in 2000 in Quincy, FL. Measurements were taken in February 2008.

When comparing the sod-based rotation to the conventional rotation in SOM and soil NO_3-N contents (Figure 4), there were no consistent differences in either trait between the two systems in previous peanut plots. In the previous cotton plots, however, SOM contents of the sod-based and conventional rotation systems were 2.1 and 1.9%, respectively, and soil NO_3-N were 8.9 and 8.5 mg kg^{-1}, respectively, averaged across the soil profile. Soil nutrient improvement from the sod-based rotation was also further confirmed by improving winter cover crop oat growth (Zhao et al., 2010). These results indicated that the sod-based rotation can improve the SOM content. Building up SOM levels is a long-term process (Katsvairo et al., 2007a). Even when the best conservation measures are practiced, SOM levels increase at a slow rate (Martin, 2003). Therefore, proper management strategies to prevent SOM depletion should be of foremost importance. Rotation of row crops with bahiagrass is a favorable and cost-effective way to increase and retain SOM (Katsvairo et al., 2007a).

Sod-based rotations help to control soil erosion and efficiently reduce economic and environmental risk (USDA-NRCS, 2004). It has been reported that incorporating bahiagrass/livestock into the peanut-cotton rotation in the southeastern USA can help sequester carbon in soils (Causarano et al., 2005). Cropping systems designed to sequester carbon could also reduce nutrient losses to the environment by 40 – 60% (Causarano et al., 2005). Overall, the soil in the sod-based rotation in Quincy, Florida contains higher NO_3-N in cotton plots or higher NH_4-N, Ca, and Zn in peanut plots, but lower P and K in all plots as compared with the conventional rotation (Table 1). Soil nutrient levels vary among minerals and depend on rotation systems. Therefore, fertilizer applications should be different between the two rotation systems.

Table 1. Soil nutrient analysis before summer crop planting in the sod-based rotation study at Quincy, FL. Sampling date was 24 April 2008 from 0 to 30 cm of soil depth before summer crop planting

Rotation	Previous crop	NO_3-N	NH_4-N	P	K	Mg	Ca	Zn	Mn
		(kg ha^{-1})							
Sod-based	Peanut	15.7	32.3	50.4	103.0	434.6	1802	6.83	29.1
Conventional	Peanut	15.1	25.0	54.3	139.4	357.3	1693	5.66	24.1
Sod-based	Cotton	20.2	34.0	62.7	137.8	317.0	1389	4.14	22.4
Conventional	Cotton	17.9	35.7	78.4	201.6	299.0	1512	4.93	24.6
$LSD_{0.05}$		1.8	3.7	5.9	22.3	58.1	106	1.13	6.9

Soil Biological Properties

There is a direct relationship between the amount of residue and the population of soil microorganisms (USDA-NRCS, 1996). When rotations are more complex and include sod crops, soil biological diversity will increase (Magdoff, 1992). Soil active organisms include bacteria, fungi, actinomycetes, protozoa, yeast, algae, earthworms and insects. Numbers of soil organisms are proportional to SOM concentrations in the upper 38 cm depth (Schnitzer, 1991). Soil fauna and their beneficial impacts on soils have been reviewed (Katsvairo et al., 2007a). A group of organisms in cropping systems used as indicators of soil quality are earthworms because they are sensitive to agricultural management practices including tillage, crop rotations, and pesticides. Jordan et al. (1997) compared earthworm densities across several crop rotations and tillage systems and concluded that tillage was the single most important factor which influenced population densities of earthworms, with no-till having the highest earthworm densities. Early studies indicated that plant residues, temperature and moisture, tillage frequency and management practices also affected earthworm population densities (Berry and Karlen, 1993). A few articles reported the effects of bahiagrass on earthworms in sod-based peanut–cotton rotation systems (Hartzog and Balkcom, 2003; Hartzog et al., 2005; Wright et al., 2004). Preliminary studies from Florida and Alabama have shown higher earthworm densities in peanut and cotton after bahiagrass when compared to a conventional rotation using strip tillage in both systems (Wright et al., 2004; Katsvairo et al., 2007a).

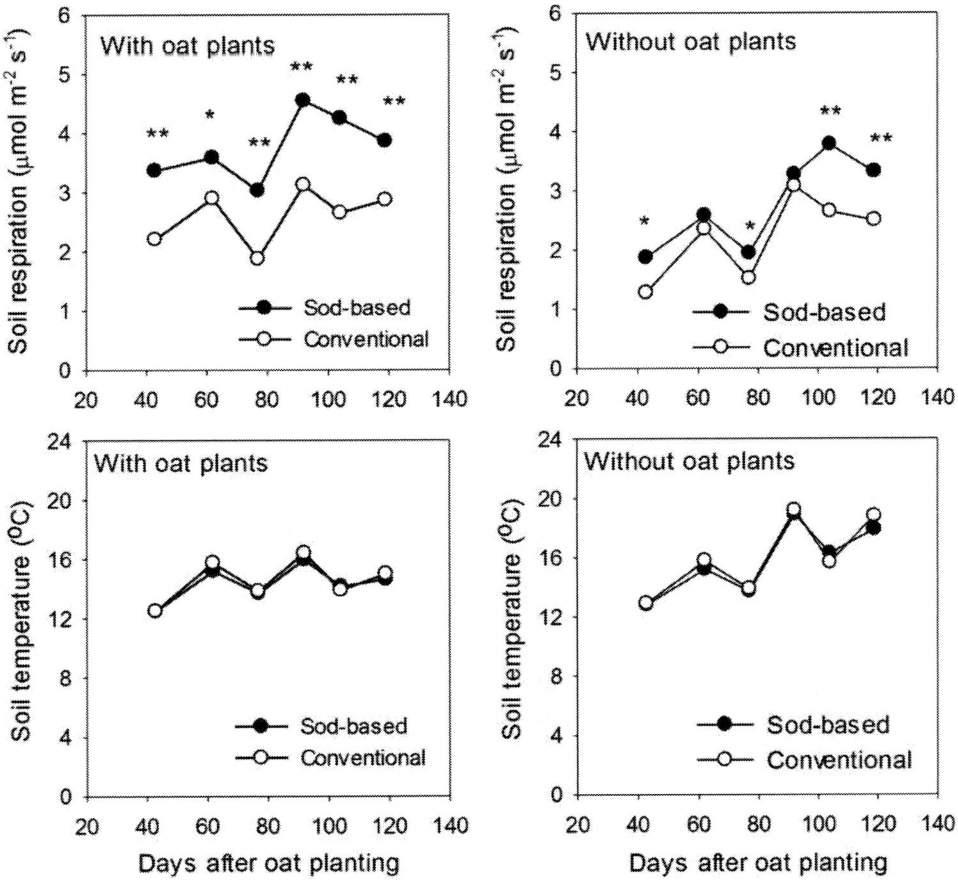

Note: * and ** indicate differences between the two rotations at $P = 0.05$ and 0.01 levels, respectively.

Figure 5. Soil respiration rates and soil temperatures measured using a LI-6400XT chamber and accessories during oat cover crop growth in the sod-based and conventional peanut-cotton rotations in 2007-2008 at Quincy, FL. Notes: Oat was planted on 7 Nov. 2007; Measurements were always set the same distance between oat rows for the with oat plants and set at the 1-m^2 away from plants on bear soil where all oat seedlings were removed immediately after emergence for the without oat plants in each plot. Soil respiration correlated with soil temperature across measurement dates with r = 0.67-0.88** in this study.

Soil respiration is one indicator of biological activity and decomposition, and it is related to aerobic microbial decomposition of SOM to obtain energy for their growth and functioning (microbial respiration), plant root and faunal respiration, and eventually from the dissolution of carbonates in soil solution. When soil respiration rates were measured in the sod-based and conventional peanut-cotton rotation systems during growth of oat cover crop from December 2007 to early April 2008, soil respiration was higher in the sod-based rotation than in the conventional rotation system, while soil temperatures did not differ between the two rotations when measured simultaneously (Figure 5). The results of soil respiration further indicated that the sod-based peanut-cotton rotation improved soil biological activity and oat root growth as compared with the conventional rotation in the southeastern USA.

Plant-parasitic nematodes in soils are important issues for crop growth and yields. Nematodes caused an approximate 12% annual loss in peanut yield and quality (Sasser and Freckman, 1987; Koenning et al., 1999). Current options for nematode management on peanut and cotton are limited to crop rotation and nematicides (Rich and Kinloch, 2007). The most common crop recommended in rotation with peanut and cotton to reduce populations of nematode is bahiagrass because it is a non-host of root-knot nematodes, effectively suppressing populations below economic thresholds for subsequent crops (Rodríguez-Kábana et al., 1994; Johnson et al., 1999). Additionally, bahiagrass reduces other soil-borne diseases and improves nutrient recycling and soil structure (Katsvairo et al., 2006). Aside from being a non-host, the release of toxic metabolites or increased presence of biological antagonists are other likely mechanisms of nematode suppression by bahiagrass (Kloepper et al., 1991; Widmer and Abawi, 2000; Chitwood, 2002; Wang et al., 2003).

Effect of bahiagrass on nematode populations in the field and their behavior under greenhouse and laboratory conditions have been investigated (Tsigbey et al., 2009). Data from these tests indicated that bahiagrass in the sod-based rotation reduced populations of spiral (*H. dihystera*), reniform (*R. reniformis*), and root-knot nematode (*M. incognita*), but not ring (*M. ornatum*) nematodes when compared to the conventional peanut-cotton-cotton rotation. These data also suggests that incorporation of bahiagrass plant residues into soil can inhibit egg production of *M. arenaria*, and reduce root galling under greenhouse conditions (Tsigbey et al., 2009). The general benefits and economics of the sod-based rotation have been summarized (USDA-NRCS, 2004). From the agricultural extension perspective, Wright et al. (2012) used a small-farm (81 hectares) economic model to compare profits of the conventional and sod-based cotton-peanut rotations. They suggested that annual profits for the sod-based and conventional rotations were \$439 and \$194 ha^{-1}, respectively. One of the main factors contributing to the improved profit potential of the sod-based rotation is the reduction in input costs because of improved soil quality, as compared with the conventional rotation (Wright et al., 2012).

Peanut in the Sod Based Rotation

Peanut Diseases

Peanut is an important row crop in the southeastern USA, and it is also a tillage intensive crop because of the incidence of soil borne and foliar diseases (Sholar et al., 1995; Cox and

Scholar, 1995; Johnson et al., 2001). Newer varieties of peanut have various degrees of resistance to spotted wilt, early and late leaf spot, white mold, and limb rot (Brenneman et al., 2003). With current integrated pest management (IPM) programs and improved disease resistant cultivars, most peanut diseases can be effectively controlled. Among these peanut diseases, spotted wilt is one of the most devastating diseases and has impacted peanut yield in recent years because it is difficult to control except for cultural practices and improved cultivars. Tsigbey (2007) found that peanut leaf spot severity in the sod-based rotation was much lower than in the conventional rotation system during the peanut growing season (Figure 6A). Tomato spotted wilt incidence can be decreased an additional 50% by strip tilling peanuts into bahiagrass as opposed to an annual cover crop thereby reducing this disease by 75% or more from conventional tillage during the growing season (Figure 6B) and across years (Figure 6C). Likewise, peanut diseases, such as leaf spot, can be delayed and have a lower rate of severity by the end of the season in a bahiagrass peanut-cotton rotation as compared to conventional rotations using conservation tillage with annual cover crops (Brenneman et al., 1995, Tsigbey, 2007; Marois et al., 2009). In addition, Sholar et al. (1995) and Taylor and Rodriguez-Kabana (1999) reported that integrating perennial grasses in crop rotation systems can effectively control peanut soil-borne diseases.

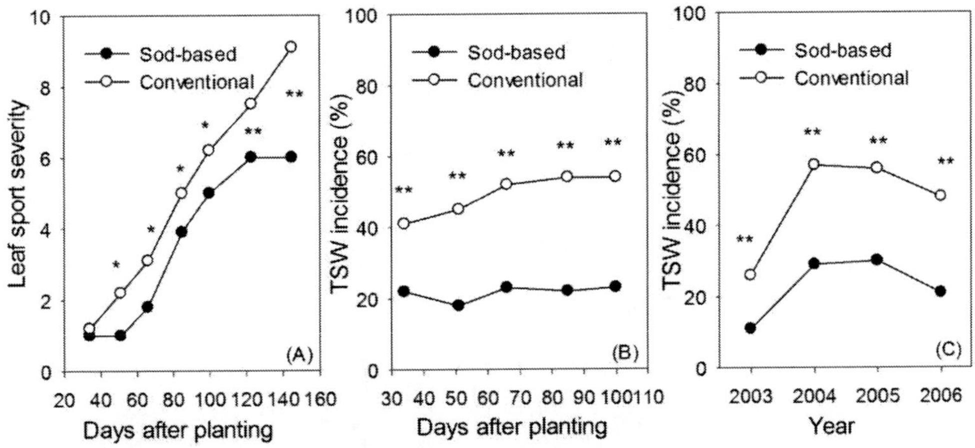

Note: * and ** indicate differences between the two rotations at $P = 0.05$ and 0.01 levels, respectively.

Figure 6. Effects of sod-based (bahiagrass-bahiagrass-peanut-cotton) and conventional (peanut-cotton-cotton) rotations on (A) peanut leaf spot severity using the 1-10 scale; (B) peanut tomato spotted wilt (TSW) incidence during the 2006 growing season; and (C) across experimental years. Data are adapted from Tsigbey (2007).

Peanut Growth

Plant canopy ground coverage was recorded in all plots of the sod-based and conventional rotations at the Quincy research site during the 2008 growing season. Peanut in the sod-based rotation always had greater ground coverage than peanut in the conventional rotation under both irrigated and non-irrigated conditions (Table 2). These results indicated that the sod-based peanut with greater ground coverage can intercept more solar radiation to produce more plant biomass as compared to peanut in the conventional rotation system.

Table 2. Peanut plant canopy ground coverage at different days after planting (DAP) for the sod-based and conventional rotation systems under irrigated and non-irrigated conditions in 2008. The rotation study was established in 2000 at the University of Florida North Florida Research and Education Center, Quincy, FL

DAP	Rotation	Irrigated	Non-irrigated	Mean
			(%)	
43	Sod-based	68.3*	45.0*	56.7*
	Conventional	63.5	40.0	51.8
51	Sod-based	80.0*	61.7**	70.9*
	Conventional	71.7	50.0	60.8
72	Sod-based	100.0*	86.3**	93.2*
	Conventional	92.4	70.7	81.6

* and ** indicate significant differences at $P = 0.05$ and 0.01 levels, respectively, between the two rotations at the same measurement date.

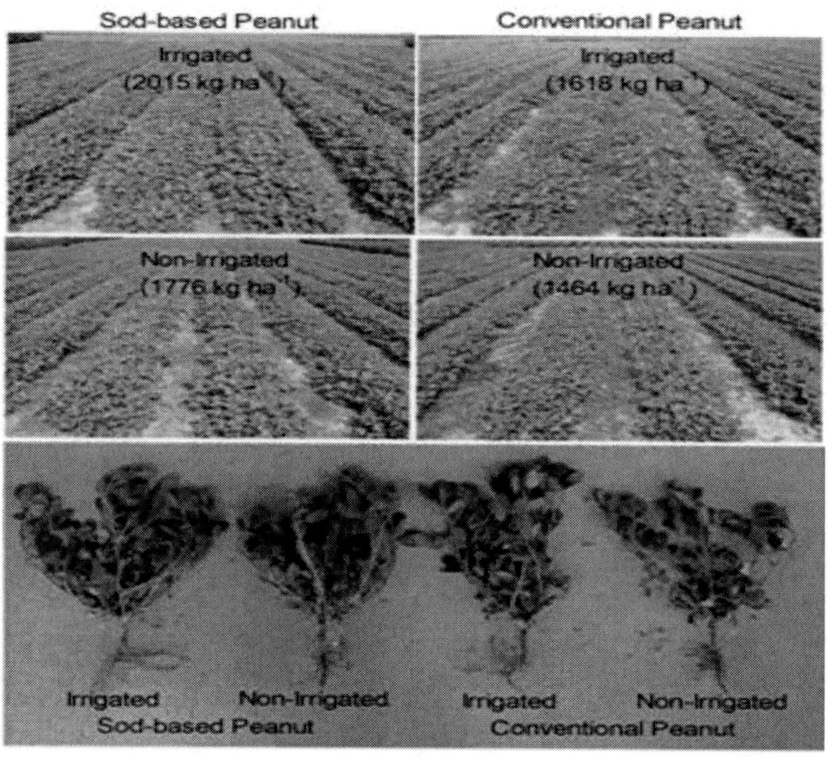

Figure 7. Peanut growth and field performance in the sod-based rotation (bahiagrass-bahiagrass-peanut-cotton) and conventional rotation (peanut-cotton-cotton) systems in Quincy, FL. Photos were taken at pigging stage on 9 July 2007 (48 days after planting). Total dry biomass data at the time are also given in the Figure.

Peanut plants in the sod-based rotation grew faster than plants in the conventional rotation with greater canopy and larger individual plant size under both the irrigated and non-irrigated conditions (Figure 7). Leaf relative chlorophyll levels (SPAD readings) have been commonly used to estimate crop plant nitrogen status and guide N fertilizer application in field crops. Averaged across irrigation treatments of the sod-based rotation study, leaf SPAD readings were not significant between the two rotations at 48 and 71 days after planting (DAP), while leaf SPAD value of the sod-based rotation (36.2) at 112 DAP was significantly lower than that of the conventional rotation (39.0) (Table 3). The lower chlorophyll level for the sod-based peanut than the conventional peanut at 112 DAP may be associated with more nitrogen moving into pods from leaves because the pod to total dry matter ratio was 0.27 for the sod-based peanut and 0.23 for the conventional peanut at 112 DAP. Total dry matter of peanut in the sod-based rotation was 23% higher than that in the conventional rotation at 48 DAP and 17% higher at 112 DAP, averaged across the irrigated and non-irrigated treatments (Table 3). Zhao et al. (2008a) reported that peanut grown in the sod-based rotation had greater leaf water potential (LWP) than peanut grown in the conventional system, especially under non-irrigated conditions. The mean LWP values of sod-based and conventional peanuts were -0.49 and -0.83 MPa, respectively, under irrigated conditions and -0.83 and -1.62 MPa, respectively, under non-irrigated conditions (Zhao et al., 2008a).

Table 3. Peanut leaf relative chlorophyll level (SPAD), total dry biomass (BM), and pod BM at different days after planting (DAP) for the sod-based and conventional rotation systems under irrigated and non-irrigated conditions in 2008. The rotation study was established in 2000 at the University of Florida North Florida Research and Education Center, Quincy, FL

DAP	Rotation	Irrigated			Non-irrigated		
		SPAD	Total BM	Pod BM	SPAD	Total BM	Pod BM
			(kg ha^{-1})			(kg ha^{-1})	
48	Sod-based	41.1	2015*	0	42.6	1776*	0
	Conventional	41.8	1618	0	42.4	1464	0
71	Sod-based	40.6	6861*	403	41.3	6303*	186*
	Conventional	39.9	6506	381	40.5	5994	117
112	Sod-based	36.8	16587**	4686**	35.5	11268**	2931**
	Conventional	40.1*	13504	3222	37.8*	10397	2358

* and ** indicate significance at $P = 0.05$ and 0.01 levels, respectively, between the two rotations for the same variable at the same measurement date.

Peanut Yield

Growers in the southeastern USA routinely use conservation tillage techniques for peanut production since it has many advantages over conventional tillage for both peanut and cotton (Pudelko et al., 1995, 1997). Strip tillage into bahiagrass shows benefits in peanut yield and quality (Katsvairo et al., 2007b). In the long-term study of sod-based and conventional rotations (both rotations with strip tillage) in Quincy, FL, the sod-based peanut had 46 to 1156 kg ha^{-1} higher pod yield than the conventional peanut in 2003-2008 under irrigated conditions and 220 to 1140 kg ha^{-1} higher yield under non-irrigated conditions, depending on years (Table 4). Averaged across years, the sod-based peanut had 15 and 20% higher pod yields, respectively, under irrigated and non-irrigated conditions.

Peanut water used efficiency (WUE) was estimated using final pod yield dividing by the sum of precipitation and amount of irrigation water during the growing season from April to

September (Zhao et al., 2008b). Overall, the non-irrigated crop had higher WUE than irrigated crop (Figure 8). The non-irrigated peanut in the sod-based rotation had the greatest, while irrigated peanut in conventional system had the least WUE, when averaged across years. The sod-based peanut had significantly greater WUE compared to conventional peanut under both irrigated (increased 15%) and non-irrigated (increased 19%) conditions ($P < 0.01$).

Table 4. Peanut pod yield responses to rotation and irrigation in the long-term sod-based peanut-cotton rotation study established in 2000 at the University of Florida North Florida Research and Education Center, Quincy, FL

Year	Rotation†	Irrigated	Non-irrigated	Mean
			(kg ha^{-1})	
2002	Sod-based	3634 a A‡	3763 a A	3698 a
	Conventional	3696 a A	3376 a B	3536 a
2003	Sod-based	3168 a A	3065 a A	3316 a
	Conventional	2461 b A	1925 b B	2193 b
2004	Sod-based	3670 a A	3681 a A	3675 a
	Conventional	2514 b B	2894 b A	2704 b
2005	Sod-based	1933 a A	2050 a A	1992 a
	Conventional	1820 a A	1830 a A	1825 a
2006	Sod-based	4594 a A	4665 a A	4630 a
	Conventional	3788 b A	3832 b A	3810 b
2007	Sod-based	4954 a A	4772 a A	4863 a
	Conventional	4908 a A	4316 b B	4612 b
2008	Sod-based	4563 a A	4006 a A	4285 a
	Conventional	3797 b A	3547 b A	3672 b
Mean	Sod-based	3788 a A	3715 a A	3752 a
	Conventional	3283 b A	3103 b B	3193 b

†The sod-based rotation was bahiagrass-bahiagrass-peanut-cotton and the conventional rotation was peanut-cotton-cotton; Details of the rotations can be found in Figure 1.

‡Means followed by the same low-case letter within a year and a column are not significant; Means followed by the same high-case letter within a row are not significant ($P > 0.05$).

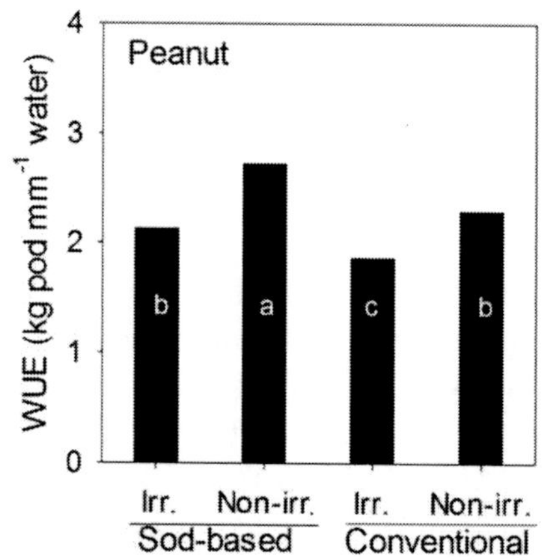

Figure 8. Water used efficiency (WUE) of irrigated (Irr.) and non-irrigated (Non-irr.) peanut under the sod-based (Bahiagrass-Bahiagrass-Cotton-Peanut) and conventional (Peanut-Cotton-Cotton) rotation systems in Quincy, FL as described in Figure 1. WUE = pod yield/(rainfall + irrigation) for the irrigated peanut and pod yield/rainfall in growing season for the non-irrigated peanut. Data with the same letter indicate the difference is not significant at $P = 0.05$ level.

Hagan et al. (2003) reported up to 34% increase in peanut yield after bahiagrass as compared with continuous peanut, while Dickson and Hewlett (1989) reported over a three-fold increase in yield for peanut after bahiagrass compared with continuous peanut. These increased yields have been attributed to reductions in disease when peanut follows a non-host crop (bahiagrass) (Elkins et al., 1977; Dickson and Hewlett, 1989; Brenneman et al., 2003). Peanut yield improvement in the long-term sod-based rotation study in Quincy, FL is also associated with the improved soil quality, reduced peanut leaf diseases, and better growth and physiological parameters as described above. The sod-based peanut had higher leaf water potential than the conventional peanut during growth under both irrigated and non-irrigated conditions (Figure 9A). Peanut kernel aflatoxin content increases under drought stress environment (Sanders et al., 1993; Arunyanark et al., 2009). Kernel aflatoxin content was very low for the sod-based peanut under irrigated and non-irrigated conditions, but kernel aflatoxin level for the non-irrigated conventional peanut was much higher (Figure 9B). These results further confirmed that sod-based rotation can improve peanut tolerance to drought stress and enhance peanut growth, yield, and quality. Zhao et al. (2009) further investigated peanut yield and kernel grade responses to timing of bahiagrass termination and tillage in a sod-based rotation. They found peanut yield and market grade characteristics were not affected when bahiagrass was terminated in spring or fall. Therefore, when using perennial grasses in sod-based rotations, farmers have a wide window from fall to spring to terminate bahiagrass for optimum peanut production.

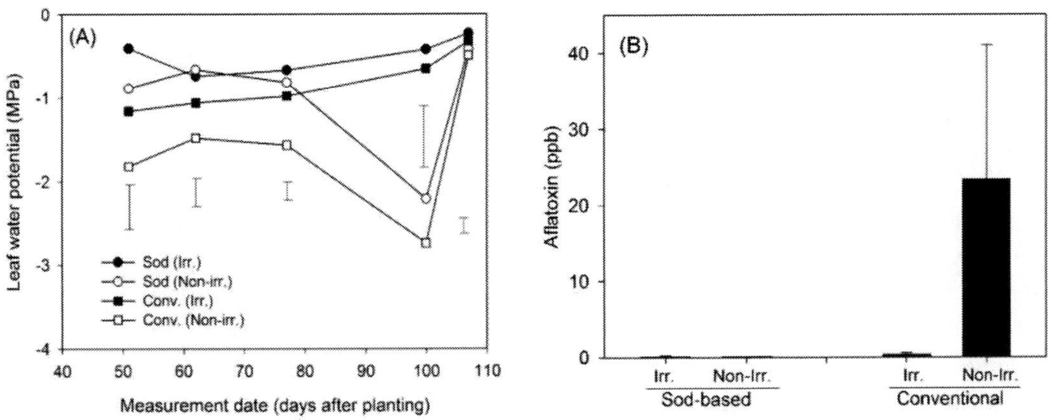

Note: vertical bars are LSD0.05 values for LWP and standard deviation for aflatoxin.

Figure 9. Comparison of (A) the daily lowest leaf water potential (LWP) during the 2007 growing season and (B) kernel aflatoxin content at harvest for the sod-based and conventional peanut under irrigated and non-irrigated conditions.

COTTON IN THE SOD BASED ROTATION

Cotton Plant Growth

Rotations of cotton with bahiagrass have been less frequently investigated compared to rotations of peanut with bahiagrass (Wright et al., 2013). Most farmers in the southeastern USA grow 2 years of cotton followed by 1 year of peanut rather than 2 years of peanut due to the increase in peanut diseases and the subsequent yield reduction. Although leaf disease pressure of cotton is much less than that of peanut in the southeastern USA, the sod-based rotation greatly reduces cotton seedling disease, such as rhizoctonia root rot, compared to the conventional rotation. Additionally, the sod-based cotton plots had significantly less weed, especially morning glory, population than the conventional cotton plots (Wright et al., 2008). Cotton hard lock, defined as incompletely opened bolls which are typically dropped from the plant prior to or at harvest, is caused by bacterial and fungal infections and is an important issue for high yield of cotton production in the region (Marois et al., 2007; Srivastava et al., 2010). In an early study in Alabama, Elkins et al. (1977) reported higher cotton yields following bahiagrass and found that it had developed a more extensive rooting system. Katsvairo et al. (2007a; 2007b) observed increased vegetative growth and more total N uptake. Katsvairo et al. (2009) further compared cotton plant height, leaf area index (LAI), relative chlorophyll level (SPAD readings), N uptake, weed densities, and residual soil nutrients in the conventional rotation versus sod-based rotation in Quincy, FL from 2000 to 2006. They found that plant height, LAI and N, P, and K uptake were generally greater for cotton in the sod-based rotation compared to the conventional cotton/peanut rotation and weed densities were reduced for cotton in the bahiagrass rotation. Cotton in the sod-based rotation in 2007 to 2009 also grew faster with great canopy coverage than cotton in conventional rotation (Figure 10).

Further measurements of plant height, the number of main-stem nodes, and leaf area index (LAI) indicated that crop rotation and N rate influenced these growth parameters significantly and sod-based cotton plants were taller with more main-stem nodes than the conventional cotton plants under both low and high nitrogen conditions (Figures 11 and 12). Starting from 65 DAP, cotton plants grown in the sod-based rotation with 95 kg N ha^{-1} were significantly taller than cotton in the conventional rotation ($P < 0.05$) (Figure 11A). The differences were much more profound during mid and late growing stages (Figure 11). At 107 DAP (about 3 weeks after first flower), plant heights of the low (0N) and high (95N) nitrogen treatments were 80 and 101 cm, respectively, for the sod-based cotton and only 71 and 88 cm, respectively for the conventional cotton. Between 80 and 110 DAP, the sod-based cotton with 95N had the greatest number of nodes, while the conventional cotton with the 0N had 3 to 4 less nodes than the sod-based 95N treatment (Figure 11B). The sod-based cotton also had 2 more nodes than the conventional cotton at the same N level at 107 DAP.

Figure 10. Cotton plants at first flowering stage (July 9, 2009) for sod-based and conventional rotations under irrigation and 95 kg N ha^{-1} conditions in Quincy, FL.

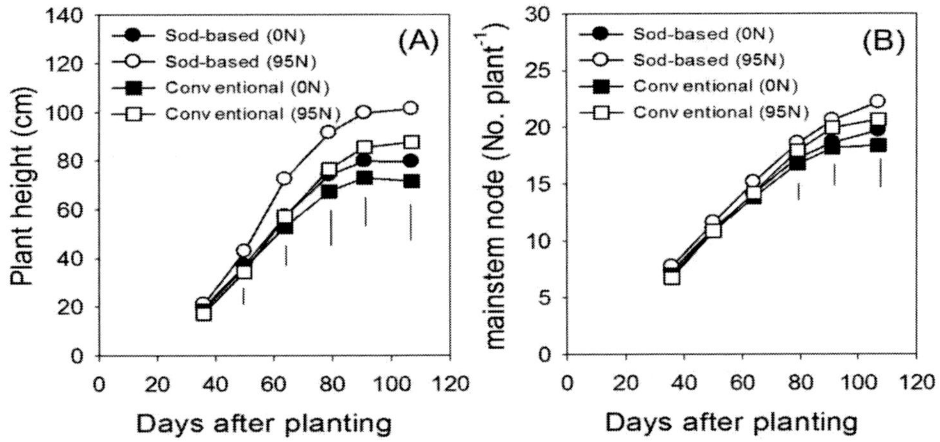

Note: vertical bars are LSD0.05 values at the specific sampling dates.

Figure 11. Changes in (A) plant height, and (B) the number of main-stem nodes during the 2007 growing season for different crop rotations and N rates under irrigated conditions. Note: all data of the conventional cotton are means of two crops (first year cotton and second year cotton) in the system in this Figure and in the following Figures and Tables.

Cotton LAI increased slowly in early growing season (from emergence to 45 DAP) and rapidly after first square stage, reached maximum value at approximately 80 to 90 DAP, and then declined during boll filling (Figure 12). Similar to plant height and number of nodes, LAI had great variation among the crop rotations and N rate treatments during flowering and boll development. Overall, LAI of the 95N-treated cotton was greater than that of the 0N treatments within a crop rotation. At the same N rate level, the sod-based cotton had greater LAI than the conventional cotton (Figure 12). Increased LAI in the sod-based rotation certainly can lead to improved interception of solar radiation contributing to improved dry matter production and radiation use efficiency.

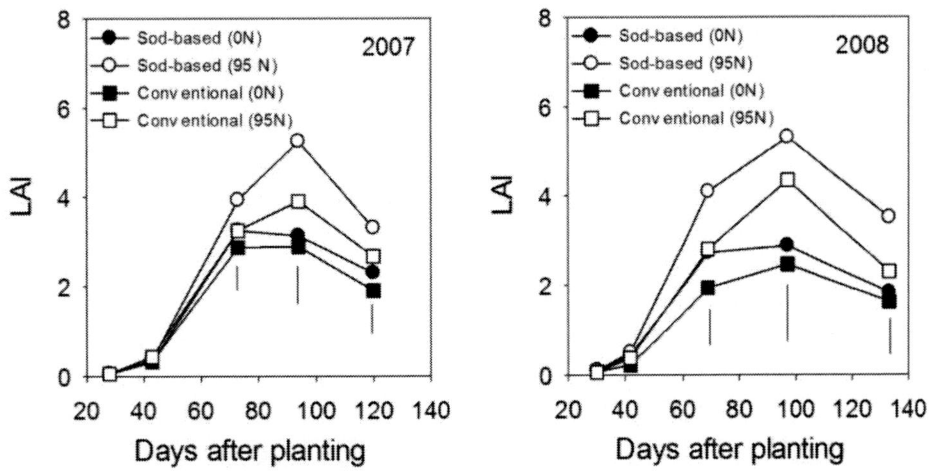

Note: vertical bars are LSD0.05 values at the specific sampling dates.

Figure 12. Changes in leaf area index (LAI) during the 2007 and 2008 growing seasons for the different crop rotations and N rates under irrigated conditions.

Katsvario et al. (2007b) investigated cotton root growth in the sod-based and conventional rotations. They found that cotton in the sod-based rotation had larger root crown diameter, total root area, total root length, and total root biomass as compared with cotton in the conventional rotation. A larger root system enables the crop to explore larger soil volume to extract more nutrients and moisture. The more extensive root growth in cotton after bahiagrass can be attributed to recolonization of the sod root channels (Katsvario et al., 2007). A recent study indicated that cattle grazing bahiagrass and winter cover crop in the sod-based rotation could further improve cotton root dimensions (Loison et al., 2012). The better cotton root growth in the sod-based rotation than in the conventional rotation results in the fast growth of above-ground plant mass.

Leaf Chlorophyll, Total N, Petiole NO_3-N, and Photosynthesis

During cotton growth, dynamics of leaf chlorophyll level (SPAD readings), leaf blade total N concentration, and petiole NO_3-N content in the sod-based and conventional rotations were determined (Figure 13). Leaf SPAD readings changed little and ranged from 46 to 51 across the measurement dates when averaged over treatments (Figure 13A). There was no statistical difference between the sod-based and conventional rotations in cotton leaf SPAD readings. Leaf total N concentration slowly declined and petiole NO_3-N level decreased sharply as plants aged (Figure 13B and 13C). The crop rotations had little effect on either leaf total N concentration or petiole NO_3-N level. In contrast to the rotations, N rate significantly affected all the three tested variables of plant N status. The high N rate (95N) treatment had greater concentrations of leaf chlorophyll, leaf blade N, and petiole NO_3-N than the low N (0N) treatment at most sampling dates throughout the growing season (Figure 13). Averaged across the crop rotations, leaf N concentrations of the 0N and 95N treatments were 5.2 and 5.3%, respectively at first square stage; 4.2 and 5.1%, respectively at first flower stage; and 3.9 and 4.7%, respectively at 3 weeks after first flower stage. Bell et al. (2003) reported that cotton leaf N concentration associated with seed cotton yield loss was 5.4% at first flower stage, 4.3% at early-flower stage, and 4.1% at mid-flower stage. According to Bell et al. (2003), leaf N concentrations of the 0N treatment for the study in Quincy, Florida were around these critical levels, but leaf N of the 95N treatment were greater than the critical levels reported by Bell et al. (2003) at all growth stages. These results suggest that no N application has risk to negatively affect cotton growth and yield, but a total amount of 95 kg N ha^{-1} seems to be too high for cotton in the sod based rotations in the southeastern USA. Therefore, refining N rate in the sod-based rotation in the region is still necessary for improving cotton growth, yield, N use efficiency, and profitability.

Cotton leaf net photosynthetic rate depended on plant growth stage and on the experimental year under irrigated conditions. Leaf photosynthetic rate had no consistent response to the rotation systems. Photosynthetic rate of the uppermost fully expanded leaves during squaring stage (mid-June) did not differ between the 0N and 95N treatments in either sod-based or conventional rotation systems (Table 5). At fruiting stage (from late July to early August), there was no difference between the two N rates within a rotation in leaf photosynthesis in 2007, the sod-based cotton had significantly higher leaf photosynthetic rate than the conventional cotton under the 95N condition. In 2008, the sod-based cotton had significantly higher leaf net photosynthetic rate than the conventional cotton under 0N condition. Plant dry matter accumulation is associated with LAI and leaf net photosynthetic rate. Overall, the response of cotton leaf photosynthesis to rotation was much smaller than the response of LAI to rotation.

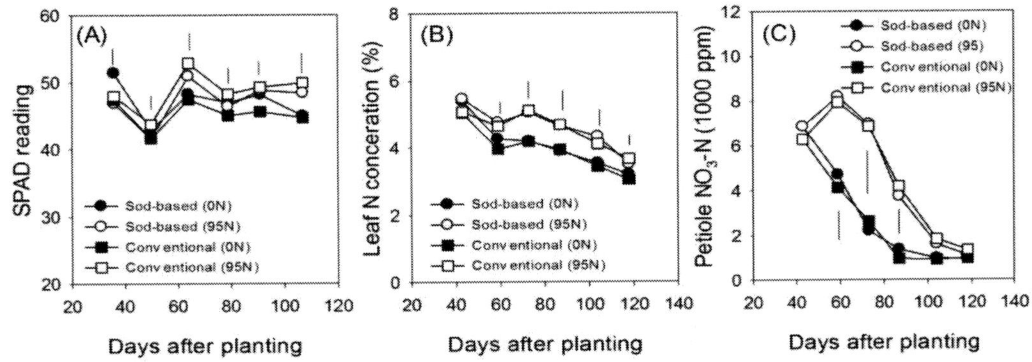

Note: vertical bars are LSD0.05 values at the specific sampling dates.

Figure 13. Changes in (A) leaf chlorophyll level (SPAD readings), (B) leaf N concentration, and (C) petiole NO_3-N content of the irrigated cotton during the 2007 growing season for the sod-based and conventional rotations and N rates of 0 and 95 kg N ha^{-1} in Quincy, FL.

Table 5. Net photosynthetic rate of uppermost fully expanded leaves for the sod-based and conventional cotton with 0 and 95 kg N ha^{-1} rates measured at squaring (mid June) and fruiting (late July- early August) stages under the irrigated condition in Quincy, FL in 2007 and 2008

Year	Growth Stage	Sod-based		Conventional		$LSD_{0.05}$
		0N	95N	0N	95N	
				(µmol CO_2 m^{-2} s^{-1})		
2007	Squaring	23.0	23.8	22.4	23.2	NS[†]
	Fruiting	25.6	27.4	24.1	23.8	2.8
2008	Squaring	30.5	31.8	28.4	29.8	NS
	Fruiting	24.8	29.2	21.9	28.6	2.6

[†]NS = not significant.

Leaf water potential (LWP) is a useful indicator of cotton plant water status. Studies have suggested that the critical value of daily lowest LWP for cotton is -1.5 MPa (Zhao et al., 1989; Oosterhuis et al., 1991; Faver et al., 1996; Zhao and Oosterhuis, 1997), and this value has been used to monitor cotton plant water deficit stress and schedule irrigation (Oosterhuis et al., 1991). The 2007 cotton growing season in Quincy, FL was a dry season with an accumulated rainfall of only 391 mm from April to September. Leaf water potential data were collected during the growing season from the irrigated and non-irrigated cotton in both the sod-based and conventional rotation systems (Figure 14). Overall, LWP of the irrigated cotton was higher than -1.5 MPa at most measurement dates and the differences between the two rotation systems in LWP were small. Averaged across measurement dates, LWP values of the sod-based and conventional cotton were -1.40 and -1.45 MPa, respectively. Under non-irrigated conditions, however, the sod-based cotton (-1.61 MPa) had significantly higher LWP than conventional cotton (-1.81 MPa) during flowering and fruiting (70 -110 DAP) (Figure 14). Similar LWP results were obtained in the 2008 growing season and in winter oat cover crop in the same study by Anguelov et al. (2009). These results indicated that the sod-based rotation improved plant water status of cotton and other crops, especially under non-irrigation conditions.

A combination of improved soil conditions after bahiagrass as described earlier led to higher total nutrient and water uptake, which contributed to the improved plant physiological characteristics and increased vegetative growth.

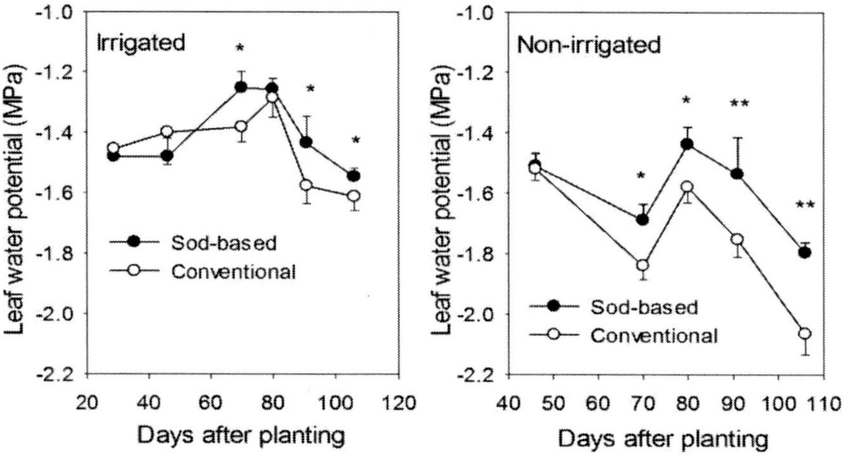

Note: vertical bars are one-side standard errors; * and ** indicate differences between the two rotations at $P = 0.05$ and 0.01 levels, respectively.

Figure 14. The lowest leaf water potential of the sod-based and conventional cotton grown under irrigated and non-irrigated conditions in Quincy, FL in the 2007 growing season. Measurements were taken between 1:00 and 3:00 pm on sunny days.

Lint Yield and Fiber Quality

The traditional peanut and cotton cropping system in the southeast region is peanut-cotton-cotton (i.e., conventional rotation) to reduce peanut diseases. Elkins et al. (1977) reported higher cotton yields following bahiagrass, but the yield improvement by rotation with bahiagrass was inconsistent across years. In a 4-year (2003 – 2006) period of the same sod-based rotation study that was compared with conventional rotation, Katsvairo et al. (2009) reported that although sod-based cotton grew faster than conventional cotton based on plant height, LAI, and total dry biomass accumulation, lint yield did not differ between the two rotation systems (Table 6). They found that cotton yield varied across the years with the greatest yield obtained in 2005 and 2006, which were also the years with the lowest plant height, LAI, and biomass, indicating a negative relationship between excessive growth and lint yield. The negative relationship between excessive growth and lint yield partially explains the lack of rotation response. There was an almost two-fold difference in lint yield between the years 2003 - 2004 vs. 2005 - 2006. The lack of yield differences for cotton in the sod-based rotation compared to cotton in the conventional rotation may have been due to excessive vegetative growth with more fruit shedding, heavy boll rot and hard-lock issues in late season. Therefore, proper management practices, such as reducing N fertilizer and irrigation water applications and optimum plant growth regulator (PIX) use for the sod-based cotton should improve cotton yield and profits in the sod-based rotation.

Irrigation was scheduled based on cotton leaf water potential in 2007 and 2008 (see Figure 13) and PIX application rate and time were adjusted based on plant growth. The optimum management practices resulted in considerable reduction in cotton fruit shedding, boll rot and hardlock diseases and substantial increase in cotton lint yields (Table 6). Clearly, the sod-based cotton had significantly higher lint yield than the conventional cotton under both irrigated and non-irrigated conditions in 2007 and 2008. Therefore, proper production management practices are needed for cotton in the sod-based rotation to obtain high yields and profits.

Table 6. Cotton lint yield responses to rotation and irrigation in a long-term sod-based peanut-cotton rotation study established in 2000 at the University of Florida NFREC, Quincy, FL

Year	Rotation†	Irrigated	Non-irrigated	Mean
			(kg ha^{-1})	
2002	Sod-based	1040 a A‡	1059 a A	1050 a
	Conventional	745 b A	716 b A	731 b
2003	Sod-based	841 a A	861 a A	851 a
	Conventional	878 a A	948 a A	913 a
2004	Sod-based	867 a A	928 a A	898 a
	Conventional	823 a A	858 a A	840 a
2005	Sod-based	1608 a A	1644 a A	1626 a
	Conventional	1600 a A	1645 a A	1622 a
2006	Sod-based	1625 a A	1497 a B	1561 a
	Conventional	1575 a A	1473 a B	1524 a
2007	Sod-based	1571 a A	1255 a B	1413 a
	Conventional	1348 b A	1153 b B	1250 b
2008	Sod-based	1658 a A	1570 a A	1614 a
	Conventional	1322 b A	1239 b A	1281 b
Mean	Sod-based	1356 a A	1259 a A	1308 a
	Conventional	1184 b A	1147 b A	1165 b

†The sod-based rotation was bahiagrass-bahiagrass-peanut-cotton; The conventional rotation was peanut-cotton-cotton; Details of the rotations can be found in Figure 1; Cotton yield of the conventional rotation are mean of 2-year means in each cycle.

‡Means followed by the same low-case letter within a year and a column are not significant; Means followed by the same high-case letter within a row are not significant ($P > 0.05$).

Table 7. Accumulated precipitation and amount of irrigation in the 2002 to 2007 growing seasons from April to September in Quincy, FL

Year	2002	2003	2004	2005	2006	2007	2008	Long-term
					(mm)			
Precipitation	640.1	729.0	922.0	932.2	436.9	391.2	759.5	762.0
Irrigation	188.0	111.8	127.0	190.5	193.0	129.5	38.1	---
Year type	Normal	Normal	Wet	Wet	Dry	Dry	Normal	---

Although irrigation is necessary for high lint yield in dry years, it is possible to use less irrigation water to reach yield goals and thus reduce production cost in the southeastern USA (Zhao et al., 2008b; 2008c). For instance, irrigation was scheduled based on LWP in 2007 (an extremely dry year, Table 7). When the daily lowest LWP, measured between 1300 and 1400 h, of cotton declined to -1.5 MPa (visually canopy leaves show slight wilt or lost tension), irrigation was provided in the irrigated plots (Figure 13). Compared to 2006 (also a dry year), 2007 had 45.7 mm less precipitation and 63.7 mm less irrigation (Table 7) during the growing season, but lint yield of irrigated cotton was equivalent (Table 6). Therefore, there is a great potential to reduce the amount of irrigation and to improve cotton yield and crop production profits even in dry years.

Cotton Water Use Efficiency

Water use efficiencies (WUE) is calculated by dividing cotton lint yield by the total amount of water (irrigation and precipitation) received during the growing season, for both irrigated and non-irrigated treatments in the sod-based and conventional rotation systems. Overall, the 2002, 2003, 2008 growing seasons were close to normal with precipitation of 641.1, 729.0, and 759.5 mm, respectively; the 2004 and 2005 growing seasons were relatively wet with 160.0 and 170.2 mm more precipitation compared to long-term average; and the 2006 and 2007 growing seasons were dry with 437 and 391 mm of rainfall, respectively (Table 7). Especially the 2007 growing season was extremely dry with only 51% of normal precipitation from April to September. The wide range of precipitation and amount of irrigation water provided during the experiment allows us to analyze crop WUE and yield responses to irrigation. Amount of irrigation in the 2002 to 2008 growing seasons for the study ranged from 38.1 to 193.0 mm (Table 7).

Table 8. Water use efficiency (WUE) under the irrigated conditions or precipitation use efficiency (PUE) under non-irrigated conditions for the sod-based and conventional cotton in the 2002 to 2008 growing seasons in Quincy, FL

Year	WUE for irrigated†		PUE for non-irrigated		Mean	
	Sod-based	Conventional	Sod-based	Conventional	Sod-based	Conventional
	(kg lint ha^{-1} mm^{-1} water)					
2002	1.26*‡	0.90	1.65*	1.12	1.46*	1.01
2003	1.00	1.04	1.18	1.30	1.09	1.17
2004	0.83	0.78	1.01	0.93	0.92	0.86
2005	1.43	1.43	1.76	1.76	1.60	1.59
2006	2.58	2.50	3.43	3.37	3.00	2.94
2007	3.02**	2.59	3.21**	2.95	3.11**	2.77
2008	2.08*	1.66	2.07**	1.68	2.07**	1.64
Mean	1.74	1.56	2.04	1.87	1.89	1.71

† WUE = lint yield/(precipitation + irrigation accumulated in the growing season) for irrigated cotton; PUE = lint yield/(precipitation in growing season) for non-irrigated cotton.

‡ The * and ** indicate that WUE or PUE is significantly different between the sod-based and conventional cotton within year at $P < 0.05$ and 0.01 level, respectively.

Zhao et al. (2009) first reported that there was the great potential to reduce irrigation water, conserve regional water resource, and improve crop WUE and profits. Compared to the conventional rotation system in 2007 and 2008, the sod-based rotation improved soil quality and other growth environment, resulting in high cotton yields (Table 6) and WUE (Table 8). In normal and wet years, there may be no need to irrigate cotton in the Southeast, but preventing cotton rank growth by applying PIX and adjusting N rate is necessary for maintaining high yield and sustainability.

Similar to lint yield response to precipitation and irrigation, WUE for irrigated cotton and precipitation use efficiency (PUE, defined as lint yield diving by amount of precipitation dring the growing season) for non-irrigated cotton were affected significantly ($P < 0.01$) by both year and irrigation treatment. The year × irrigation interaction was also significant ($P < 0.01$). The differences in WUE or PUE between the two rotations were smaller compared to the year effects. Water use efficiency of irrigated cotton varied greatly among years and ranged from 0.83 to 3.02 kg lint ha^{-1} mm^{-1} water for the sod based rotation and from 0.78 to 2.59 kg lint ha^{-1} mm^{-1} water for conventional rotation. Precipitation use efficiency of non-irrigated cotton ranged from 1.01 to 3.43 kg lint ha^{-1} mm^{-1} water for the sod based rotation and from 0.93 to 3.37 kg lint ha^{-1} mm^{-1} water for conventional rotation (Table 8). In drier years (2006 and 2007), irrigation significantly ($P < 0.05$ to 0.01) improved cotton WUE as compared with non-irrigated cotton (Table 8). Benefits of water-saving irrigation for cotton production in the southeastern USA region are to reduce not only irrigation cost, but also PIX application and the pressure of diseases and insects. The WUE (PUE) of sod-based cotton was 12% higher than that of conventional cotton under irrigated conditions and 9% higher than conventional cotton under non-irrigated conditions averaged across years. In 3 of 7 years, the sod-based cotton had significantly greater WUE and PUE than the conventional cotton under both irrigated and non-irrigated conditions (Table 8).

CONCLUSION

This chapter provided a comprehensive review on the sod-based (bahiagrass-bahiagrass-peanut-cotton) and conventional (peanut-cotton-cotton) rotation systems. Results indicated that the sod-based rotation along with proper production management practices substantially improved soil quality, conserved natural resources, depressed crop diseases, reduced applications of pesticides and nematicides and other environmental risk, improved crop growth and yields (especially peanut yield), and increased long-term sustainability and profitability. In the southeastern USA, there is the great potential to further improve peanut and cotton yields and profits by refining field management practices, including reducing irrigation and N-fertilizer and pesticide applications in the sod-based rotation. Integration of livestock into the sod-based rotation can further improve sustainability and profitability of the system, but increased management skills and additional new equipment are required to match changed farming practices. Therefore, training and transferring these technologies to growers require a team of scientists, including agronomists, soil scientists, entomologists, plant pathologists, weed scientists, animal scientists, sociologists, economists, and extension specialists. The scientists need to work together for further enhancing long-term economic value of row crop production by increasing yields while decreasing production costs and for extending these technologies to growers in the southeastern USA, especially in the rural communities that are still dependent upon farm production and other natural resources.

References

Anguelov, G., D. Wright, J. Marois, and D. Zhao. 2009. Soil-aggregate stability and leaf water potential under conservation tillage and sod-based crop rotations in a sequence of dry and wet years. p. 122-128. *In:* M.S. Reiter (ed.) A Multidisciplinary Approach to Conservation. Proc. 31st Southern Conserv. Agric. Syst. Conf., Melfa, VA. 20-23 July 2009. Available at: http://www.ag.auburn.edu/auxiliary/nsdl/scasc/Proceedings/2009/proc2009.html (Accessed on 30 March 2015).

Arunyanark, A., S. Jogloy, S. Wongkaew, C. Akkasaeng, N. Vorasoot, G.C. Wright, R.C.N. Rachaputi, and A. Patanothai. 2009. Association between aflatoxin contamination and drought tolerance traits in peanut. *Field Crops Res.* 114:14-22.

Bell, P.F., D.J. Boquet, E. Millhollon, S. Moore, W. Ehelhar, C.C. Mitchell, J. Varco, E.R. Funderburg, C. Kennedy, G.A. Breiten-beck, C. Craig, M. Holman, W. Baker, and J.F. McConnell. 2003. Relationships between leaf-blade nitrogen and relative seedcotton yields. *Crop Sci.* 43:1367-1374.

Belvins, R.L., W.W. Frye, M.G. Wagger, and D.D. Tyler. 1994. Residue management strategies for the southeast. p. 63-76. *In:* J.L. Hartfield and B.A. Stewart (eds.) Crop Residue Management. Lewis Publishers, Boca Raton, FL.

Berry, E.C., and D.L. Karlen. 1993. Comparison of alternative farming systems: II. Earthworm population density and species diversity. *Am. J. Alternative Agric.* 8:21-26.

Boman, R.K., S.L. Taylor, W. Raun, G.V. Johnson, D.J. Bernardo, and L.L. Singleton. 1996. The Magruder plots, a century of wheat in research in Oklahoma. Agronomy 96-1. *Oklahoma Agric. Exp. Stn., Stillwater.*

Boquet, D.J., R.L. Hutchinson, and G.A. Breitenbeck. 2004. Long-term tillage, cover crop, and nitrogen rate effects on cotton: plant growth and yield components. *Agron. J.* 96:1443-1452.

Brenneman, T.B., D.R. Sumner, R.E. Baird, G.W. Burton, and N.A. Minton. 1995. Suppression of foliar and soilborne peanut diseases in bahiagrass rotations. *Phytopathology* 85:948-952.

Brenneman, T.B., T. Timper, N.A. Minton, and A.W. Johnson. 2003. Comparison of bahiagrass, corn, and cotton as rotational crops for peanut. Proc. Sod-based Cropping Systems Conf. Quincy, FL. Feb. 20-21, 2003. p. 59-65.

Bruns, H.A. 2012. Concepts in Crop Rotations. p. 25-48. *In*: G. Aflakpui (ed.) Agricultural Science. ISBN: 978-953-51-0567-1, InTech, Available at http://www.intechopen.com/books/agricultural-science/conceptsin-crop-rotation. (Accessed on 5 April 2015).

Campbell, R.B., D.C. Reicosky, and C.W. Doty. 1974. Physical properties and tillage of Paleudults in the Southeastern Coastal Plains. *J. Soil Water Conserv.* 29:220-224.

Causarano, H.J., A.J. Franzluebbers, D.W. Reeves, J.N. Shaw, and M.L. Norfleet. 2005. Potential for soil carbon sequestration in cotton production systems of the southeastern USA. Available at http://www.spcru.ars.usda.gov/publicationsall.htm. (Accessed on 7 Oct. 2015).

Chitwood, D.J. 2002. Phytochemical based strategies for nematode control. *Ann. Review Phytopathol.* 40:221-249.

Cox, F.R., and J.R. Sholar. 1995. Site selection, land preparation, and management of soil fertility. p. 7-10. *In:* H.A. Melouk and F.M. Shokes (eds.) Peanut Health Management. *Am. Phytopathol. Soc.,* St. Paul, MN.

Crookston, R.K. 1995. The rotation effect in corn: A summary of fifteen years of field research in Minnesota. *Minnesota Agric. Exp. Stn.* no. 22-180. Univ. of Minnesota, St. Paul, MN.

Dickson, D.W., and T.F. Hewlett. 1989. Effects of bahiagrass, and nematicides on *Meloidogyne arenaria* on peanut. *Suppl. J. Nematol.* 21, 4S:671-676.

Doran, J.W., and T.B. Parkin. 1994. Defining soil quality for a sustainable environment. p. 1-45. *In:* J.W. Doran, D.C. Coleman, D.F. Bezdicek, and B.A. Stewart (eds.) Special Publication No. 35, American Society of Agronomy, Madison, WI.

Elkins, C.B., R.L. Haaland, and C.S. Hoveland. 1977. Grass roots as a tool for penetrating soil hardpans and increasing crop yields. p. 21-26. Proc. 34th Southern Pasture and Forage Crop Improv. Conf., Auburn University, Auburn, Alabama.

Faver, K.L., T.J. Gerik, P.M. Thaxton, and K.M. El-Zik. 1996. Late season water stress in cotton: II. Leaf gas exchange and assimilation capacity. *Crop Sci.* 36:922-928.

Ferrell, J.A., G.E. MacDonald, and B.J. Brecke. 2006. Weed management in cotton. Available at http://edis.ifas.ufl.edu/WG003. (Accessed on 14 March 2014). University of Florida, IFAS, Gainesville, FL.

Field, T.G., and R.E. Taylor. 2002. Beef production management decisions. Pearson Education. Prentice Hall, Upper Saddle River, NJ. 07458.

Fisher, M. 2011. Sod-based rotations: New system could increase profits, reduce risk, and conserve resources. Crop and Soil magazine, Sept. – Oct. 2011, pp. 10. https://www.agronomy.org/ files/publications/crops-and-soils/sod-based-rotations.pdf.

Franzluebbers, A.J. 2007. Integrated crop–livestock systems in the southeastern USA. *Agron. J.* 99: 361-372.

Franzluebbers, A.J., J.A. Stuedemann, H.H. Schomberg, and S.R. Wilkinson. 2000. Soil C and N pools under long-term pasture management in the Southern Piedmont USA. *Soil Biol. Biochem.* 32:469-478.

Gale, W.J., C.A. Cambardella, and T.B. Bailey. 2000. Root-derived carbon and the formation and stabilization of aggregates. *Soil Sci. Soc. Am. J.* 64:201-207.

Gamble, A.V., J.A. Howe, C.W. Wood, D.B. Watts, and E. van Santen. 2014. Soil organic carbon dynamics in a sod-based rotation on Coastal Plain soils. *Soil Sci. Soc. Am. J.* 78:1997-2008.

Gardner, J.C., and D.B. Faulkner. 1991. Use of cover crops with integrated crop–livestock production systems. p. 185-191. *In:* W.L. Hargrove (ed.) Cover Crops for Clean Water. *Soil and Water Conserv. Soc.,* Ankeny, IA.

Gates, R N. 2003. Integration of perennial forages and grazing in sod-based crop rotations. Proc. Sod-based Cropping Systems Conf. Quincy, FL. Feb. 20-21, 2003. p. 7-13.

Girma, K., S.L. Holtz, D.B. Arnall, B.S. Tubana, and W.R. Raun. 2007. The Magruder plots. *Agron. J.* 99:1191-1198.

Hagan, A.K., L.H. Campbell, J.R. Weeks, M.E. Rivas-Davila, and B. Gamble. 2003. Impact of bahiagrass, cotton, and corn cropping frequency on the severity of diseases of peanut. Proc. Sod-based Cropping Systems Conf., Quincy, FL. Feb. 20-21, 2003. p. 46-58.

Hartzog, D.L., and K. Balkcom. 2003. Initiation of the Tri-State sod-based rotation project - An overview. Proc. Sod-based Cropping Systems Conf., Quincy, FL. Feb. 20-21, 2003. p. 7-13.

Hartzog, D.L., T.W. Katsvairo, D.L. Wright, J.J. Marois, and P.J. Wiatrak. 2005. Earthworm activities in integrated livestock peanut/cotton cropping systems. ASA-CSSA-SSSA Annual Meetings Abstracts [CD-ROM]. ASA, CSSA, and SSSA, Madison, WI.

Hesterman, G.A., C.C. Sheaffer, and E.I. Fuller. 1986. Economic comparisons of crop rotations including alfalfa, soybean, and corn. *Agron. J.* 78:24-28.

Johnson, W.C., T.B. Brenneman, S.H. Baker, A.W. Johnson, D.R. Sumner, and B.G. Mullinix, Jr. 2001. Tillage and pest management considerations in a peanut–cotton rotation in the Southeastern Coastal Plain. *Agron. J.* 93:570-576.

Johnson, A.W., N.A. Minton, T.B., Brenneman, G.W. Burton, A.K. Culbreath, G.J. Gasco, and S.H. Baker. 1999. Bahiagrass, corn, cotton rotations, and pesticides for managing nematode diseases, and insects on peanut. *J. Nematology* 31:191-200.

Jordan, D., J.A. Stecker, V.N. Cacnio-Hubbard, F. Li, C.J. Gantzer, and J.R. Brown. 1997. Earthworm activity in no-tillage and conventional tillage systems in Missouri soils: A preliminary study. *Soil Biol. Biochem.* 29:489-491.

Kashirad, A.J., G.A. Fiskell, V.W. Carlisle, and C.E. Hutton. 1967. Tillage pan characterization of selected Coastal Plain soils. *Soil Sci. Soc. Am. Proc.* 31:534-541.

Katsvairo, T.W., and W.J. Cox. 2000. Economics of cropping systems featuring different rotations, tillage, and management. *Agron. J.* 92:485-493.

Katsvairo, T.W., J.R. Rich, and R.A. Dunn. 2006. Perennial grass rotation: An effective and challenging tactic for nematode management with many other positive effects. *Pest Manag. Sci.* 62:793-796.

Katsvairo, T.W., D.L. Wright, J.J. Marois, D.L. Hartzog, J.R. Rich, and P.J. Wiatrak. 2006. Sod-livestock integration into the peanut-cotton rotation. *Agron. J.* 98:1156-1171.

Katsvairo, T.W., D.L. Wright, J.J. Marois, D.L. Hartzog, K.B. Balkcom, P.P. Wiatrak, and J.R. Rich. 2007a. Cotton roots, earthworms, and infiltration characteristics in sod–peanut–cotton cropping systems. *Agron. J.* 99:390–398.

Katsvairo, T.W., D.L. Wright, J.J. Marois, D.L. Hartzog, and J.R. Rich. 2007b. Performance of peanut and cotton in a bahiagrass cropping system. *Agron. J.* 99: 1245-1251.

Katsvairo, T.W., D.L. Wright, J.J. Marois, J.R. Rich, and P.P. Wiatrak. 2009. Comparative plant growth and development in two cotton rotations under irrigated and non-irrigated conditions. *Crop Sci.* 49:2233-2245.

Kloepper, J.W., R. Rodríguez-Kábana, J.A McInroy, and D.J. Collins. 1991. Analysis of populations and physiological characterization of microorganisms in rhizospheres of plants with antagonistic properties to phytopathogenic nematodes. *Plant Soil* 136:95-102.

Koenning, S.R., C. Overstreet, J.W. Noling, P.A. Donald, J.O. Becker, and B.A. Fortnum. 1999. Survey of crop losses in response to phytoparasitic nematodes in the United States for 1994. *J. Nematology* 31:587-618.

Loison, R., D. Rowland, W. Faircloth, J.J. Marois, D.L. Wright, and S. George. 2012. Cattle grazing affects cotton root dimensions and yield in a bahiagrass-based crop rotation. *Crop Management* doi:10.1094/CM-2012-0925-02-RS.

Long, F.L., and C.B Elkins. 1983. The influence of roots on nutrient leaching and uptake. p. 335-352. *In:* R.T. Lowrance, L. Asmussen, and R. Leonard (eds.) Nutrient Cycling in

Agricultural Ecosystems. University of Georgia. *College of Agric. Exp. Stations,* Spec. Pub. 23.

Magdoff, F. 1992. Building Soils for Better Crops. University of Nebraska Press, Lincoln and London. p. 90-102.

Marois, J.J., D.L. Wright, J.A. Baldwin, and D.L. Hartzog. 2002. A multi-state project to sustain peanut and cotton yields by incorporating cattle in a sod based rotation. p. 101-107. *In:* E. van Santen (ed.) Proc. 25th Annual Southern Conserv. Tillage Conf. for Sustainable Agric., Auburn, AL. June 24-26, 2002.

Marois, J.J., D.L. Wright, B. Leite, D. Mailhot, and E. Osekre. 2007. Hardlock of cotton: historical review and perspectives. World Cotton Res. Conf.-4. 10-14 Sep. 2007, Lubbock, TX. Available at http://wcrc.confex.com/wcrc/ 2007/techprogram/P1837.HTM. (Accessed on 6 March 2015).

Marois, J., D. Wright, F. Tsigbey, J. Rich, and G. Anguelov. 2009. Impact of sod-based rotation on peanut diseases using conservation technology. p. 58-67. *In:* M.S. Reiter (ed.) A Multidisciplinary Approach to Conservation. Proc. 31st Southern Conserv. Agric. Syst. Conf., Melfa, VA. 20-23 July 2009. Available at: http://www.ag.auburn.edu/ auxiliary/ nsdl/scasc/Proceedings/2009/Marois.pdf. (Accessed on 26 March 2015).

Martin, H. 2003. Transition to Organic Farming. Available at www.omafra. gov.on.ca/english/crops/organic/transition.htm (Accessed on 27 Apr. 2015). Ontario, Canada.

Milchumas, D.G., W.K. Lauenroth, J.S. Singh, and C.V. Cole. 1985. Root turnover and production by 14C dilution: Implications of carbon partitioning in plants. *Plant Soil* 88:353-365.

Mitchell, C.C., D.P. Delaney, and K.S. Balkcom. 2008. A historical summary of Alabama's old rotation (circa 1896): The world's oldest, continuous cotton experiment. *Agron. J.* 100:1493-1498.

Oosterhuis D.M., R.A. Ball, and R.E. Hampton. 1991. Plant indicators of water stress for improved irrigation scheduling. p. 33-43. *In*: D.M. Oosterhuis, (ed.) Proc. 1991 Cotton Res. Meeting. University of Arkansas Agricultural Experiment Station, Special Report 144.

Pudelko, J.A., D.L. Wright, and J. Spitalniak. 1997. Aspects of strip till and conservation tillage in the Southeast USA. *J. Rocz. AR Pozn*. CCLXXXV, Roln. 48:85-89.

Pudelko, J.A., D.L. Wright, and P.J. Wiatrak. 1995. No tillage research review. *Polish Acad. Sci.* 66:154-162.

Puget, P., and L.E. Drinkwater. 2001. Short-term dynamics of root- and shoot-derived carbon from a leguminous green manure. *Soil Sci. Soc. Am. J.* 65:771-779.

Ranney, R.W. 1969. An organic carbon-organic matter conversion equation for Pennsylvania surface soils. *Soil Sci. Soc. Am. Proc*. 33:809-811.

Raun, W.R. 2006. Long term fertility experiments at Oklahoma State University. Available at http://nue.ofstate.edu.Long_Term_Experiments. htm. (Accessed on 29 May 2014).

Reddy, C.K, E.Z. Nyakatawa, and D.W. Reeves. 2004. Tillage and poultry litter application effects on cotton growth and yield. *Agron. J.* 96:1641-1650.

Reeves, D.W. 1997. The role of soil organic matter in maintaining soil quality in continuous cropping systems. *Soil Tillage Res*. 43:131-167.

Rhoads, F. 2002. Florida cotton production guidelines: water requirements for cotton. University of Florida, IFAS Extension. Available at http://edis.ifas. ufl.edu. (Accessed on 26 Feb. 2015).

Rich, J.R., and R.A. Kinloch. 2007. Peanut Nematode Management. University of Florida IFAS: Gainesville. Available at http://edis.ifas.ufl. edu/NG016. (Accessed on 12 Oct. 2014).

Rodríguez-Kábana, R., N. Kokalis-Burelle, D.G. Robertson, P.S. King, and L.W. Wells. 1994. Rotations with coastal bermudagrass, cotton, and bahiagrass for management of *Meloidogyne arenaria* and southern blight in peanut. *J. Nematology* 26:665-668.

Sanders, T.H., R.J. Cole, P.D. Blankenshi, and J.W. Dorner. 1993. Aflatoxin contamination of peanut from plant drought stressed in pod or root zones. *Peanut Sci.* 20:5-8.

Sasser, J.N. and D.W. Freckman. 1987. A world perspective on nematology: the role of the society. p. 7-14. *In:* J.A Veech and D.W. Dickson (eds.) Vistas on Nematology Society of Nematology, Hyattsville, MD.

Schnitzer, M. 1991. Soil organic matter and soil quality, Technical-Bulletin-Agriculture-Canada-No. 1991-1E. p. 33-49.

Sholar, R.E., R.W. Mozingo, and J.P. Beasley, Jr. 1995. Peanut cultural practices. p. 354-382. *In:* H.E. Pattee and T.H. Stalker (eds.) Advances in Peanut Science. *Am. Peanut Res. and Educ. Soc.,* Stillwater, OK.

Srivastava, P., D.J. Mailhot, B. Leite, J.J. Marois, D.L. Wright, and R.L. Nichols. 2010. *Fusarium verticillioides* associated with hardlock of cotton. *Current Microbiol.* 61:79-84.

Stanger, T.F., J.G. Lauer, and J. Chavas. 2008. The profitability and risk of long-term cropping systems featuring different rotations and nitrogen rates. *Agron. J.* 100:105-113.

Tanaka, D.L., J.M. Krupinsky, M.A. Liebig, S.D. Merrill, R.E. Ries, J.R. Hendrickson, H.A. Johnson, and J.D. Hanson. 2002. Dynamic cropping systems: An adaptable approach to crop production in the Great Plains. *Agron. J.* 94:957-961.

Taylor, C.R., and R. Rodriguez-Kabana. 1999. Optimal rotation of peanuts and cotton to manage soil-borne organisms. *Agric. Syst.* 61:57-68.

Tsigbey, F.K., J.R. Rich, J.J. Marois, and D.L. Wright. 2009. Effect of bahiagrass (*Paspalum notatum* Fluegge) on nematode populations in the field and their behavior under greenhouse and laboratory conditions. *Nematropica* 39:111-119.

Tsigbey, F.K. 2007. Management of Major Peanut (*Arachis hypogaea* L.) Diseases Using Bahiagrass (*Paspalum notatum* Fluegge) Rotation. A Ph.D. Dissertation, University of Florida. (http://ufdcimages.uflib.ufl.edu/ UF/E0/02/14/49/00001/tsigbey_f.pdf).

USDA-NRCS. 1996. Conservation Crop Rotation Effects on Soil Quality. Soil Quality-Agronomy Technical Note, No. 2. pp. 8. Available at http://www.nrcs.usda.gov/Internet/ FSE_DOCUMENTS/nrcs142p2_053 269.pdf. (Accessed on 27 March 2015).

USDA-NRCS. 2003. Soil Compaction: Detection, Prevention, and Alleviation. Soil Quality-Agronomy Technical Note. No. 17. pp. 7. Available at http://www.nrcs.usda.gov/ Internet/FSE_DOCUMENTS/nrcs142p2_ 053258.pdf. (Accessed on 27 March 2015).

USDA-NRCS. 2004. Sod-based rotations: A proven old practice to improve soil productivity. Soil Quality- Agronomy Technical Note. No. 18. pp. 7. Available at http://www.nrcs. usda.gov/Internet/FSE_DOCUMENTS/nrcs142p2_053292.pdf. (Accessed on 27 March 2015).

Wang, K.-H., B.S. Sipes, and D.P. Schmitt. 2003. Enhancement of *Rotylenchulus reniformis* suppressiveness by *Crotalaria juncea* amendment in pineapple soils. *Agric. Ecosyst. Environ.* 94:197-203.

Wedin, D.A., and D. Tilman. 1990. Species effects on nitrogen cycling: A test with perennial grasses. *Oecologia* 84:433-441.

Widmer, T.L., and G.S. Abawi. 2000. Mechanism of suppression of *Meloidogyne hapla* and its damage by green manure of Sudan grass. *Plant Dis.* 84:562-568.

Wright, D.L., T.W. Katsvairo, J.J. Marois, and P.J. Wiatrak. 2004. Introducing bahiagrass in peanut/cotton cropping systems-effects on soil physical characteristics. ASA-CSSA-SSSA International Annual Meetings (October 31 - November 4, 2004), Seattle, WA. Abstract No. 3637.

Wright, D.L., J.J. Marois, C. Fraisse, and D. Dourte. 2012. Sod-based rotation: A management option for climate variability and change. Available at http://agroclimate.org/management/Sod-based-rotation.pdf. (Accessed on 16 Aug. 2015).

Wright, D.L., J.J. Marois, C. Mackowiak, D. Zhao, G. Anguelov, and C. Lamb. 2013. Sod-based/livestock/row crop integration: An improved conservation farming system. *In:* J. Bergtold, G. Hawkings, K. Iversen, and R. Raper (eds.) Conservation Tillage Systems: Production, Profitability, and Stewardship. Beltsville, MD: SARE Outreach (in press).

Wright, D.L., J.J. Marois, M. Vargas, and P.J. Wiatrak. 2003. Perennial grasses, soil organic matter and crop yield. Proc. Sod-based Cropping Syst. Conf., Quincy, FL. Feb. 20-21, 2003. p. 34- 39.

Wright, D., J.J. Marois, D. Zhao, and C. Mackowiak. 2008. Overview of the sod based rotation using conservation techniques. p. 79-83. *In:* D.M. Endale (ed.) Proc. 30th Southern Conserv. Agric. Syst. Conf. and 8th Ann. Georgia Conserv. Prod. Syst. Trng. Conf., Tifton, Georgia, July 29-31, 2008. Available at: http://www.ag.auburn.edu/auxiliary/nsdl/scasc/. (Accessed on 20 Feb. 2015).

Zentner, R.P., C.A. Campbell, V.O. Biederbeck, P.R. Miller, F. Selles, and M.R. Fernandez. 2001. In search of a sustainable cropping system for the semiarid Canadian Prairies. *J. Sustain. Agric.* 18:117-136.

Zentner, R.P., D.D. Wall, C.N. Nagy, E.G. Smith, D.L. Young, P.R. Miller, C.A. Campbell, B.G. McConkey, S.A. Brandt, G.P. Lafond, A.M. Johnston, and D.A. Derksen. 2002. Economics of crop diversification and soil tillage opportunities in the Canadian prairies. *Agron. J.* 94:216-230.

Zhao, D., and D.M. Oosterhuis. 1997. Physiological responses of growth chamber-grown cotton plants to the plant growth regulator PGR-IV under water-deficit stress. *Environ. Exp. Bot.* 38:7-14.

Zhao, D., D.L. Wright, and J.J. Marois. 2009. Peanut yield and grade responses to timing of bahiagrass termination and tillage in a sod-based crop rotation. *Peanut Sci.* 36:196-203.

Zhao D., D.L. Wright, J.J. Marois, C. Mackowiak, and T.W. Katsvairo. 2008a. Yield and water use efficiency of cotton and peanut in conventional and sod-based cropping systems. *In:* D.M. Endale (ed.) Proc. 30th Annual Southern Conserv. Agric. Syst. Conf., Tifton, Georgia, July 29-31, 2008. Available at: http://www.ag.auburn.edu/auxiliary/nsdl/scasc/. (Accessed on 29 Jan. 2015).

Zhao, D., D. Wright, J. Marois, and C. Mackowiak. 2008b. Cotton growth and yield responses to nitrogen rate in the sod-based peanut-cotton rotations. Proc. Beltwide Cotton Conf., National Cotton Council, Memphis, TN. p. 93-99.

Zhao, D., D. Wright, J. Marois, and C. Mackowiak. 2008c. Reducing irrigation and improving cotton yield and profitability in the southeast USA. Proc. Beltwide Cotton Conf., National Cotton Council, Memphis, TN. p. 37-43.

Zhao, D., D.L. Wright, J.J. Marois, C.L. Mackowiak, and M. Brennan. 2010. Improved growth and nutrient status of an oat cover crop in sod-based versus conventional peanut-cotton rotations. *Agron. Sustain. Dev.* 30:497-504.

Zhao, D., Y.Z. Xu, and X. Xu. 1989. Effects of water stress during flowering and fruiting on the development and physiological characteristics of cotton plants. *J. Northwestern Agric. Univ.* 17 (2):79-85.

In: Crop Rotations
Editor: Bao-Luo Ma
ISBN: 978-1-63484-496-3

Chapter 5

LAND USE PRACTICES, CROPPING SYSTEMS AND CLIMATE CHANGE VULNERABILITY TO MOUNTAIN AGRO-ECOSYSTEMS OF NEPAL

***Kalidas Subedi*[1,*], *Netra B. Chhetri*[2] and *Tika B. Karki*[3]**
[1]Agriculture and Agri-Food Canada,
Central Experimental Farm, Ottawa, Canada
[2]Schools of Geographical Sciences and Urban Planning and Consortium for Science, Policy and Outcomes, Arizona State University, AZ, US
[3]Nepal Agriculture Research Council (NARC), Kathmandu, Nepal

ABSTRACT

Characterized by fragile geo-ecology, marginality, inaccessibility, and subsistence livelihoods, the land resources in the Middle Mountain region of Nepal are intensively cultivated beyond their carrying capacity. Lack of off-farm employment opportunities combined with limited productive lands has forced the mountain communities to eke out their living through intensive cultivation of crops to fragile mountain slopes. Cropping systems vary considerably with land types, elevation, slope, aspect, seasonal water availability, soil types, and their fertility. Multiple cropping, in combination with several crop rotations are predominant to safeguard food supply and meet dietary requirements of households. Land management practices such as terracing, traditional agro-forestry practices and intercropping are some of the best examples that have been developed by the farmers by their ingenuity to cope with the harsh and fragile mountain ecosystems. The mountain cropping systems have to face numerous natural and human-induced challenges, including land degradation and loss of agro-biodiversity, leading to food insecurity and unsustainable livelihoods. This region is highly vulnerable to environmental degradation and climate change is seen as a risk multiplier. Meeting the ever-growing food demands while sustaining land productivity and maintaining resiliency at the farm level is the major challenge faced by the mountain farmers. This

* The views expressed by the author in this article are his own and the organization bears no responsibilities for them.

chapter outlines the typical characteristics of the mountain cropping systems and discusses how they are coping with the ongoing natural and socio-economic dynamics. The vulnerability of the fragile mountain agro-ecosystems to climate change and its impacts on land use and cropping systems are discussed. The mountains agriculture seems to be highly vulnerable to climate change effects particularly erratic rainfall events and droughts, and land degradation due to soil erosion, landslides, flash floods, and siltation leading to loss of productive lands, crops failure and food insecurity. Effective measures to cope with such impacts in the mountain agro-ecosystem are suggested. Increased public awareness about the climate change effects, building adaptive capacity to cope with such effects, sustainable soil and water conservation practices, and resilient cropping practices (drought tolerant crop/varieties, change in crop rotations, and water use efficiency) seem to be the key strategies to be adopted to cope with the climate change effects in the fragile mountain agro-ecosystems.

Keywords: land use, cropping systems, crop rotations, climate change adaptation, and mountain agro-ecosystem

1. Introduction

Characterized by physical isolation, poor mobility, vulnerability to risks, and biophysical diversity (Jodha, 1992), the crop-livestock integrated systems of the mountain region are extremely sensitive to climate variability and change (Chhetri et al., 2013). Harsh climate, rough terrain, poor soils and short growing season often lead to low agricultural productivity and food deficits (Kurvits, et al., 2014). This is further compounded by the subsistence nature of the system, fragmented and small-sized farms, poor technical know-how, land degradation, and erratic climatic events (Subedi and Dhital, 2007). Rain-fed agriculture is dominant in the mountain region of Nepal, thus productivity is associated with the seasonal rainfall patterns.

Nepal's oblong 147,181 square km of land mass is located between 26° 22' to 30° 27' north latitude and 80° 14' to 88° 12' east longitude, between India and China, which falls within the Hindu Kush Himalaya region. It extends 885 km in the east-west direction and has a non-uniform mean width of 193 km from the north to south. It is a landlocked country, bounded on the east, south, and west by India and on the north by the Tibetan autonomous region of the People's Republic of China.

The elevation starts at about 70 m above the sea level in the south, adjoining the Indo-Gangetic Plain to the highest peak on earth, the Mt. Everest (8848 m) in the north. The elevation, slopes and aspects of mountains create difference in micro-climatic conditions, natural vegetation, land types, and cropping systems. Nepal's diverse terrain is comprised of five distinct physiographic regions including the flat plains, or the *Terai*, in the southern part of the country, rising to the middle hills or the Siwaliks, and to even higher elevations categorized sequentially as Middle Mountains, High Mountains, and Himalayas (Figure 1), the latter forming the highest mountain ranges in the world. Each of these regions represents a well-defined geographic area with distinct geomorphology, climate, and hydrological characteristics that are significantly different from each other. However, for the purpose of planning agricultural development, the country has traditionally been categorized into three ecological regions: the *Terai* (flat plains and southern portions of the Siwaliks), Hills (northern portions of the Siwaliks and middle mountains), and Mountains (high mountains and high Himal).

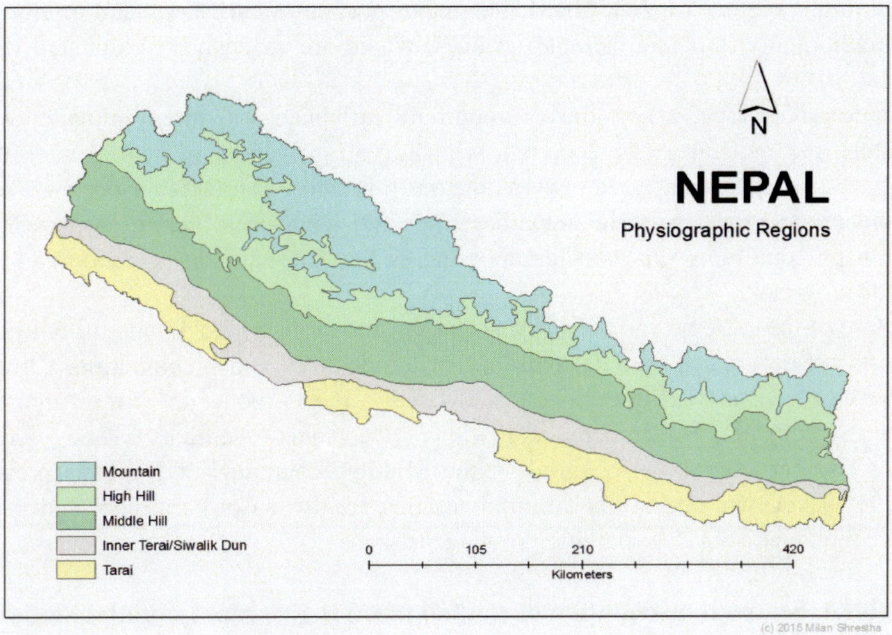

Figure 1. Five physiographic regions of Nepal.

As the Middle Mountains are intensively cultivated, densely populated and highly vulnerable to environmental degradation due to natural and human-induced causes, for the purpose of this Chapter, only this region is taken into consideration. This chapter outlines the biophysical and socio-economic characteristics of Middle Mountain agro-ecologies, current land use practices, cropping systems, predominant crop rotations and discusses how the adverse climatic conditions brought by the global warming impact to the mountain agriculture.

2. Biophysical Characteristics of the Middle Mountains

The Middle Mountain (Middle hills and High hills as shown in Figure 1 above) is a wider belt of land aligned east to west in the middle part of Nepal, bordered by the *Mahabharat* range in the south and the high mountains (Himalayan region) in the north. Intercepted by several north-to-south flowing rivers and gorges, the altitudes ranges from deep river-basins (≤ 300 m) and valleys to higher mountain ridges (≥ 2500 m). This region is characterized with rugged mountain terrain with variable elevations, slope angles, soil depths, and aspects resulting in diverse climate and vegetation. The lands are steep with shallow soils, and dissected and predominately (≤ 75%) un-irrigated (rain-fed). Over 80% percent of the land in Nepal is mountainous with rugged topography and steep to very steep slopes gradients (Shrestha et al., 2004), over 66% of the countries' land area falls under >30% slopping land and 12.7% is with 8 to 30% slope, and almost 19 million people inhabit in such marginal areas (Pratap, 2003). The Middle Mountain region of Nepal occupies about 42% of the country's total land area (147,181 km^2) and 43% of 26.8 million population, with the highest population density of 126/km (CBS, 2012). Of the of total land area, only about 21% is cultivated in this region (MoAD, 2013). The major portion of cultivated land in this region is worked traditionally into innumerable terraces, which are extensively cultivated (Shrestha, 1992).

Altitude and slope aspect have important influence on micro-climatic variations (temperature and moisture) of a location, which creates diversity in natural vegetation and suitability of crops particularly. In general, the north or north-east facing aspects are relatively cooler and retain more moisture than the south and south-west facing slopes. Therefore, micro-climatic conditions vary considerably and as a result, natural vegetation and cropping systems differ greatly.

Because of the extreme variation in topography and altitude, the mountains terrain, slopes and aspects create greater spatial variability of temperature and precipitation. Climatically, there are four seasons, as pre-monsoon (March –May), monsoon (June-August), post-monsoon (September –November), and winter (December- February). The precipitation pattern of five representative locations in the Middle Mountains of Nepal is presented in Table 1. However, wet and hot and humid summer (rainy season) and cool and dry winter season are the two distinctly separable growing seasons.

Table 1. Seasonal distribution of rainfall (mm) in five representative stations (from east to west) in the Middle Mountains of Nepal (Average of 1971-2000). The numbers in parenthesis are percentage of total annual rainfall

Station	Pre-monsoon	Monsoon	Post-monsoon	Winter	Total
Dhankuta (931 m)	183 (18)	723 (72)	65 (6)	38 (4)	1009
Kathmandu (1336 m)	203 (14)	1126 (78)	65 (5)	46 (3)	1440
Pokhara (827 m)	550 (14)	3127 (79)	195 (5)	79 (2)	3951
Dailekh (1402 m)	182 (10)	1504 (82)	55 (3)	96 (5)	1837
Dadeldhura (1848 m)	201 (15)	1004 (73)	48 (3)	131 (9)	1384

Source: CBS (2013).

The most outstanding feature of Nepal's climate is the monsoon precipitation, which is characterized by two distinct phases: the "wet" and the "dry." The wet phase (June-September) refers to the summer season, when warm and moist winds enter the country from the southeast. Diurnal temperatures and amount of precipitation vary greatly within a short vertical distance. Over 75 percent of the annual precipitation occurs during this phase (Webster, 1987; Shrestha, 2000). The variation in the pattern of rainfall from east to west is substantial and is further accentuated by the diverse terrain within each physiographic belt (Lang and Barros, 2002; Kansakar et al., 2004), creating many micro-regions with differing agricultural conditions. The amount of monsoon rainfall decreases substantially as it moves to the northwestern part of the country (Lang and Barros, 2002). Not only the amount of summer monsoon becomes less, the number of days with rainfall decreases as the monsoon circulation progresses toward the western part of the country, creating variable climatic regimes for rice cultivation (Kansakar et al., 2004). The dry phase (December-January) is the period when the direction of the winds reverse to bring cool and dry air from the northwestern part of the country (Webster, 1987). While precipitation is comparatively less during this time, winter rain tends to be more concentrated in the western part of the country than in the east.

As the rain-fed agriculture predominates in the mountain region, abnormally wet or dry monsoons have been directly responsible for frequent famines in certain parts of the country. Generally lower than normal or spatially variable monsoon rains are considered to be a cause of concern. Using the database developed by the Indian Institute of Tropical Meteorology for the years 1871–1997, Parthasarathy et al. (1994) reported 21 major droughts and 19 flood years when precipitation was at least +10% below or above normal in India. Though neighbouring Nepal lacks such a database, by virtue of lying in the same monsoon path, such findings hold true. Impending climate change may increase the intensities of these extreme climatic events (Kripalani and Ashwini, 1997) and the effect of such extreme events on agriculture should not be underestimated.

Over 80% of the population in these areas depends on agriculture as the primary source of livelihood. Because of limited land, intensive cultivation in steep slopes is common (Figure 2). Only about 26% of the agricultural land in this region is irrigated (CBS, 2012). Therefore, crop production depends primarily on seasonal rainfall and is thus prone to droughts and unreliable weather. The mountain agriculture has distinct bio-physical characteristics. The degree of diversity, fragility, marginality, human adaptation and inaccessibility are directly linked to factors such as elevation, slope angle, slope orientation, and exposure (Jodha, 1992). Small and scattered land holdings, slopping and shallow soil depths, marginal lands, rainfall-fed farming is the characteristic of a typical mountain farm. Farmers are in their desperate bid to maintain their livelihoods in such challenging environment. There exists a close interrelationship between crops, livestock and forests to fulfil the livelihood needs of resource-poor farmers and maintain ecological stability (Baul et al., 2013).

Figure 2. Typical terraced land in the Middle Mountain of Nepal showing the land use system.

Land resources in the mountains are undergoing degradation (Thapa and Paudel, 2002; Maskey et al., 2003; Acharya and Kafle, 2009). Degradation of productive agricultural lands is contributed by both natural conditions and directly or indirectly through human activities. Fragile geographical formations and heavy seasonal rainfall combined with several human activities such as deforestation, cultivation in sloppy lands, excessive tillage practices, over-grazing of animals, and improper infrastructure development and maintenance such as roads and urban development are considered as the key factors contributing the land degradation (Acharya and Kafle, 2009; Subedi et al., 2015). Soil fertility decline due to soil erosion and nutrients losses through leaching is a serious problem in the hills of Nepal (Subedi et al., 1989; Tripathi et al., 2003; Acharya et al., 2007). The production potential of land is reduced, which leads to further encroachment of forest and marginal lands and the intensification of cropping practices further depletes the fertility of soils (Maskey et al., 2003). Pratap (2003) analysed the indicators for unsustainable agriculture in the upland farming in the Hindu-Kush Himalayas during the period from 1954 to 1991 and reported that the cultivation in steep slopes (>30%) has increased by 10 to 15%, and soil erosion in the sloping lands increased by 20 to 30%.

In addition to loss of productive soils, decline in soil fertility and consequential decline in agricultural productivity is a growing concern in the mountain agriculture. Soil fertility problems associated with human-induced nutrient depletion are wide spread and the process of soil nutrient depletion is a potentially serious threat to world food security and sustainable agriculture (Tan et al., 2005). Thapa, (1996) and Neupane and Thapa (2001) have observed that rates of removal of plant nutrients from farm lands usually exceed the rates of their replenishment. Agricultural expansion into marginal lands and intensification in irrigated agriculture (triple annual crop rotations) are leading to rapid soil nutrients depletion (Schreier et al., 2005; Subedi et al., 2015).

3. Socio-Economic Characteristics of the Middle Mountains

Productivity of land not only depends on the biophysical characteristics of land but also on socioeconomic parameters of specific environment (Pratap, 2003). The typical mountain farming in Nepal has a number of socio-economic characteristics as follows, which make unique farming conditions and challenges for improved productivity.

I. *Small and fragmented land:* Small holdings, fragmented parcels and rain-fed lands (>75%) with low production potentials are typical characteristics of Nepalese land holdings. As arable land is limited and any expansion of cultivated land is at the expense of the forest, which is inherently unsustainable (Subedi and Dhital, 2007). Agricultural land holding in this region is very small: about 45% of the population owning less than 0.5 ha of land and are highly fragmented with about 4 parcels per holding (CBS, 2011). A rapidly expanding population (1.3% per annum) and urbanization has further reduced the average size of farming land. Limited lands are also encroached for urbanization and non-agricultural use such as construction of roads.

II. *Subsistence/labor intensive agriculture*: Agriculture in the mountains is predominately at the subsistence level with no mechanization. The labor intensive production practices lead to drudgery especially on women.

III. *Mixed and complex farming systems:* incorporating crops, livestock, and agro-forestry components is a common characteristic of mountain farming systems. The farming systems are complex and there is an interdependency of crop, livestock and forestry to each other. Livestock and crops also serve as complementary investments, with crops providing feed for the livestock while the livestock provides manure to crops.

IV. *Limited access to inputs and market:* Farmers in the remote mountain areas have limited access to markets, quality inputs, roads, and institutional credits. Therefore, poor, land-less, marginal farmers and women are further marginalized when it comes to accessing improved agricultural practices/technology (Subedi and Dhital, 2007). Lack of access to productive inputs and market make the farming as low-external input and organic manure-based. Agricultural value chains in the region are poorly developed.

V. *Traditional knowledge-based farming:* Farmers lack both productivity and business skills. While farmers are eager to increase their productivity, they frequently lack knowledge in crop management, crop and varietal selection, proper storage techniques, irrigation methods, and other agro-techniques. National agricultural research system is also not efficient in generating location specific technologies. Extension services are rarely available and those available are mainly focussed to the pro-rich or to influential farmers in the more accessible areas (i.e., road-sides); the farmers in the remote mountain areas are often isolated from modern agricultural technologies. As a result, they have poor technical knowhow leading to low productivity.

VI. *Food insecurity:* Because of limited agricultural lands with low production potentials, rainfall dependent cultivation, and low external inputs crop productivity is low, thus majority of the households in the mountain region are with food-insecurity (Subedi and Dhital, 2007; Krishnamurthy et al., 2013). Food production is rarely sufficient to meet the household requirements, and it is the major spending item for smallholder households.

VII. *Lack of off-farm employment*: Since income generative opportunities are rare especially in the rural areas and the majority of the farmers are poor with low cash income. Because of the lack of other off-farm employment opportunities, disguised employment is common phenomenon.

VIII. *Changing farming population:* There appears a new socioeconomic dimension in the country. There is an ever growing trend of youth outmigration in Nepal, resulting in rural population composed of elderly, women, disabled, children, and physically less active population. One of the greatest social challenges to farming in the Hindu Kush Himalaya region is from the outmigration of labour (Kurvits et al., 2015). Moreover, farming is not viewed as an attractive future by the young generation (Subedi et al., 1989; Acharya et al., 2007; Kurvits et al., 2015). The outmigration of youths is the root cause of shortage of agricultural workers leading to declining productivity. There is also a growing trend of land abandonments in the mountain region in recent years.

Out-migration of the able-bodied workforce has burdened already-overworked women, creating greater gaps in response to climate and other changes. Absent a radically different model of innovation and development for the agriculture sector, the coupled crop-livestock livelihoods system - once considered sustainable and in harmony with local ecological systems - is at risk. This phenomenon, commonly known in the region as the increasing *feminization of agriculture* coupled with the challenge of *climate variability and change,* is weakening the entire socio-ecological systems. These dynamics not only have adversely impacted the resilience of crop-livestock systems but have opened up forward-backward linkages (including market interactions) leading to an increased vulnerability at a range of scales.

4. Land Use Systems

As the livelihood of the majority of the population (≥75%) is dependent on agriculture and farming lands are limited, land becomes precious natural resources in Nepal. Land is the wealth and foundation of livelihood (food, income and employment). Despite a principal source of livelihood of majority of the people, agricultural land is highly unequally distributed across geographical regions and within households. The bottom 45% of the agricultural household operate only 12% of the total agricultural land area, while 5% occupy 27% of the total agricultural land (Sharma, 1999). Although there are distinct geographical and biophysical characteristics associated with biophysical conditions of the region, a coupled crop-livestock based livelihood systems characterizes them all.

Agricultural production in the Middle Mountains is determined by various factors including altitude, rainfall, slope, and aspect. A typical household possesses small, dispersed parcels of land in combination with a few numbers of livestock including small ruminants. Farmers in the mountains of Nepal have developed and practiced complex farming systems with close integration of crop, livestock and common pool resources (e.g., forest, rangeland). Demand for subsistence production has been met by the expansion of rain-fed agriculture (Brown and Shrestha, 2000). Agricultural lands are generally surrounded by forests, mostly managed by the communities. Crop residues contributes almost one third of the total livestock feed while rest is derived from communal forests and rangelands (Pilbeam et al., 2000).

Mountain farmers have adopted multiple strategies in their desperate bid to maintain their livelihoods in the face of an ever shrinking land base and dwindling crop yields (Pratap, 2003). Terracing in the steep hill slopes is an indigenous land management practice and as a means of extending cultivation in marginal lands. Because of the scarcity of flat lands, steep slopes are reclaimed by means of terracing (e.g., Figure 2) and shifting cultivation. The main purposes of terracing are (i) to conserve soil loss from the sloppy lands; (ii) make the land easy for agricultural operations, and (iii) to hold the runoff during the monsoon season for rice planting. Although terracing was found to be more costly to establish but have higher long-term financial returns than other soil and water conservation measures (Mishra and Rai, 2014). There exists a negative correlation between the terrace width and slope (Pandit and Ballad, 2004). Depending on the water availability, terraces are either levelled or slopping (up to 20%). Generally, the outward sloped terraces are common in the mountain slopes. While cultivation in the hill slopes without terracing can be found in certain part of the region, which further aggravates the problem of land degradation, maintaining the terrace is a labor-intensive endeavour. In addition to terracing, other indigenous practices such a contour bunding, vegetative barriers such as planting trees along the terrace risers and fence are common practices to protect/conserve soil are also common. Agricultural lands in the Middle Mountains of Nepal are categorized in different types based on the availability of water for irrigation, slope gradient, soil fertility/productivity, and aspects. The following are the major land types:

I. *Khet:* Levelled terraces with bunds, seasonally or year-round irrigated paddy lands. *Khet* lands are situated mostly in the foot-hills, river basins and valleys, which are irrigated with gravitational canal. In the hills slopes, this is created through the construction of bunds around the edge of levelled terrace so that uniform depth of water is maintained throughout the terrace (Figure 2). The flat *Khet* lands in the valleys and river-basins are called *Phant* while the hillside narrow terraces with bunds for the retention of seasonal irrigation are called as *Tari Khet.* The term *Khet* land refers to a type of land where water is held for considerable period of time by earth bonds to create favourable conditions for rice (*Oryza sativa*) cultivation. Therefore, rice based cropping patterns are predominant in the *Khet* lands. The irrigated lands are considered as prime lands and depending on the altitude and water availability, up to three crops are gown in a year in *Khet* lands.

II. *Bari:* The outward slopping, freely draining rain-fed lands are locally called as *Bari* lands. *Bari* lands become predominant as the elevation increases and where there is no possibility of canal irrigation. Traditionally, maize (*Zea mays*) based crop rotations are predominant in *Bari* lands. If irrigation is feasible, farmers become eager to convert *Bari* into *Khet* land and go for rice cultivation (Subedi et al., 1991).

III. *Tars: Tars* are also a type of *Bari* lands. *Tars* are large, flat rain-fed lands that are left-over by big rivers. They have greater potentials of converting into *Khet* lands if irrigation can be provided.

IV. *Khoriya*: Sloppy lands prepared after slash and burn such as in shifting cultivation. *Khoria* are used for planting upland rice, legumes such as ricebean (*Vigna umbellate*), horsegram (*Macrotyloma uniflorum*), blackgram (*Vigan mungo*) and raising nurseries for rice and finger millet (*Eleusine coracana*). One or two years of cultivation is followed by fallows for regeneration. This system is in a diminishing trend in recent years.

V. *Kharbari*: marginal lands protected for collection of thatch and grasses. Generally such lands are too steep or rocky, less fertile and unsuitable for cultivation of crops or conversion into terrace.

The *Bari* and *Khet* are the two major land types in the Middle Mountains. The *bari* land constitutes about 64% of cultivated land and over two-third of this lies in the middle mountains (Carson, 1992). The irrigated lands are considered as precious lands. Where water is available for irrigation and slope permits for bench terracing, farmers convert their *Bari* and *Tars* lands to *Khet* and paddy rice is the most preferred crop (Subedi et al., 1991; Pandit and Ball, 2004; Shrestha et al., 2004).

5. Cropping Systems

Cropping systems is defined as cropping patterns used on a farm and their interaction with farm resources and other farm enterprises. Cropping systems evolve based on climate, soil, water availability, and farmers' priorities for the crop commodities. Traditionally, cropping systems in the Middle Mountains of Nepal have been relied on the close integration of forestry, livestock and crop production. A typical farmer grows several types of food crops, fruit trees and vegetable crops around the homestead, keeps both large and small ruminant animals, backyard poultry, and plant fodder trees along fences, terrace risers and in the marginal lands. Crops may include cereals, food legumes, vegetables, spices such as ginger (*Zingiber officinale*) and turmeric (*Curcurma domestica*). Home garden may composed of a few fruit trees, seasonal vegetables, and herbaceous plants. The use of tree-forage and fodder from forest areas and terrace risers as animal feed ensures the flow of nutrients from common pool resources to agricultural land (Pilbeam et al., 2000).

Multiple cropping (i.e., growing two or more crops at the same time on the same piece of land or in a given growing season) is the most common practice adopted by the subsistence mountain farmers as a means of meeting their multiple needs. Confronted with problems of increasing food demand, small landholdings, and the lack of non-farming opportunities, farmers need to ensure maximum possible harvest to meet their food-requirements (Thapa, 1996; Thapa and Paudel, 2002). Multiple cropping is a strategy of securing more produce by tapping limited resources (Thapa, 1996) as well as buffer to avoid a complete crop failure due to disease/pests or unusual climate. In addition to meet their food and fibre demands, multiple cropping is also adopted to capture the niche endowed by mountain ecosystems. The crop combinations vary greatly with altitude, growing season, aspects and land types. Knowingly or unknowingly, the multiple cropping systems provide greater soil cover, which is important in terms of soil and water conservation.

Various forms of multiple cropping systems are practiced to accommodate more than one crop in the system in a given piece of land, within a given period of time or growing season. The aim of multiple cropping is diversifying crops and growing more than one crop at a time or within a given time. This system is adopted to produce more foods, better economic returns and utilize resources available during the growing season (i.e., land, water and solar radiation) more efficiently so as to increase farm productivity per unit land. The most common types of cropping systems in the mountains of Nepal are as follows:

i. *Sequential cropping:* The sequential cropping system refers where more than one crop is grown in a piece of land in a sequence, within a given time frame. The second crop is planted once the previous crop is harvested. This system of crop rotation is denoted by (-) sign. For example, rice-wheat-maize (3 crops in a year) or maize-potato-barley (3 crops in two years).
ii. *Mix-intercropping:* This is a kind of intercropping in which, two or more crops are grown simultaneously in the same field. This system of intercropping is denoted by (+) sign. The companion crops are planted at the same time without row arrangements. For example, wheat + peas, millet + legumes. In this system of intercropping, more emphasis is given to the main crop and the plant density of the main crop is not reduced to accommodate the companion crop. Generally, the companion crops are harvested at different times as the crops do not mature simultaneously. Using combine harvester (where mechanization is available) is a limitation in such intercropping system.
iii. *Row or strip-intercropping:* Two or more companion crops are grown together in different rows or strips arrangements in a given land. Crops are sown in different

rows without affecting the population of the main crop when sown as sole crop. The main objective of row-intercropping is to utilize the space left between two rows of the main crop. This system of intercropping is also denoted by (+) sign. For example, maize + soybean. Generally, the companion crops are harvested separately as the crops mature at different times.

iv. *Relay-intercropping:* This is a type of intercropping in which two or more crops are grown simultaneously during the part of the life cycle of each crop. The second crop is planted once the first crop is established but before harvest so that part of the companion crops share the land at least part of their growing season. This system of intercropping is denoted by (/) sign. For example, maize/finger millet, rice/lentil, where lentil (*Lens culinaris*) seeds are broadcasted under rice crop before its harvest, lentils grow and establish following rice harvest. Relay inter-cropping system is practiced primarily to accommodate two crops in a limited growing season, where two crops in a sequence is not possible to grow.

v. *Agro-forestry (tree/crop) system:* Although the agro-forestry system is not generally considered as an intercropping system, in the context of mountain cropping systems, we think that this system should be included under this topic. Growing certain shade-loving annual and perennial crops such as vegetables, ginger, turmeric, herbs, coffee under the shade of fruit trees (e.g., mandarin) and fodder trees is a common practice in the mountains of Nepal, adopted to better utilize the scarce land resources. Mountain farmers also utilize their marginal lands by growing a number of multipurpose trees such as fodder trees in the bunds, terrace risers, and along the farm fence.

The advantages of crop rotations and inter-cropping systems are well documented elsewhere. Various indexes are used to assess the benefits of these systems. Some of the more common indices are described below:

(i) Multiple Cropping Index (MCI):

The MCI is defined as the ratio of total area cropped in a year to the land area available for cultivation, expressed in percentage. It is calculated as follows:

$$MCI = \frac{\sum_{i=1}^{n} ai}{A} \text{x } 100$$ (Dalrymple, 1971)

Where, "n" is the total number of crops, "a" is area occupied by i^{th} crop and "A" is total land area available for cultivation. The MCI is also referred as cropping intensity (CI).

(ii) Land Equivalent Ratio (LER):

The LER is defined as the relative land area under sole crop that would be required to produce the equivalent yield under a mixed or intercropped condition at the same level of management. This is the most common index to measure the benefits of intercropping system. The LER is calculated as follows:

$$LER = \frac{\text{YA}mix}{\text{YA}mono} + \frac{\text{YB}mix}{\text{YB}mon}$$ (Willey and Osiru, 1972)

Where, Y_A and Y_B are the yields of component crops "A" and "B" under intercropping conditions (mix) and as a pure stand (mono). An LER value of 1.0 indicates no difference in yield between the sum of intercrops and the sole crop.

(iii) Monetary Equivalent Ratio (MER):

The monetary equivalent ratio (MER) measures the economic advantage of intercropping over the sole crop that has the largest economic return (Adetiloye and Adekunle, 1989). It provides the economic or monetary value of the intercropping against the sole crop monetary value. The MER is calculated using the following equation:

$$MER = \frac{\text{ra+rb}}{\text{Ra}}$$ (Adetiloye and Adekunle, 1989)

Where "ra" and "rb" are monetary returns from component crops "a" and "b," respectively, under intercropping situation and "Ra" is the highest sole crop monetary return.

6. Crop Rotations

Viewed as increased utilization of arable land for higher production, crop rotation is a systematic approach of growing crops in a defined sequence on the same piece of land, for a given period of time. Within a given crop rotation sequence, there may be a sole crop or intercrops during a growing season. Crop rotation has several benefits on soil fertility, disease/pest management, nutrients and water use efficiencies, diversifying farm income, and economic returns. In the Mountain region of Nepal, crop rotations are associated with change in land-use and shortening of fallow period so as to maximize the land productivity from limited available land.

Choice of crops and their rotations are determined by various factors such as type of land, elevation, availability of seasonal water, soil fertility conditions, and household and/or market demands. In the mountains of Nepal, temperatures (as determined by elevation and aspect) and seasonal water availability, inputs and labor availability are the major determinants of the choice of crops and their rotations. The major crop rotations in different land types and altitude range in the Middle Mountains of Nepal are summarized in Table 2.

In the lower hills below approximately 1000 m above sea level (valley bottom and foot-hills), where seasonal water availability is not limited for irrigation, up to three crops are grown in a year. At up to 1800 m altitude, generally two crops are grown in a year and above this range, growing two crops in a year is difficult; therefore, generally three crops are grown in a two-years rotation. Irrespective of biophysical conditions associated with different regions of Nepal, crop rotation implies higher frequency of cultivation with higher inputs of labor and possibly other inputs.

It is apparent from Table 2 that the upper regions of mountains which are almost entirely rain-fed, maize-based crop rotations are predominant in the *bari* lands and rice-based rotations in the *khet* lands. The listed crop rotations are only based on the major crops (rice, maize, finger millet and wheat). Among various crop rotations, the maize/millet intercropping is the predominant cropping system used by the mountain farmers throughout the Middle Mountains. There are numerous other minor crops such as horticultural crops, spices and herbs like ginger, cardamom *(Amomum* sp.) and coffee (*Coffea arabica*) cultivated under fruit trees or under agro-forestry system. In the upper mountains, some minor crops such buckwheat (*Fagopyrum esculentum*), grain amaranths (*Amaranthus* sp.), proso millet (*Panicum maliacum*) and foxtail millet (*Setaria italic*) are also cultivated in rotations.

Table 2. Land types and major crop rotations in the Middle Mountains of Nepal

Land type	Elevation (m)	Major Crop Rotations*
Bari	>2300	Maize or finger millet-potato-barley (3 crops in two years) Potato-wheat or barley-maize or finger millet (2 years rotation) Beans-wheat (1 year) Beans-proso or foxtail millet (1 year)
	1600-2200	Maize/finger millet-fallow Maize-mustard (oilseed rape) Maize/finger millet - mustard Maize-wheat or wheat + peas Maize- barley + peas Maize-potato Maize-buckwheat
	1000-1500	Maize/millet-fallow Maize+soybean-mustard Maize-potato or vegetables Maize –wheat+mustard Maize-wheat+ peas
	300-900	Maize/millet-mustard Maize-millet + legumes** Maize+soybean-whaet Maize-vegetables/potatoes
Tars	300-700	Upland rice-blackgram Upland rice-blackgram + niger Maize-soybean-mustard Maize-blackgram or cowpeas Upland rice-sesame Maize-blackgram or soybean
Khet	1600-2200	Rice-fallow Rice-potatoes Rice-buckwheat
	1000-1500	Rice- wheat-fallow Rice-potato-fallow Rice- maize+dry beans or cowpeas Rice- vegetables-maize Rice-fallow- foxtail millet or figer millet
	<1000	Rice-fallow-maize Rice-wheat-maize Rice-wheat- rice Rice-fallow-rice Rice-wheat-fallow Rice-vegetables-maize Rice-potato or vegetables-maize

* The symbols +, / and – are for mixed cropping, relay-intercropping, and sequential cropping, respectively.

** Grain legumes such as blackgram, cowpeas (*Vigna unguiculata*), soybean (*Glycine max*), horsegram (*Macrotyloma uniflorum*) and ricebean (*Vigna umbellata*) are generally mixed with finger-millet.

Source: Compiled from Subedi et al. (1989; 1990); Subedi (1991); Karki (2001); Raut et al. (2011).

Relay cropping of finger millet with maize is a pre-dominant crop rotation in the *Bari* lands of Middle mountains. This is a unique example of cereal-cereal inter-cropping, developed by farmer's ingenuity to accommodate two important crops in the same piece of land within a limited growing period (Subedi, 2001). At and above 1000 m elevation, maize and millet in a sequence are not possible to grow because flowering time of millet coincides with cold temperatures. The major reasons given by the farmers for adopting the maize/millet relay intercropping system are to (i) accommodate two major staple crops in a limited growing season, (ii) effortless land preparation, (iii) better utilize residual soil moisture and nutrients, and (iv) better labor distribution. A typical maize and millet relay intercropping system is illustrated in Figure 3.

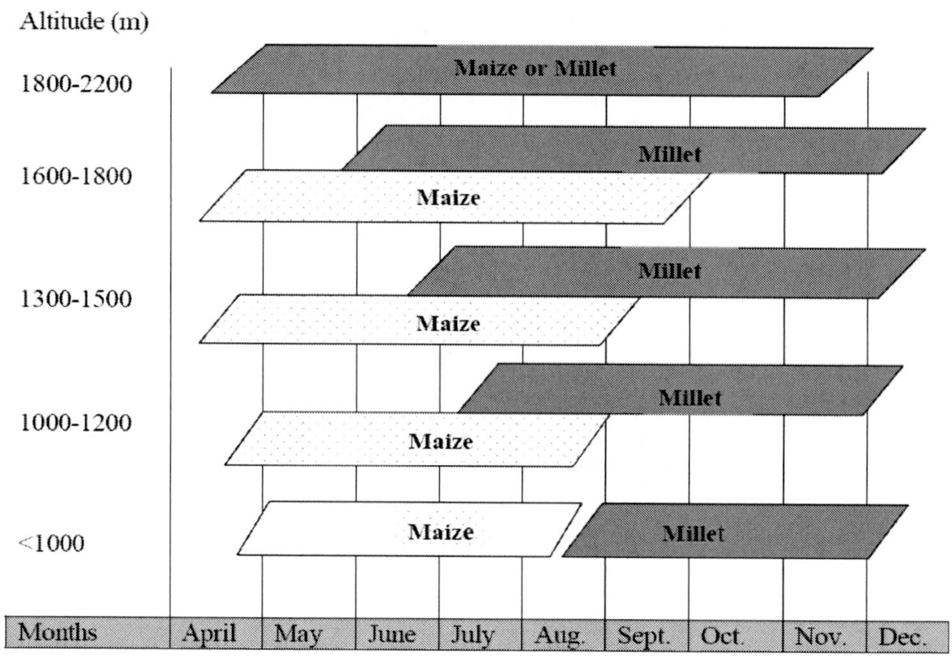

Figure 3. Timing of maize and finger millet planting and duration of overlapping in the maize/finger millet relay-intercropping system at different altitudes in the Middle Mountains of Nepal (After Subedi, 2001).

As the figure illustrates, maize and finger millet are grown separately in sequence at lower altitudes (<1000 m) and at higher altitudes (>1800 m). Occasional relay systems are also observed at 700-1000 m. As the altitudes rises from 1000 m, the overlapping period becomes longer and the competition between the two component crops increases and *vice versa* (Subedi et al., 1991; Subedi, 2001).

Traditionally, farmers in the mountains include different types of grain legumes in rotations. Soybean, blackgram, cowpeas, field beans (*Phaseolus vulgaris*), and ricebean are the warm season legumes while lentil, chickpea (*Cicer arietinum*), and field peas (*Pisum sativum*) are the three important cool season legumes. The warm season legumes are either intercropped with maize or finger millets, or grown as sole crop in rotation. They are also commonly planted on the bunds of rice fields and in terrace risers. The cool season legumes are also either intercropped with wheat and barley or grown separately following rice or maize.

The mix-cropping and crop rotations by the mountain farmers are the traditional cropping systems adopted since long time. Mountain farmers have continuously changed their farming systems over time to cope with their needs and opportunities. Change in land use intensification is characterized by changes in cropping patterns, fertilizers use, irrigation and mechanization (Chhetri and Easterling, 2010; Raut et al., 2011). Recently, some new crops have been introduced in the crop rotations, especially in the lower elevation areas with access to market, technology, and assured irrigation. For example, the introduction of commercial vegetables and potatoes in the rice-based systems has prompted innovation of new cropping patterns. Increasing market demand, access to inputs and technologies (e.g., off season vegetable production) accompanied by the availability of irrigation has been a major driver of the change in traditional crop rotations. On-going social changes such as migration of able-bodied labor force to non-farm sector and increasing urbanization in the productive lands have also been the major factors inducing changes in traditional crop rotations or abandment of some crops or rotations. While the new cropping patterns are emerging, the staple crops such as rice (in *Khet*) and maize (in *Bari*) are still predominant across the middle mountains of Nepal.

7. Climate Change: Trend and Potential Impacts on Mountain Agriculture

Climate change referrers to the change in the state of climate that can be identified by changes in the mean and/or variability of its properties and that persists for an extended period, typically decade or longer (IPCC, 2007). Changes in climate are driven by natural processes (or external forcing), or by anthropogenic changes in the composition of the atmosphere or in land use. As a result, climate change is expected to bring an increase in the frequency, intensity, spatial extent, and duration of weather and climate extremes (Lavell et al., 2012). Recent report of the Intergovernmental Panel on Climate Change (IPCC) shows that over the last 50 years, extreme events have been on the rise in most regions of the world (Field et al., 2012).

The most outstanding feature of Nepal's climate is the monsoon precipitation, which is characterized by two distinct phases: the "wet" and the "dry." The wet phase (June-September) refers to the summer season, when warm and moist winds enter the country from the southeast. Over 75 percent of the annual precipitation occurs during this phase. The annual cycle of the monsoon determines the practice of agriculture in Nepal's mountain region. Abnormally wet or dry monsoons have also been directly responsible for frequent famines in certain parts of the country.

Shrestha et al. (1999) showed a general warming trend in Nepal. The temperature differences are most pronounced during the dry winter season, and least when the monsoon peaks. The study also showed significant warming in the higher elevations of the Hills and Mountains in the western half of the country compared to the lower elevations in the south. Unlike temperature, there is no evidence of change in aggregate precipitation (Shrestha,

2000). However, studies conducted to examine stream flow show an increase in number of flood days in certain rivers (Shakya, 2003).

Based on the observed trends of change in temperature as reported by Shrestha et al. (1999), Chhetri and Easterling (2010) reveal a negative association between the amount of rainfall and general trends of warming. For example, the western mountain region of country received lower than average rainfall and exhibited a higher degree of warming compared to the central and eastern regions, which are comparatively wetter. Theoretically, if this trend continues in the foreseeable future, the drier regions of the country will become even drier due to the projected increase in temperature. For farmers, such a prognosis imposes further challenges in their effort to ensure better food security.

The Organization for Economic Co-operation and Development (OECD) has assessed the change in average temperature and precipitation in Nepal using over a dozen general circulation models (Agrawala et al., 2003). There is a significant and consistent increase in temperatures projected for Nepal for 2030, 2050, and 2100. While the study also projects an overall increase in precipitation, mostly during the monsoon season, it is not clear whether the existing rainfall patterns will remain the same. It is also not apparent how these changes will affect the timing and period of monsoon rainfall. The potential increase in monsoon precipitation in an area that has already experienced heavy rainfall may lead to more flooding. Conversely, areas with low monsoon rainfall may be subject to drier conditions in the future. In short, it is difficult to make predictions with any degree of certainty.

7.1. Vulnerability of Mountain Agriculture to Climate Change

Because the mountain region is naturally fragile, they are susceptible to accelerated soil erosion due to rain with downstream effects. Lands are highly prone to mass movements including landslides, avalanches, debris flows and flooding, due to high and erratic rainfalls that can lead to disasters. With the persistent pressure of population growth, Nepal's mountains are intensively cultivated, and are highly susceptible to the impacts of climate change. Evidence of climate change, such as general warming, receding snowline, prolonged drought, and unpredictable rainfall patterns, has been well documented in the mountain region of Nepal (MoE, 2010). By analyzing the time series of Indian monsoon, Rajeevan et al. (2008) reveal an increasing trend of the extreme rain events between 1901 and 2005. The study also reveals a stronger trend of the extreme rain after 1950. Likewise, Sen Roy (2009) also found widespread increases in heavy precipitation events across India, mostly in the high-elevation regions of the northwestern Himalaya as well as along the foothills of the Himalaya extending south into the Indo-Ganges basin.

Farmers in the mountains of Nepal face a range of socio-economic pressures, including population growth, a low level of technology usage, and an exodus of able-bodied labor forces (mainly males) to other economic sectors, that compound the impacts of climatic changes. The changing climate is an additional burden to the poor people in the mountains who are already living in the poverty, are vulnerable and excluded with prediction of additional risks to livelihoods and further inequity in the future (Gentle and Maraseni, 2012). Since the agricultural production is heavily reliant on monsoonal rain, any changes in the monsoonal patterns, their intensity and frequency can impact on crops to be grown, crop-rotations to be followed and ultimately agricultural production. Poor and marginalized households were more vulnerable to the climate change impacts (SAGUN, 2009).

7.2. Indications of Climate Change in Nepal

Nowhere in the world is the need of addressing possible consequences of climate change are more pressing than the mountains of Nepal as it already having a pronounced impact on agriculture, water resources, energy, health, and biodiversity sectors. It is also important to note that observed indicators of climate change is not uniform across space and time. For example warming is not uniform across the country, with higher increases observed in high altitude regions (Shrestha and Aryal, 2011). Annual precipitation data shows a general decline in pre-monsoon precipitation in far- and mid-western Nepal, while there is a general trend of increasing pre-monsoon precipitation in the rest of the country (Shrestha et al., 1999). This can have severe impacts on the agriculture system (Malla, 2008) in general and crop rotations in particular. Recent extreme climatic events such as longer periods of drought, shorter but more intense periods of rainfall and variable monsoon rainfall are some of the climate change impacts in Nepalese mountains. Several studies (Sagun, 2009; Gentle and Maraseni, 2012; Krishnamurthy et al., 2013; Baul et al., 2013) have reported increasingly erratic rainfall and unpredictable onset of monsoon seasons, prolonged drought, landslides, storm, glacial retreat, are the major effects of climate change in Nepalese mountains. Himalayan glacier melt and retreat have also been documented (Bolch et al., 2012), with an increased risk in glacial lake outburst flooding (MoE, 2010; Shrestha et al., 2004). The potential increase in monsoon precipitation in an area that has already experienced heavy rainfall may lead to more flooding. Conversely, areas with low monsoon rainfall may be subject to dryer conditions in the future.

There are also perceptions of the farmers that climate has changed in Nepal. In a recent survey with farmers in the 148 households in the Middle Mountains of Nepal, Baul et al. (2013) reported that almost all farmers perceived that summers are becoming hotter and longer while 81% of the interviewed farmers responded that winters are becoming warmer. Similarly, a focussed group discussion and key informant interviews with communities has reported decreasing and erratic trend of pre-monsoon and monsoon rainfalls in recent years (Gentle and Maraseni, 2012; Devkota, 2014). In a case study conducted in three ecological zones of Nepal, SAGUN (2009) indicated that temperatures increased at all sites and rainfall patterns were altered such as delayed monsoon, erratic rainfall, shorter rainfall duration and reduced winter rainfall. The major climate risks in the study sites were droughts, landslides, floods, and riverbank erosion, fire and hail stones. These changes indicate that unpredictable climate variability will be a major obstacle for agricultural production.

Farmers' experiences on such climate changes with erratic pre-monsoon and winter rainfall was also justified by meteorological data (Gentle and Maraseni, 2012; Baul et al., 2013; Devkota, 2014). Krishnamurthy et al. (2013) observed no clear trends of precipitation across the country, but generally minor or no decreases in precipitation in the western Nepal, an increase of up to 10% annual rainfall in eastern Nepal. The majority of this increase is due to more intense monsoon precipitation, resulting in up to a 20% increase in rainfall in the summer months. While a decrease in post-monsoon rainfall in the winter months in the western region leads to droughts in the western mountains. Although there is a spatial and temporal variability in rainfall, Krishnamurthy et al. (2013) found that the rainfall over the whole country has decreased since 1960 but there is a high inter-annual variability. The decrease has largely been due to a decline in mean precipitation during the dry season (December to February).

As predicted by Dixit (2015), the observational records also show a general warming trend in Nepal. The mean temperature in Nepal has reported to be increased by 1.8°C in the last 32 years (Malla, 2008), and the increase was greater in the Himalayas (0.08°C/year) than in the *Terai* region (0.04°C/year). The general circulation models and regional circulation models have indicated an increase in temperature across Nepal (overall temperatures have increased by around 1.5°C over the period of 1975 to 2009), due to increase in atmospheric greenhouse gas concentration (Krishnamurthy et al., 2013). Based on the model they used, the temperatures in Nepal are expected to increase by 1.2 to 1.4°C by 2030 compared to the 2000 baseline.

7.3. Impacts of Climate Change on Crop Production

In Nepal, climate change many manifest in the form of rainfall irregularity, intensity, prolonged drought, flood, drying of stream, heat waves, and extreme and unpredictable weather patterns. The impacts on water resources include small, gradual changes in climate change-related hazards, such as glacial lake outbursts, landslides, debris flows, and floods that will specifically impact the Nepali Himalayas (Nyaupane and Chhetri, 2009). Any changes such as delayed monsoon, prolonged drought or intense rainfall events will have significant negative consequences on existing cropping systems and of agricultural production. Such impacts may be intense at high elevations and in regions with complex topography as is in the Nepal's mid-hills (Dixit, 2015).

Current practice of agriculture in the mountains of Nepal is as such constrained by incidence of natural disasters: floods, droughts, landslides, intense rains, hailstorms and cold and heat waves (Selvaraju et al., 2014). Projected scenarios of climate change suggest that climate conditions in Nepal will worsen, which may imply an increase magnitude and intensity of climatic extremes (Malla, 2008; Selvaraju et al., 2014). Karki et al. (2009) reported that the climate change is already threatening Nepal's food security, human habitats, water resources, and tourism sectors seriously. Overall, climate change is projected to cause food production to fall, with lower yields with major crops (Cameron, 2014). A recent study by Challinor et al. (2014) showed that global warming of 2^0C will be detrimental to crop yield from the 2030 onward. The major impacts of climate change in the mountain agriculture in Nepal seem to be as follows:

i. Land Degradation

Land degradation entails a loss of productive capacity of land and is often induced by human activities, which has environmental, economic, and social consequences. The extent of land degradation depends on the geology of the land, rainfall patterns, and land use practices such as tillage, crop rotations, and cultural practices. The land degradation is a crucial issue in the mountains of Nepal because of the geology and steepness of the slope (Shrestha et al., 2004). Loss of productive topsoil through soil erosion during the monsoon season is the major driver of land degradation. While soil erosion is a natural process, it is accelerated by human activity, especially cultivation of crops in delicate hill slopes. Increased intensity of rainfall, as has been observed in recent decades, might significantly increase the potential for soil erosion in the mountains of Nepal. A study by Yang et al. (2003) estimated that nearly 60 percept of 0.38 mm ha^{-1} per year of global soil erosion is induced by human activity. In Nepal, approximately 240 million cubic meters of topsoil are eroded annually (Maskey et al., 2003) and the declining trend of agricultural productivity in the mountains is attributed to the erosion of topsoil causing soil fertility to decline.

The geologically young mountain slopes of the Hindu Kush Himalaya region are highly susceptible to soil erosion (Bartlett et al., 2010). They are also prone to landslides when exposed to unfavourable climatic factors such as intense rainfalls. Such fragile land types combined with inappropriate farming practices can lead to severe land degradation. While the rate of soil erosion vary significantly depending on the types of sediment source, land cover, topography, climate, and land management practices (Ghimire et al., 2013), intensive monsoon rainfall generating more runoff and less infiltration into the ground is an important driver of soil erosion in the mountains of Nepal. If unabated, the erosion of topsoil can have a detrimental effect on crop production and food security of the large number of smallholder farmers whose livelihoods is derived from growing crops in the rain-fed mountain slopes (UNEP, 2015). Gardner et al. (2000) estimated that soil loss through erosion in the Middle Mountains of Nepal varied from 2 to 105 t ha^{-1} yr^{-1}. Additionally, Ghimire et al. (2013) estimated soil losses from various types of erosion in a degraded catchment of the Siwalik Hills such that landslide has the greatest erosion (26 t ha^{-1}) followed by sheet (16 t ha^{-1}) and gully erosion (14 t ha^{-1}). They also observed a total sediment loss of as high as 64 t ha^{-1} annually, within a catchment in the study area. The extent of soil loss due to erosion in various land use types and crop rotations is summarized in Table 3.

As summarized in the table, the bench-terraced rice fields (*Khet* land) and densely forested hill slopes have minimal soil loss while the outward slopping terraces have the highest level of soil loss. Increased deforestation, overgrazing, and construction of roads and irrigation canals without proper consideration of the fragility of the slope are some of the major human-induced factor of soil degradation.

Studies have revealed that as much as 60-90% of annual soil loss is associated with first couple of major pre-monsoon storms (Schreire et al., 2005; Aterya et al., 2006; Tiwari, 2009). This is also the beginning of crop growing season with little or no ground cover on agricultural fields and degraded sites. This leads to preferential removal of soil's organic carbon and clay contents (Yang et al., 2003), raising a serious issue of nutrients depletion in the mountains of Nepal (Acharya et al., 2007). Carson (1992) estimated that a 5 t ha^{-1} soil loss is equivalent to a loss of 75 kg ha^{-1} of OM, 3.8 kg ha^{-1} of N, 10 kg ha^{-1} of K and 5 kg ha^{-1} of P in the mid-hills of Nepal. Another important aspect of the soil loss is the sedimentation in the plains and water reservoirs. Transfer of sediments from the hills to the downstream (valleys, river basins, reservoirs and the plains) through land surface, gullies, landslides, cutting of river banks, and rivers changing courses makes the fertile lands unproductive. Structural damage by floods carrying sediment loads is another significant factor causing land degradation.

ii. Drought

For the successful crop production in a rain-fed environment, rainfall at regular interval is a must. Prolonged and frequent droughts can have severe consequences to agricultural production thus the livelihoods of smallholder farmers. The severity of drought will be even more pronounced in the rain-fed agriculture, typical as of the mountains of Nepal. Climate change is expected to increase the frequency and the magnitude of dryness, increasing the risk of crop failure and food insecurity (Gentle and Maraseni, 2012). A prolonged drought also causes a drying of springs, depletion of ground water, decrease in natural recharge, and reduction in river flow. Lower-than-average rainfall can impact the country's agricultural productivity and food security in profound ways (Chhetri et al., 2013). Lower-than-average rainfalls can have profound impact the region's agricultural productivity and food security. In the mid- and far-western regions of Nepal, for example, less than 50% of the normal rainfall in the winter season of 2009 reduced wheat and barley production by 14% and 17%, respectively (MoAC/WFP/FAO, 2009). As a result, 43 of 75 districts were reported to be food-deficient in the same year. Therefore, the impacts of climate change on crop yield vary spatially and temporally, with impacts becoming greater as the rainfall become erratic.

Table 3. Estimated soil loss from different land use systems in the Middle Mountains of Nepal

Land type	Altitude range (m)	Cropping patterns	Annual rainfall (mm)	Average soil loss (t ha^{-1} yr^{-1})	References
Bari lands (labelled terraces)	600-1850	Maize/finger millet	1591-3524	2.5-3.4	Tripathi et al. (2003)
	1200-2000	Maize/Finger millet	3100–3600	2.5-5.0	Gardner et al. (2000)
	1500	Maize+soybean	1455-2143	11.1-16.6	Aterya et al. (2006)
Bari land (Outward sloping	782-1201	Maize/Finger millet	1304	17.7-32.0	Shrestha (1997)
	680-1850	Maize/Finger millet		2.7-12.9	Gardner and Gerrard (2003)
Khet lands	782-1201	Rice-wheat	780	0.2-0.3	Shrestha (1997)
Grazing lands				0.8-8.1	
Degraded forests				0.5-2.5	
Dense Forests				0.3	

iii. Increased Incidence of Diseases/Pests

Warm temperatures and humid conditions are generally considered to be favourable for disease and pest progression in crop and livestock. As temperature rise, vector borne diseases of crops and livestock can increase in the areas where they were not present traditionally. Rising land temperatures, change in precipitation patterns, and increased frequency and intensity of extreme heat undermine natural regulation of pest and disease, while increasing the ranges of various pests, thus expected increasing damage of crop (Cameron, 2014). Additionally, changes in mean temperature during the crop-growing season in the mountains may alter the cropping patterns and may also increase the incidence of diseases and pests. Studies on aphids and moths have shown that increasing temperatures can allow insects to reach their minimum flight temperatures sooner, aiding in increased dispersal capabilities (Woiwod, 1994).

iv. Shift in Traditional Knowledge and Biodiversity

Climate change effect especially the rise in temperatures is likely to cause shifting in traditional crop zones and crop phenology (e.g., early flowering and maturing). Altered cropping practices such as planting and harvesting times have been observed due to climate change effects. There are reports of extinction of natural vegetation (Malla, 2008) and threat to pollinating insects on a global basis (Cameron, 2014) because of the climate change effects.

8. ADAPTATION TO CLIMATE CHANGE

The sensitivity of crop production to climate makes agriculture highly vulnerable to the risks associated with climate change especially in the fragile mountain agro-ecosystems. In general, adaptation may lessen crop yield losses due to climate change or may improve yield in regions where beneficial climate change occurs. For the purpose of this chapter, we define adaptation as efforts to prepare for and withstand current or future impacts at the intersection of social and environmental changes. Hence, adaptation refers to all responses that may be used to reduce severity or actions designed to take advantage of new opportunities that may arise as a result of climate change. Adaptation strategies may range from short-term fixes to incremental change or transformation of whole systems in question. Three main objectives of adaptation are: (i) reduce vulnerability, (ii) enhance resiliency, and (iii) innovation of technology on demand. Adaptation can be pursued at all levels of governance and scales of action, from on-farm innovation that reduces the deleterious effects of climate change to climate forecasting to avoid loss of crop.

Historically, farmers across the mountains of Nepal have shown a strong capacity for adaptation to social and climatic stimuli by devising a wide range of technologies and social strategies to cope with variable climatic conditions. They have learned to thrive in a wide range of climatic conditions, spanning from extreme cold to hot and from very dry to a humid climatic conditions. It is therefore, reasonable to expect that farmers in the mountains of Nepal respond to new crop growing environments brought about by climate change the same way they have responded in the past. To cope with these constraints, farmers must increase their adaptive capacity and become more resilient. Following adaptive strategies can be incorporated:

i. *Conservation Agriculture:* Conservation agriculture (CA) is a concept for optimum use of local resources to achieve acceptable level of crop production while concurrently conserving crop biodiversity and the environment. If applied appropriately, the CA is expected to increase natural biological processes above and below the ground. According to Hobbs et al. (2008), some of the principles of CA include: i) minimal soil disturbance, ii) conservation of soil through enhancing permanent soil cover, iii) focus on crop rotations as a land management practices, and iv) minimal use of external inputs such as agrochemicals and inorganic fertilizers. Minimum soil disturbance (e.g., no-till or reduced tillage), proper management of crop residues, crop rotation as well as integrated nutrient management are essential for successful CA. When no-till practice is properly managed, the soil structure and its biological life will improve; runoff, erosion, and labor will decrease. At the same time, a well-covered soil will also increase the rate of infiltration of the surface runoff. Atreya et al. (2006) showed that reduced tillage could be a viable option for minimizing soil and nutrient losses without sacrificing economic yield losses. The CA reverses degradation process, improves soil quality, reduces production costs and helps achieve high productivity (Karki and Shrestha, 2014).

Without a proper knowledge and approach to CA, planting a crop without tillage will have problems and the farmers will likely fail, blaming no-till, not the lack of management (UNL, 2014). The CA practices are also considered climate smart agriculture as it helps to sequester carbon, improve water management and substantial increase in crop yields. Farmer's centred participatory approach (with focus on public-private partnership) that accelerates the technology generation and adoption of the CA by smallholder farmers should be the focus of adaptive agriculture in the Middle Mountain region of Nepal.

Ground cover and cultivation activities appeared to be the most important factors affecting soil erosion. Montoro et al. (2000) observed a marked reduction of runoff and sediment yields with light mulching of straw. With the reduced tillage practice, surface runoff was reduced by 7-11% and soil loss by 18-28%. Gardner and Gerrard (2003) recommended that maintenance of some form of ground cover is advisable if runoff and erosion are to be minimized. In addition to the ground cover, reduced tillage practices that minimally disturb the soil surface seem to be desired soil conservation practices. Atreya et al. (2006) reported that in a maize based cropping system in the hills of Nepal, total annual soil loss from the conventional and reduced tillage were 16.6 and 11.1 t ha^{-1}, respectively, and concluded that reduced tillage could be a viable option for minimizing soil and nutrient losses without sacrificing economic yields. A well-managed CA is also considered resilient.

ii. *Sustainable soil management*: Land use practices in the mountains should be better targeted to respond to climate change effects. Sustainable land management practices that prevent soil loss while improve land productivity should be promoted. Researches have demonstrated that sustainable interventions such as soil and water conservation practices have very positive results in reducing the impacts of climate change such as soil erosion. Acharya et al. (2008) investigated the efficacy of a combination of legumes, mulch and strip in controlling nutrient losses in surface runoff and leachate. Low input strip crop technologies were effective in soil and water conservation through the sieve-barrier effect, while increasing farm income and hence potential to maintain the overall sustainability of land system. Similarly, Mishra and Rai (2014) in the mountains of Sikkim showed that agro-forestry and vegetative barriers are the most favourable practices for soil and water conservation. Other sustainable land management system include planting more forage/fodder trees in marginal lands to conserve soil, stall feeding system of livestock raring and abandoning of free-grazing system (Subedi et al., 2015).

iii. *Climate adaptive farming:* in the context of emerging trend of adverse growing conditions due to climate change, climate-smart and resilient crop production practices should be adopted. Such practices include enhanced understanding of growing season, improved crop rotation systems, adaptive water management techniques and higher quality weather forecasts (Cameron, 2014). Adaptive water management techniques include enhancing storage and access to irrigation water, more efficient water delivery systems, and efficient irrigation technologies such as drip irrigation. Development of heat and drought tolerant crop varieties, strategic planting to avoid such effects, and improved water use efficiency are some the other important strategies to avoid crop failure from moisture and heat stress. Similarly, understanding and better use niche-based agricultural practices should be adopted to cope with the climate threats.

iv. *Water harvesting techniques:* Water harvesting has ancient roots and still forms an integral part of many farming systems worldwide. They imply the collection and storage of rainy season precipitation that would have otherwise seeped into the soil or run off into stream channels. For smallholder farmers in the mountains of Nepal, even a small volume of stored water for supplemental irrigation can significantly improve crop yield. Building upon China's long history of rainwater harvesting techniques, scientists in the Gansu Province of China have since the mid 1980s developed a newer approach to dry land agriculture (termed rainwater harvesting agriculture) to cope with erratic rainfall. They are promoting small water harvesting tanks to collect surface runoff for supplemental irrigation in wheat and maize. Research shows a significant wheat yield increase (average 35%) in areas with supplemental irrigation from such water harvesting tanks compared to areas with no such irrigation (Li et al., 2000). In regions with inadequate and/or unreliable rainfall, such water harvesting techniques can ease the constraints of water scarcity and help improve crop yields. Improvement of existing community water ponds and building of such ponds to store the monsoon rain could also serve as source of water during dry seasons.

v. *Crop management strategies:* strategies such as multiple cropping, crop diversification (planting different crops at a time to avoid total failure of harvest), and conservative cropping (growing traditional varieties or combinations of early and late mature varieties that copes better in harsh climate) are regarded as the most important strategies to avoid complete crop failure due to uncertain climate in developing countries. Multiple cropping, commonly observed in many traditional farming systems in Nepal is not only considered as a means to avoid risk of crop failure, it is pivotal in achieving yield stability, maintaining soil fertility, and attaining a constant supply of human food and animal feed (Subedi, 1998). These are evidences of deliberate choices by the farmers to safeguard a minimum supply of food during periods of climatic uncertainty but are seldom recognized as climate adaptation. By growing a range of crops with different climatic response characteristics, farmers in climatically sensitive regions iron out fluctuations in crop yields under a broad spectrum of seasonal conditions. Case studies from the African Sahel suggest that farmers have adapted to changes in rainfall regimes by switching to different crop varieties and by using a range of water saving techniques such as bench terrace construction, mulch farming, multiple cropping, and use of organic manure. Research shows that through more efficient use of nutrients, soil moisture, and light, yields from multiple cropping are relatively higher than the proportional area planted with a single crop (Tiffin and Mortimore, 2002).

vi. *Capacity Building:* Farmers in the mountain areas of Nepal are generally less aware of the impacts of climate change on agriculture sector. Training farmers and communities about what climate change is, how it is experienced, what are its consequences and what are the community-based adaptations strategies will help them to strengthen their capacity to cope with disasters through improved land-management skills. Building of adaptive capacity of people is important in a situation like that of Nepalese mountains, which will make them aware of the impacts of climate change and prepare to cope with such threats.

CONCLUSION

The smallholder farmers in the Middle Mountain region of Nepal have several challenges for their subsistence, where limited land resources are intensively cultivated with different rotations and intercropping practices for their livelihoods. Traditionally, multiple cropping systems are practiced with different crop rotations that are adapted to better utilize available growing season, to fit with soil fertility conditions and water availability, and to meet food and feed requirements of the households. As lands in the Middle mountains of Nepal are naturally fragile and highly prone to ecological degradation; climate change is seen as a risk multiplier. Climate change is projected to alter crop-growing conditions for farmers globally; the Mountain region of Nepal is no exception.

Despite farmers have been trying their efforts to cope with such challenging conditions, there are growing evidences of climate change effects facing by the agriculture sector in Nepal and especially so in the middle Mountain region. The major risks associated with climate change seem to be extreme weather events such as prolonged droughts and erratic rainfall resulting in excessive soil loss through surface erosion, landslides, floods and siltation, and damage of physical infrastructures. The projected changes seem to cause degradation of land and water resources, shift in crop rotations, resultant in low agricultural productivity, with concomitant increase in food insecurity.

Land resources in the middle Mountain are highly prone to degradation and farmers' current land use practices are not adequate to sustain crop productivity and protect lands from further degradation. It is high time to plan and promote sustainable land management practices suitable for the mountain agro-ecosystems. Appropriate land management technologies and a range of actions are required to enhance the resilience of crop production conditions that increase land productivity by reducing the potential threat from climate change and land degradation. More wakefulness campaigns, reliable weather forecast system, capacity building of farmers to make the required adjustments to climate change effects, promotion of conservation agriculture practices such as minimized soil disturbance, reduced-tillage, proper management of crop residues, and appropriate crop rotations seem to be important. Other sustainable soil conserving practices suitable for the mountain regions can be planting of more perennial crops such as tea, coffee, cardamom, fruits and fodder trees in the sloppy lands, preventing free-grazing of animals in sloppy lands. Developing and adapting resilient cropping practices such as use of more drought tolerant crop/varieties in the rotation and adjusting planting time so as to avoid extreme weather events are equally important. Infrastructures development and their maintenance in the fragile mountain agro-ecosystems should also be given specicial consideration so as to minimize the impacts on land degradation.

REFERENCES

Acharya, A.K., and N. Kafle. 2009. Land degradation issues in Nepal and its management through agroforestry. *J. Agrl. Environ.* 10:115-123.

Acharya, G.P., M.A., McDonald, B.P. Tripathi, R.M. Gardner, and K.J. Mawdesley. 2007. Nutrient losses from rain-fed bench terraced cultivation systems in high rainfall areas of mid-hills of Nepal. *Land Degrad. Dev.* 18:486-499.

Acharya, G.P., B.P. Tripathi, R.M. Gardner, K.J. Mawdesley, and M.A. McDonald. 2008. Sustainability of slopping land cultivation systems in the mid-hills of Nepal. *Land Degrad Dev.* 19:530-541.

Adetiloye, P.O., and A.A. Adekunle. 1989. Concept of monetary equivalent ratio and its usefulness in the evaluation of intercropping advantages. *Trop. Agri.* 66:337-341.

Agrawala, S., V. Raksakulthai, M. van Aalst, P. Larsen, J. Smith, and J. Reynolds. 2003. Development and Climate Change in Nepal: Focus in Water Resources and Hydro Power. COM/ENV/EPOC/DCD/DAC 1/FINAL, OECD, Paris.

Atreya, K., S. Sharma, R.M. Bajracharya, and N.P. Rajbhandari. 2006. Applications of reduced tillage in hills of central Nepal. *Soil Tillage Res.* 88:16-29.

Bartlett, R., L. Bharati, D. Pant, H. Hosterman, and P. McCornick. 2010. Climate change impacts and adaptation in Nepal. Colombo, Sri Lanka. International Water Management Institute. 35 pp. IWMI Working Paper 139.

Baul, T.K., Ullah, K.M.A., Tiwari, K.R., and M.A. McDonald. 2013. People's local knowledge of climate change in the middle-hills of Nepal. *Indian J. Tradit. Know.* 12:585-595.

Bolch, T., A. Kulkarni, A. Kääb, C. Huggel, F. Paul, J.G. Cogley, H. Frey, J.S. Kargel, et al. 2012. *Science* 336: 310-314.

Brown, S., and B. Shrestha. 2000. Market-driven land-use dynamics in the middle mountains of Nepal. *J. Environ. Mgt.* 59:217-225.

Cameron, E. 2014. Climate Change: Implication for Agriculture. Key Findings from the Intergovernmental Panel on Climate Change Fifth Assessment Report. p. 15. www.cisl.com.ac.uk/ippc.

Carson, B. 1992. "The Land, The Farmer and The Future- A Land Fertility Management Strategy for Nepal." ICIMOD Occasional Paper No. 21, International Centre for Integrated Mountain Development (ICIMOD), Kathmandu, Nepal.

CBS. 2011. Statistical Yearbook of Nepal 2009. National Planning Commission, Central Bureau of Statistics, Kathmandu, Nepal.

CBS. 2012. Irrigation. Agricultural Monograph. Central Bureau of Statistics (CBS), Nepal. pp. 120-128. Accessed at http://cbs.gov.np/wp-content/uploads/2012/Agriculture/Agricultural%20Monograph%20Final/Chapter08.pdf. 13 May 2015.

CBS. 2013. Environment Statistics of Nepal 2013. Central Bureau of Statistics (CBS), Nepal. Accessed at http://cbs.gov.np/wp-content/uploads/2014/07/Environment%20Statistics%20of%20Nepal%202013.pdf.

Challinor, A.J., J. Watson, D. Lobell, M. Howden, D. Smith, and N. Chhetri. 2014. A meta-analysis of crop yield under climate change and adaptation. *Nature Climate Change,* 4:287-291.

Chhetri, N., and W.E. Easterling. 2010. Adapting to climate change: retrospective analysis of climate technology interaction in rice based farming systems of Nepal. *Ann. Assoc. Am. Geogr.* 100:1156-1176.

Chhetri, N., M. Subedi, and S. Ghimire. 2013. Niche-based responses in addressing the climatic constraints to farm production: Analogues to climate-change adaptation in Nepal, Climate and Dev. 5:174-181.

Dalrymple, D.F. 1971. Survey of multiple cropping in less developed nations. United States Department of Agriculture, Washington D.C. FEDS, 108 p.

Devkota, R.P. 2014. Climate change: trends and people's perception in Nepal. *J. Environ. Prot.* 5, 255-265. http://dx.doi.org/10.4236/jep.2014.54029.

Dixit, A. 2015. Climate Change in Nepal: Impacts and Adaptive Strategies. World Resources Institute. Accessed at: http://www.wri.org/our-work/project/world-resources-report/climate-change-nepal-impacts-and-adaptive-strategies. Accessed on-line on 06 April 2015.

Field, C.B., V. Barros, T.F. Stocker, D. Qin, D.J. Dokken, K.L. Ebi et al. 2012 (ed.). A Special Report of Working Groups I and II of the Intergovernmental Panel on Climate Change (IPCC). Cambridge, UK, and New York, NY, USA: Cambridge University Press, 231-290.

Gardner, R.A.M., and A. J. Gerrard. 2003. Runoff and soil erosion on cultivated rained terraces in the middle hills of Nepal. *Appl. Geography*. 23:23-45.

Gardner, R.A.M., K. Mawdesley, B.P. Tripathi, S. Gaskin, and S. Adams. 2000. Soil erosion and nutrients loss in the middle hills of Nepal (1996-1998). ARS Lumle, Soil Science Division, NARC and Queen Marry and Westfield College. QMWC London.

Gentle, P., and T.N. Maraseni. 2012. Climate change, poverty and livelihoods: adaptation practices by rural mountain communities in Nepal. *Environ. Sci. Pol.* 21:24-34.

Ghimire, S.K., D. Higaki, and T.P. Bhattarai. 2013. Estimation of soil erosion rates and eroded sediment in a degraded catchment of the Siwalik hills, *Nepal. Land.* 2:370-391.

Hobbs, P.R., K. Sayre, and R. Gupta. 2008. The role of conservation agriculture in sustainable agriculture. *Phil. Trans. R. Soc. B.* 363:543-555.

IPPC, 2007. PCC Fourth Assessment Report: Climate Change 2007. https://www.ipcc.ch/publications_and_data/ar4/syr/en/mains1.html.

Jodha, N.S. 1992. Mountain perspective and sustainability: A framework for development strategies. P. 41-81. In: N.S. Jodha, M. Banskota and T. Pratap (Eds.). Sustainable Mountain Agriculture. Oxford and IBH Publishing Co. Pvt. Ltd., New Delhi, India.

Kansakar, S.R., D.M. Hannah, A.J. Gerrard, and G. Rees. 2004. Spatial pattern in the precipitation regimes of Nepal. *Int. J. Climatol.* 24:1645-1659.

Karki, L.S. 2001. Maize (*Zea mays L.*) in the Far-Western hills of Nepal. *In*: N. R. Rajbhandary, J.K. Ransom, K. Adhikari, and A.F.E. Palmer, editors, Sustainable Maize

Production Systems for Nepal. Proceeding of the Maize Symposium from 3-5 December 2001, CIMMYT, and Kathmandu, Nepal. *p*. 246-250.

Karki, M., P. Mool, and A. Shrestha. 2009. Climate change and its increasing impact in Nepal. *The Initiation*. 2009:30-37.

Karki, T. B., and J. Shrestha. 2014. Conservation agriculture: significance, challenges and opportunities in Nepal. *Adv. Plants Agric. Res*. 1:1-3.

Kripalani, R.H., and A. Kulkarni. 1997. Rainfall variability over Southeast Asia – connections with Indian monsoon and ENSO extremes: new perspectives. *Int. J. Climatol.* 17:1155-1168.

Krishnamurthy, P.K., C. Hobbs, A. Mathiassen, S.R. Hollema, R.J. Choularton. K. Pahari, and M. Kawabata. 2013. Climate risk and food security in Nepal- analysis of climate impacts on food security and livelihoods. Working Paler No. 48, CGIAR Research Program on climate Change, Agriculture and Food Security (CCAFS), Copenhagen, Denmark. Accessed at: www.ccafs.cgiar.org.

Kurvits, T., B. Kaltenborn, S. Nischalke, B. Karky, M. Jurek, and T.H. Aase. 2014. The last straw: food security in Hindu Kush Himalayas and additional burden of climate change. p. 60. A study report jointly prepared by CICIR, GRID and ICIMOD. Assessed on 24 May 2015 at http://www.grida.no/publications/last-straw/.

Lang, J.T., and A.P. Barros. 2002. An investigation of the onsets of the 1999 and 2000 monsoons in Central Nepal. *Mon. Weather Rev*. 130:1299–1316.

Lavell, A., M. Oppenheimer, C. Diop, J. Hess, R. Lempert, J. Li, R. Muir-Wood, and S. Myeong. 2012. Climate change: new dimensions in disaster risk, exposure, vulnerability, and resilience. In: C.B. Field, V. Barros, T.F. Stocker, D. Qin, D.J. Dokken, K.L. Ebi, M.D. Mastrandrea, K.J. Mach, G.-K. Plattner, S.K. Allen, M. Tignor, and P.M. Midgley, editors, Managing the Risks of Extreme Events and Disasters to Advance Climate Change Adaptation. A Special Report of Working Groups I and II of the Intergovernmental Panel on Climate Change (IPCC) (pp. 25-64). Cambridge, UK, and New York, NY, USA: Cambridge University Press.

Li, F., S. Cook, G.T. Geballe, and W.R. Burch Jr. 2000. Rainwater harvesting in agriculture: an integrated system for water management on rainfed land in China's semiarid areas. *AMBIO*. 29:477-483.

Malla, G. 2008. Climate change and its impact on Nepalese agriculture. *J. Agr. Environ*. 9:62-70.

Maskey, R.B., B.P. Sharma, and M. Joshi. 2003. Human dimensions in sustainable land management in degraded land areas of Nepal. Paper presented at the Open Meeting of the global environmental Change Research community, Montreal Canada, 16-18 October 2003.

Mishra, P.K, and S.C. Rai. 2014. A cost-benefit analysis of indigenous soil and water conservation measures in Sikkim Himalaya, India. *Mt. Res. Dev*. 34:27-35.

MoAC/WFP/FAO, 2009. 2008/09 Winter Drought in Nepal: Food Security Assessment. Ministry of Agriculture and Cooperatives (MoAC), World Food Program (WFP) and Food and Agricultural Organization (FAO). (http://documents.wfp.org/stellent/groups/public/documents/ena/wfp202832.pdf).

MoAD, 2013. Statistical Information on Nepalese Agriculture 2012/13 (2069/070). Agri-business promotion and Statistics Division, Ministry of Agriculture Development (MoAD). Government of Nepal, Kathmandu, Nepal.

MoE. 2010. National Adaptation Programme of Action to Climate Change. Ministry of Environment, Kathmandu, Nepal www.napanepal.gov.np/pdfreports/NAPA_Report.pdf) accessed: 02/15/2011.

Montoro, A, J.A. Regel, J. Querejeta, E. Diaz, and V. Castillo. 2000. Three hydro seeding revegetation techniques for soil erosion control on anthropic steep slopes. *Land Degrad Dev.* 11: 315-325.

Neupane, R.P., and G.B. Thapa 2001. Impact of agro-forestry intervention on soil fertility and farm income under subsistence farming system of the middle hills of Nepal. *Agr. Ecosys. Env.* 84:157-167.

Nyaupane, G. P., and N. Chhetri. 2009. Vulnerability of climate change on nature-based tourism in the Nepalese Himalaya. *Tourism Geogr.* 11:95-119.

Pandit, K., and M.K. Balla. 2004. Indigenous knowledge of terrace management in Paundu Khola watershed, Lamjung district, Nepal. Him. J. *Sci.* 2:33-36.

Parthasarathy, B., A.A. Munot, and D. R. Kothawale. 1994. All Indian monthly and seasonal rainfall series: 1871–1993. *Theor. Appl. Climatol.* 49:217-224.

Pilbeam, C.J., B.P. Tripathi, D.P. Sherchan, P.J. Gregory, and J. Gaunt. 2000. Nitrogen balance for households in the mid-hills of Nepal. *Agr. Ecosyst. Environ.* 79:61-72.

Pratap, T. 2003. Mountain agriculture, marginal lands, and sustainable livelihoods: challenges and opportunities. In: T.Ya, and P. M. Tulachan, editors, Mountain agriculture in the Hindu Kush-Himalayan region: Proceedings of an International Symposium held in Kathmandu, Nepal on 21-24 May, 2001. ICIMOD, Kathmandu, Nepal. p. 7-16.

Rajeevan, M., J. Bhate, and A.K. Jaswal. 2008. Analysis of variability and trends of extreme rainfall events over India using 104 years of gridded daily rainfall data. *Geophysical Res. Let.* 35: L18707, doi:10.1029/2008 GL035143, 2008.

Raut, N., B.K. Sitaula, J.B. Aune, and R.M. Bajracharya. 2011. Evaluation and future direction of intensified agriculture in the central mid-hills of Nepal. *Int. J. Agric. Sust.* 9:537-550. https://dox.doi.org/10.1080/14735903.2011. 609648.

SAGUN. 2009. Climate change impacts on livelihoods of poor and vulnerable communities and biodiversity conservation: A case study on Banke, Bardia, Dhadhing and Rasuwa Distrists of Nepal. Strengthening Actions for Governance in Utilization of natural Resources (SAGUN) Program, CARE Nepal, Kathmandu, Nepal. Accessed at: htpp://www.carenepal.org.

Schreier, H., B. Sandra, and P.B. Shah. 2005. The Jhikhu Khola Watershed Project in the Nepalese Himalayas: Approaches used and Lessons Learned. Institute for Resources and Environment, University of British Columbia and: International Centre for Integrated Mountain Development (ICIMOD). Accessed at: http://idl-bnc.idrc.ca/dspace/bitstream/ 10625/ 22526/ 1/121920.pdf.

Selvaraju, S., K.R. Regmi, C. Adhikari, B. Gautam, R. Dhakal, N. Pandey, B. Pokhrel, A. Deo, and N. Rana. 2014. Managing climate risks and adaptation to climate change in the agriculture sector in Nepal. Food and Agriculture Organization of the United Nations, Rome.

Sen Roy, S. 2009. A spatial analysis of extreme hourly precipitation patterns in India. *Int. J. Climatol. 29:*345-355.

Shakya, N.M. 2003. Hydrological Changes Assessment and its Impact on Hydropower Project of Nepal. In: Draft Proceeding of Consultative Workshop on Climate Change Impacts and Adaptation Options in Nepal's Hydropower Sector with a Focus on

Hydrological Regime Changes Including GLOF. Department of Hydrology and Meteorology and Asian Disaster Preparedness Centre, 5-6 March, 2003, Kathmandu, Nepal.

Sharma, S. 1999. Land tenure and poverty in Nepal. Paper presented in WDR-2000 Consultation meeting organized by the World Bank, April 4-6, Dhaka.

Shrestha, R.P. 1992. Agro-ecosystem of the mid-hills. In: J.B. Abington (Ed.). Sustainable Livestock Production in the Mountain Agro-ecosystem of Nepal. Food and Agriculture Organization (FAO), Rome.

Shrestha, D.P. 1997. Assessment of soil erosion in the Nepalese Himalayas. A case study in Likhu Khola valley, middle mountain region. *Land Husb.* 2:59-80.

Shrestha, M.L. 2000. Interannual variation of summer monsoon rainfall over Nepal and its relation to Southern Oscillation Index. *Meteorol. and Atmos. Phys.* 75:21-28.

Shrestha, A.B., and R. Aryal. 2011. Climate change in Nepal and its impact on Himalayan glaciers. *Reg. Environ. Change.* 11: 65-77.

Shrestha, A., B.K. Bishwakarma, and R. Allen. 2014. Climate smart management options for improving the soil fertility and farm productivity in the middle hills of Nepal. *Univ. J. Agri. Res.* 2:253-263. DOI:10.13189/ ujar.2014.020705.

Shrestha, A.B., C.P. Wake, P.A. Mayewski, and J.E. Dibb. 1999. Maximum temperature trends in the Himalaya and its vicinity: an analysis based on the temperature record from Nepal for the period 1971-1994. *J. Climate.* 12:2775-2789.

Shrestha, D.P., J.A. Zinck, and E. Van Ranst. 2004. Modelling land degradation in the Nepalese Himalaya. *Catena.* 57:135-156.

Subedi, K.D. 1998. Profitability of Barley and pea mixed intercropping in the subsistence farming systems of the Nepalese hills. *Expl. Agric.* 34:405-474.

Subedi, K.D. (editor). 1992. Problems and Potentials of Upland Rice Production in the Western Hills of Nepal: Report of a Topical *Samuhik Bhraman*. LARC Working Paper No. 92/19, Lumle Agricultural Research Centre, Nepal.

Subedi, K.D. 2001. Maize and finger millet relay intercropping system in the hills of Nepal: issues for sustainability. In: N.R. Rajbhandary, J.K. Ransom, K. Adhikari, and A.F.E. Palmer editors, Sustainable Maize Production Systems for Nepal. Proceeding of the Maize Symposium from 3-5 December 2001, CIMMYT, Kathmandu, Nepal. p. 170-174.

Subedi, K.D., and B.K. Dhital. 2007. Food security, livelihood and Nepalese Agriculture: Challenges and Potentials. Proceedings of Unfolding Futures: Nepalese Economy, Society and Politics. 5-7 October, 2007. Ottawa Canada. http://cffn.ca/category/ publications/unfolding-futures.

Subedi, K.D., B.K. Dhital, and B.P. Tripathi. 2015. Sustainable soil fertility management practices for Nepal. In: Adhikari, A.P. et al. (Eds.). Sustainable Livelihood Systems in Nepal. IUCN/CFFN, Kathmandu, Nepal (in press).

Subedi, K.D., G.B. Gurung, D.R.S. Paudel, K.J. Gurung, D.B Gurung, and J.B. Gurung. 1989. Traditional Methods of Maintaining Soil Fertility in the Mid and High Hills (1200–2100 m asl) of the Western Development Region of Nepal: Problems and Potential. Working Paper No. 3. LARC, Nepal.

Subedi, K.D., B.P. Tripahi, B.K Gyawali, T.B., Gurung, K. Adhikari, R.J. Khadka et al. 1991. Maize/Millet Relay Inter-cropping System in the Western Hills of Nepal (Kaski, Syangja and Tanahun Districts): Working Paper No. 15/91. Lumle Agricultural Research Centre (LARC), Nepal.

Tan, Z.X., R. Lal, and K.D. Wiebe. 2005. Global soil nutrient depletion and yield reductions. *J. Sust. Agri.* 26:123-145.

Tejada, M., J.L. Gonzalez, A.M. Garcia-Martinez, and J. Parrado, 2008. Effect of different green manures on soil biological properties and maize yield. *Biores. Tech.* 99:1758-1767.

Thapa, G.B. 1996. Land use, land management and environment in a subsistence mountain economy of Nepal. *Agric. Ecosys. Environ.* 57:57-71.

Thapa, G.B., and G.S. Paudel. 2002. Farmland degradation in the mountains of Nepal: A study of watershed with and without external intervention. *Land Degrad Dev.* 13:479-4963.

Tiffin, M., and M. Mortimore. 2002. Questioning desertification in dryland sub Saharan Africa. *Nat. Res. Forum.* 26:218-233.

Tiwari, K.R. 2009. Runoff and soil loss responses to rainfall, land use, terracing and management practices in the middle mountains of Nepal. *Acta Agr. Scand.* 59: 179-207.

Tripathi, B.P., S.P. Shrestha, G.P. Acharya, R. Gardner, K. Mawdesley, S. Gaskin, and S. Adams. 2003. Assessment of soil and nutrients losses from rain-fed upland (bari) terraces in the western hills of Nepal. In: T.Ya and P. M. Tulachan, editors, Mountain Agriculture in the Hindu Kush-Himalayan Region: Proceedings of an International Symposium held in Kathmandu, Nepal on 21-24 May, 2001. ICIMOD, Kathmandu, Nepal. p. 49-54.

UNEP. 2015. Managing Fragile Ecosystems: Sustainable Mountain Development. Accessed at http://www.unep.org/Documents. Multilingual/Default.Print.asp?DocumentID=52& ArticleID=61. On 09 February, 2015.

UNL. 2014. Crop Watch. 2014. Accessed at http://cropwatch.unl.edu/ tillage/notillman.

Webster, P.J. 1987. The elementary monsoon. In: Fein and P. L. Stephens, editor. Monsoons. p. 3-32. John Wiley.

Willey, R.W., and D.S.O. Osiru. 1972. Studies on mixtures of maize and beans (*Phaseolus vulgaris*) with special reference to plant population. *J. Agri. cultural Sci.* 79: 519-529.

Woiwod, I.P., and R. Harrington. 1994. Flying in the face of change: the Rothamsted insect survey. Long-term experiments in agricultural and ecological sciences. In: R.A. Leigh and A.E. Johnston, editors. CAB International, London, UK. p. 321-337.

Yang, D., S. Kanes, T. Oki, T. Koike, and K. Musiake. 2003. Global potential soil erosion with reference to land use and climate changes. *Hydrol. Process.* 17:2913-2918.

In: Crop Rotations
Editor: Bao-Luo Ma

ISBN: 978-1-63484-496-3

Chapter 6

CROP ROTATION SYSTEMS AND THEIR ECOLOGICAL IMPACTS IN THE LOESS PLATEAU OF CHINA

You-Cai Xiong[1,*], Feng-Min Li[1], Muhammad Ashraf[2], Nudrat Aisha Akram[3], Sabeeh ur Rasool Sabir[1], Fei Mo[1] and Xiao-Ling Wang[4]

[1]School of Life Sciences, State Key Laboratory of Grassland Agroecosystem, Lanzhou University, Lanzhou, China
[2]Pakistan Science Foundation, Islamabad, Pakistan
[3]Department of Botany, Government College University, Faisalabad, Pakistan
[4]Henan University of Science and Technology, Luoyang, Henan, China

ABSTRACT

The Loess Plateau of northwest China is geographically dominated by hilly and gully topography. Intense rainfall during the summer months eroded the bare topsoil from the hill tops to the valleys, which led to serious soil and water loss, land degradation and social poverty. Since the 1990s, a micro-field rain-harvesting farming technology has been extensively used, i.e., the technology of ridge-furrow with plastic mulching at field scale. This technology has led to a significant increase in grain yield and water-use-efficiency in major staple crops such as maize, wheat and potato. However, the soil quality started to decline with 3-5 years of application on this technology, due to significant negative relationships between soil microbial biomass carbon (MBC) and soil organic carbon (SOC) or soil mineral nitrogen (MN). The C/N ratio was also decreased with the prolonged planting years, suggesting that soil fertility tended to decrease. In this case, an integrated model of stratified strategies for ecosystem management (SSEM) has been proposed to conserve endangered ecosystems in the semiarid and arid regions. Based on the SSEM model, the forage-based crop rotation system has led to considerable improvement in agricultural cropping patterns, and could be used together with alfalfa planting at the perspective of sustainable management under global climate change. This

* Corresponding author address: xiongyc@lzu.edu.cn.

chapter provides a comprehensive review of the forefront of the ridge-furrow planting with plastic film mulching system and the application of the stratified strategies for ecosystem management in the Loess Plateau of northwest China.

Keywords: alfalfa, dry soil layer, forage and crop rotation, productivity, sustainability

1. Introduction

The Loess Plateau locates at the upper and middle regions of the Yellow River, with an area of some 640,000 km² in northwest China (Figure 1). The Loess is formed by wind deposition on the land surface over a long period of time and is recognized as "the most highly erodible soil on earth" (Chen et al., 2007). The Loess Plateau and its dusty soils cover almost whole areas of Shanxi and Shaanxi provinces, and partial areas of other five provinces including Gansu, Henan, Qinghai, Ningxia, and Inner Mongolia (Li, 1989).

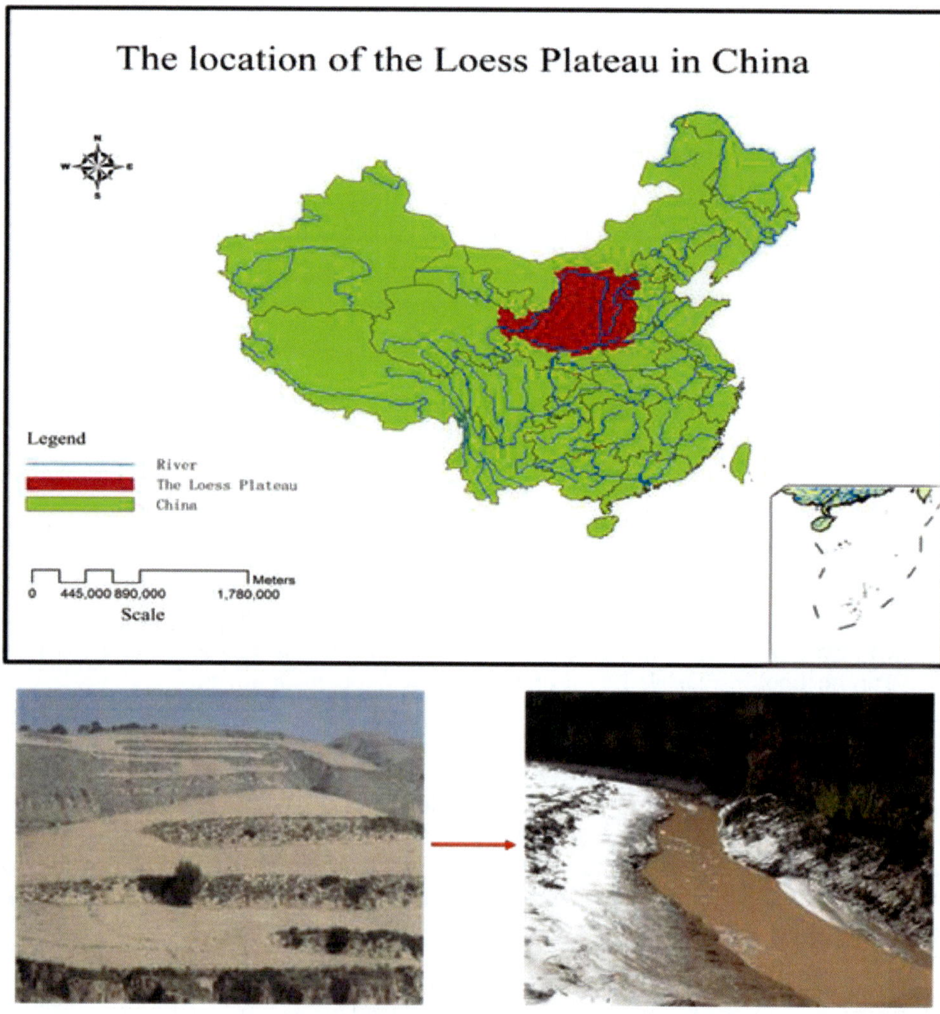

Figure 1. Geographic location of the Loess Plateau in China and the soil erosion.

This region is strongly affected by monsoon climate, with low and variable rainfall and more than 60% of annual precipitations that occur from July to September. Frequent drought is among the most serious climatic events (Li and Gong, 2002). During the early growth stages of spring-planted crops, soil moisture tends to decrease substantially, due to a doubling amount of surface soil evaporation over rainfall during the same period (Tao et al., 1993). In this case, the soil cannot supply sufficient water for crop growth and development. Therefore, grain yield per unit area is largely reduced due to the misplacement between rainy season and crop water demand. Li et al. (1999) reported that growth of spring wheat was largely affected by temporal water deficiency, especially during the periods of seed germination and seedling establishment. Considering the seasonal characteristics of rainfall in the Loess Plateau, summer-planted crops such as potato and maize are endowed with a great potential to fully utilize natural rainfall througout the period of growth (Ma, 1991; Xie et al., 2005).

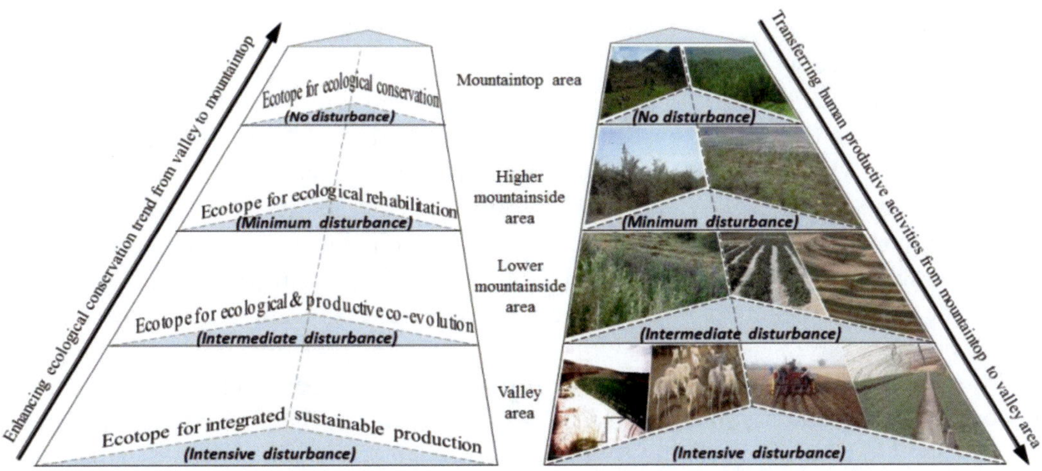

Figure 2. An integrated model of Stratified Strategies for Ecosystem Management (SSEM) in the Loess Plateau. Note: There are four ecotopes according to vertical stratified principle due to hilly and gully terrain characteristics in the Loess Plateau (Zhao et al., 2012).

From the late 1970s, rapid growth of local population has led to a speedy deterioration in the average arable land per capita in the region. Local farmers have been forced to convert more and more marginal lands into croplands for grain crop production. Particularly, increasing steep sloping lands were reclaimed for food crop production. Subsequently, the magnitude and scope of soil erosion tended to be expanded year by year, and soil fertility kept declining. The deterioration of soil system and vegetation cover threatened the sustainability of agricultural system (Li and Xu, 2002).

The Loess Plateau is a typical hilly and gully ecosystem. To achieve a sustainable management paradigm, the ecotope concept proposed by Sørensen is employed to establish the basic framework. In this chapter, we incorporate conceptual framework of ecotope into the theoretical system of arid and semiarid agroecological management on the Loess Plateau (Yang et al., 2005; He et al., 2007; Wang et al., 2009). The term "ecotope" is defined as the smallest ecologically-distinct landscape features in a landscape mapping and classification system (Ellis, 2008). A general approach is to stratify landscapes into distinct ecological unit

(Bastian et al., 2003). Actually, the criteria of ecotope identification are loosely defined. A realistic ecotope is generally referred to a specific ecosystem mapping and classification scale. Each ecotope consists of stratified sub-landscape, involving in the interaction between biotic and abiotic factors, including vegetation, soil, hydrological process and others (Sørensen, 1936; Tansley, 1939; Costanza et al., 1997; Dobson et al., 1997; Assessment 2005).

On the basis of conceptual framework of ecotope, we proposed an integrated model that displays how to manage fragile arid and semiarid agricultural ecosystem on the Loess Plateau (Wang et al., 2014), through the Stratified Strategies for Ecosystem Management (SSEM) (Zhao et al., 2012). The SSEM is theoretically interlinked with the perspectives of the 'dryland development paradigm (DDP)' proposed by Reynolds et al. (2007) and the 'coupled natural and human systems (CNHS)' of Liu et al. (2007). The model of SSEM was established at multiple scales from individual organisms into a global ecosystem, and across the dimensions from natural to social ecosystems (Wang et al., 2012). The model aims to improve both vegetation cover and agricultural productivity to bring greater economic benefits to local farmers, and ultimately help to enhance the sustainability of the ecosystem. It provides a practical approach to significantly improve the field productivity and water use efficiency of maize and alfalfa by developing a double ridge-furrow and plastic-mulching technology in lowland ecotopes, together with the appropriate adoption of crop-fallow rotation systems accompanied with conservation tillage. Particularly, the alfalfa-crop rotation has led to a considerable improvement in agricultural cropping patterns, and promoted the upgrade of agriculture and animal husbandry in the arid and semiarid rainfed agricultural areas of the Loess Plateau (Jia et al., 2006ab; Jiang et al., 2006; Liu et al., 2009; Jia et al., 2009; Zhou et al., 2009).

Since the 1990s, an alfalfa-grain-crop rotation system has been extensively used in the terraced fields of the Loess Plateau. In the 2000s, the Terraced Field Construction Program (TFCP) started to be implemented as a strategy of ecological and productive co-evolution. To restrain the soil and water loss, the terracing of the hillsides has been rapidly expanded due to its productive and ecological effectiveness in this area. The terraces are generally located in the lower hillside areas, i.e., lowland ecotope, with less than 25° slope, accounting for almost 40% of the total cropland area. Over the past 40 years, soil water storage and sediment trapping in terraced fields were increased to 237 m^3/ha and 112 t/ha, respectively, i.e., 82% of the runoff water and 84% of the sediment from the sloping land was preserved and trapped (Table 1). Soil quality of level terraced fields was significantly improved, in comparison with that of sloping fields. Soil organic matter and total nitrogen in terraced fields were increased to 13.0 g/kg and 79.6 mg/kg respectively, significantly greater than those of sloping lands (Table 1). In addition, terracing contributed to the obvious increases in total phosphorus and available potassium, but a decrease in soil bulk density (Table 1). On the other hand, alfalfa displayed higher economic output than that of wheat crops (Table 1). Therefore, terrace construction in conjunction with alfalfa-led legume crop rotations demonstrated a great potential in the SSEM model of the Loess Plateau (Figure 2).

The SSEM conceptual diagram was constructed with the aim to harmonize the relationship between ecological conservation and agricultural production. As shown in Figure 2, the hilltop and upper sloping lands serve as an ecotope for ecological conservation and rehabilitation, with zero disturbance of human activities. The lower hillside and valley areas are the ecotopes for ecological and productive co-evolution with intermediate and intensive disturbance to carry out integrated sustainable production. As such, human activities would be gradually transferred from the hilltops to the valleys (Figure 2). In summary, the goal of implementing the SSEM model is to boost ecological restoration in large areas (mainly the fragile and infertile hilltop areas and steep sloping lands), and extend the integrated agricultural production pattern in small but potentially productive areas (mainly the lowland and flat valley lands surrounded by the mountains). Alfalfa and forage crop-based rotation systems may act as an ecosystem engineer to cope with increasing challenges caused by global change and variability in arid and semiarid agroecosystem management program of the Loess Plateau.

Table 1. Differences in ecological and economic performances between sloping land and terraced croplands in the Loess Plateau

Parameter	Sloping land	Terraced cropland
Soil and water conservation		
Runoff amount (m^3 ha^{-1})	290.2 ± 94.5	—
Sediment amount (t ha^{-1})	133 ± 49.6	—
Water storage (m^3 ha^{-1})	237 ± 77.2	—
Water storage efficiency (%)	—	81.66
Sediment trapping (t ha^{-1})	—	112.2 ± 41.8
Sediment trapping efficiency (%)	—	84.34
Soil quality and field productivity		
SOM (g kg^{-1})	12.97 ± 1.4	10.69
TN (mg kg^{-1})	79.62 ± 7.9	70.06
TP (mg kg^{-1})	138.1 ± 18.5	131.38
AK (mg kg^{-1})	71.2 ± 4.4	68.87
BD (g cm^{-3})	1.12 ± 0.01	1.32
Crop economic output (CNY)	740.02 (wheat)	1442.93 (alfalfa)

Notes: SOM, soil organic matter; TN, total nitrogen; TP, total phosphorus; AK, available potassium; BD, bulk density. The data are extracted from Wang (1998), Jiao et al. (1999), Wu et al. (2003) and Kang et al. (2005).

Table 2. Grain yield and WUE of maize in integrated rainwater-harvesting system in the Loess Plateau

Treatments	Size design	2006 (dry year)		2007 (wet year)	
		Grain yield (kg/ha)	WUE (kg/ha/mm)	Grain yield (kg/ha)	WUE (kg/ha/mm)
CK	Flat planting	170	0.4	536	1.3
Ridge-furrow with plastic mulching	40-50 cm optimum ridge width	1150	3.1	6130	16.6

The data are extracted from from Jia et al. (2006, 2009), Liu et al. (2009), and Zhou et al. (2009).

2. Critical Challenges in the Rainfed Farming

In the rainfed Loess Plateau agricultural region, the long-term average of annual total precipitation ranges from 300 to 550 mm. Since the 1980s, various rainwater-harvesting techniques have been developed and extended in the Loess Plateau. Particularly, the cultivation technique named ridge-furrow with plastic mulching system has been extended to a large area, which has ensured the massive increase in grain yield per unit area (Table 2). Currently, this technique has been used in maize, potato, wheat, alfalfa and other cash crops, and the planting areas of major staple food crops were increased dramatically (Figure 3). Taking Gansu province as an example, the planting area has risen up to 1.2 million hectares for maize, and 45,000 hectares for potato.

Figure 3. Some photos showing the micro-field rain-harvesting farming system (potato, wheat, maize and alfalfa) in the Loess Plateau.

From the perspectives of technical design, the ridges are used to prevent runoff, and the furrow serves to collect rain water for use by the crops planted in the furrow. The width ratio of ridge to furrow varied with crop type as well as the trend in rainfall amount and air temperature. In most areas of the Loess Plateau, the optimal width ratios between ridge and furrow are 60 cm: 40 cm for potato and 60 cm: 60 cm for maize, to achieve the maximum yields. Among different mulching materials, plastic sheet is the most widely used, mainly because it displays the advantage of preventing evaporation and substantially improving crop yields with reasonablly low cost. Previous studies showed that plastic mulching would increase soil surface temperature and minimize evaporation from soil surface, leading to better grain yield and dry matter production. For those areas where maize is not suitable for planting due to shortage of enough crop heat units (Dwyer et al., 1999), such as in high elevation areas, maize has now become the major and profitable crop in the region. In virtue of the application of ridge-furrow plastic-mulching technology, the production of forage maize has provided sufficient high-quality elite fodder for local livestock industry. In areas where annual mean air temperature is less than 5°C, grain yield for the plastic-mulched maize was up to 11 times greater than that of traditional flat planting system (Table 2). The data showed that this technique increased the root residues in the soil, and accordingly enhanced population size and biological activities of soil microorganisms. In addition, growth period of the crop was shortened by 7-15 days as a result of application of this technology in the Loess Plateau (Liu et al., 2009; Zhou et al., 2012). However, after application of this technique for 3-5 years, the risk and unsustainability of soil quality decline and related environmental problems have been increasing.

As is well known, the important indicators of soil health are nirogen and carbon. The crop residues are rich in carbon and nitrogen, and when they are returned into the field, it takes some time to decompose in the soil system (McCallum et al., 2000). To some extent, the favourable soil temperatures with adequate soil moisture under the plastic mulching led to an accelerated rate of soil organic matter decomposition (Wang et al., 2005). The microorganisms tended to consume available form of nitrogen from the soil, which frequently ends up binding available nitrogen and resulted in insufficient supply of the available nitrogen to the crop (Carter et al., 1991; Gan et al., 2013).

It was reported, in soil database of China 2005, that the soil can be deficient of nitrogen when total nitrogen content was below 2 g kg^{-1}. Soil mineralized nitrogen (MN) tended to decrease with prolonged utilization of plastic mulching. Existing evidence showed that MN was negatively correlated with MBC, and increased microbial activity restricted the formation of MN. In this case, the competition for soil nitrogen emerges between plants and soil microorganisms (Zhou et al., 2012). Mulching cultivation lowers the SOC content due to an increased microbial activity and excessive consumption of soil nutrients within a few years, though the content of MBC increased gradually (Malhi et al., 2003; Zhou et al., 2010). At the same time, mineralized nitrogen tended to decrease with the increase in soil microbial biomass carbon. There existed significantly negative correlations between MBC and SOC, and between MBC and MN (Figure 4). This phenomenon would directly affect the increase in yield and efficiency for the succeeding crops. On the other hand, the ridge and furrow plastic mulching system increased the absorption and utilization of soil available phosphorus by plants due to the increased soil temperature, resulting in lower AP level in soil (Liu et al., 1999). For example, Zhao et al. (2009) reported that the soil AP layer in maize fields was reduced when soil moisture was improved.

As is well known, the ratio of SOC to TN (C/N) is a characteristic of soil organic matter. The soil organic matter tended to decompose rapidly under the condition of ridge-furrow mulching, and the lowered C/N ratio accelerated the supply of SOC accordingly, which would further lowered the C/N ratio. In this case, the effective strategies and practices to increase SOC and C/N ratio would have to be adopted in order to restore soil quality and reduce N loss (Zhou et al., 2012). Alfalfa may act as an ecological engineer in optimizing the nutrient balance and supplying more mineral nitrogen in the Loess Plateau.

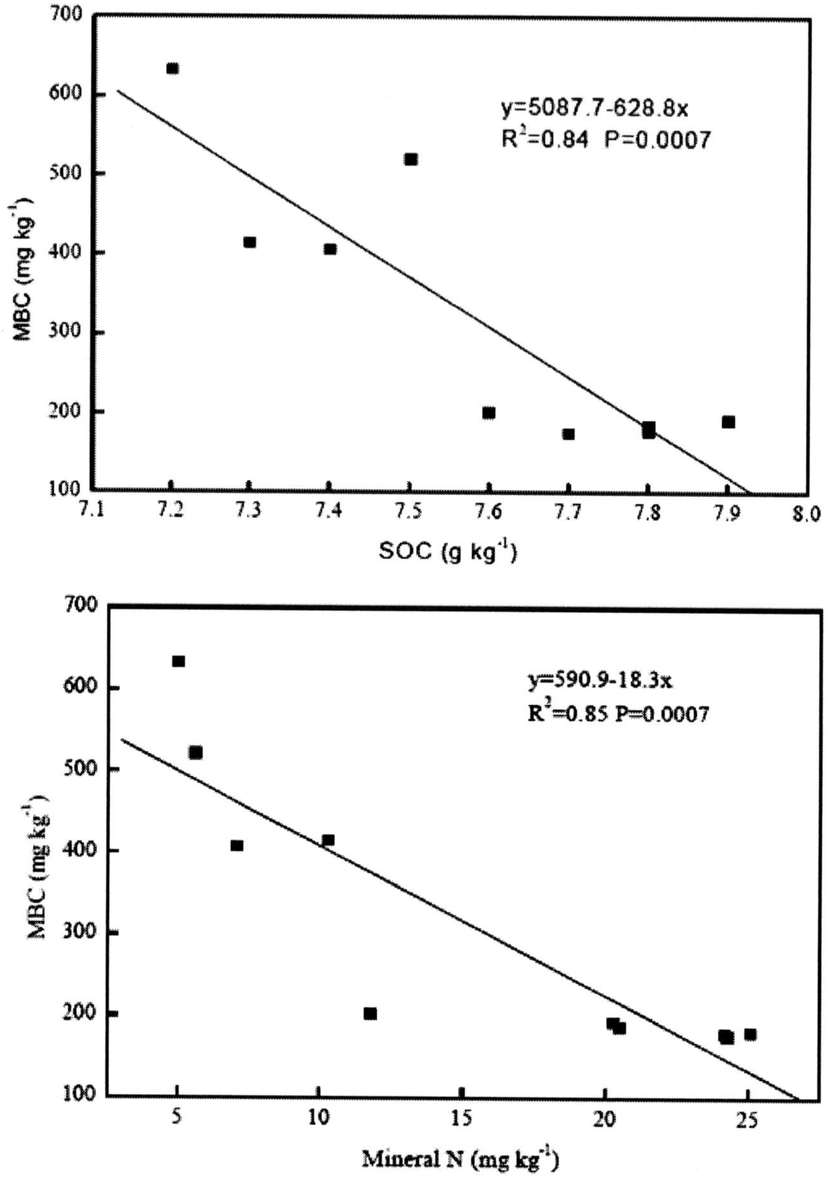

Figure 4. Relationship between microbial biomass carbon (MBC) and soil organic carbon (SOC) or soil mineral N (MN).

3. Establishment of Alfalfa-Based Crop Rotation Systems

3.1. Advantages of Alfalfa Grassland

Alfalfa is endowed with several reasons. It has been planted in the Loess Plateau for a long time to reduce soil erosion and increase soil fertility. It is also used as a primary species in a series of large vegetation restoration campaigns, such as the "Grain to Green" (Jun et al., 2014). Under conventional farming conditions, the productivity of alfalfa was remained at a relatively low level. Since the ridge-furrow plastic-mulching farming technique was introduced to this area, the planting area and forage production of alfalfa have increased significantly. As a legume-forage crop, alfalfa is able to fix atmospheric nitrogen into the available form of soil nitrogen through symbiotic nitrogen fixation. As a perennial crop, it can grow well in marginal land without disturbance to the soil. The root residues are remained in the soil for years, which can be converted into soil organic carbon. As an elite pasture, it has much higher protein content but less carbohydrate level in the straw than graminaceous crops such as maize. According to the nutrient demand of livestock, alfalfa overcomes the defect of low protein content in graminaceous crops. From the perspective of labor demand, alfalfa field management requires very low labor input. The aboveground forage can be harvested twice or three times each year for a duration of up to ten years. Therefore, alfalfa displays a great potential to restore soil fertility and economic return.

3.2. Effects of Alfalfa-Based Crop Rotation System on Soil Quality and Water Conservation

When alfalfa field was established using the micro-field rain-harvesting technique, both SOC and STN contents would increase greatly (Jia et al., 2006). Also, adoption of this technique substantially increased both the forage yield and water use efficiency (WUE) during the initial 5 years, as compared to those with conventional cultivation (Jia et al., 2006). However, there is a critical constraint in the available rainfall in the arid and semiarid Loess Plateau. Increasing the WUE of a crop is the first question to be addressed, because it is widely recognized as a critical paramter to judge the output of arable land. Alfalfa is a highly water-consuming crop compared with other crops (Blad and Rosenberg, 1976). Actually, WUE of alfalfa was generally lower in both conventional and water-harvesting cultivation systems than that of conventional crop rotation system (non-alfalfa system). However, from an economic point of view, alfalfa is still an ideal forage crop regarding available rainwater resource utilization in arid and semiarid areas, compared with spring wheat, pea, potato and maize, because of its advantages of absolute forage yield to meet the demand of animal husbandry. Due to the attributes of perennial species, the duration to cover the land surface is much longer than annual crops (Figure 2). The prolonged coverage time can help to reduce wind erosion efficiently in spring and autumn in the Loess Plateau (Hu et al., 2002). Importantly, average annual biomass production of alfalfa can achieve the same level as wheat in over 10 years of planting, and in the meantime soil total nitrogen was 2–3 times higher in alfalfa than in wheat (Shen et al., 2004). In this case, introduction of rain-harvesting technique provides a great potential for sustainable land use in the water-limited environments and ensures the extension of alfalfa planting.

Introduction of alfalfa to grain crop system is an effective strategy to increase the soil quality, particularly to improve the concentration of mineral nitrogen. Among those nutrient components of soil, organic C is a representative parameter, including light fraction and heavy fraction carbon. As a major component of soil organic C, the light fraction C pool is a derivative from freshly-added organic materials (e.g., plant residues) that can remain un-decomposed over a short period of time (Mueller et al., 1998). Light fraction organic C is more labile than gross organic C, and is a delicate parameter of changes in SOC in response to land management and environmental strains (Malhi et al., 2003). In alfalfa-crop rotation system and conventional farming system, the accumulation of soil organic matter resulted in producing high soil light fraction C and N contents, but this was not the case in constant alfalfa system. The lowered SOC/STN ratio might mean an acceleration of soil organic matter decay in alfalfa (Li et al., 2003; Jia et al., 2006). A high input of organic matter would also increase soil light fraction C (SLC) and N (Graham et al., 2002). Unfortunately, we observed that organic matter input in alfalfa caused a decrease in SLC and SLN contents. This was a case which occurred very infrequently. On the other hand, microbial community contains a high proportion of carbon, i.e., microbial biomass C. Changes in microbial biomass C acts as a critical indicator to assess the effect of management practices on soil biological and biochemical properties (Powlson et al., 1987; Nannipieri et al., 1990; Carter, 1991).

Microbial biomass C in the continuous alfalfa and alfalfa-crop rotation systems proved to be higher than that of the conventional system, suggesting that introducing alfalfa into crop rotation system could improve soil biochemical properties and microbial activity. The MBC/SOC ratio is another sensitive paramter to indicate the effectiveness of organic C conversion into microbial C and the extent of soil C loss during the process of organic matter decomposition (Sparling, 1992, 1997). For degraded soils, the MBC/SOC ratio generally tends to accelerate deterioration at a higher rate than soil organic matter does. Therefore, the high MBC/SOC ratio in continuous alfalfa culture and alfalfa-crop rotation systems appears to be advantageous towards soil biological activity. For soils with lower SOC, however, high MBC/SOC ratio and soil biological activity generally suggest that the decay of soil organic matter be accelerated, which would be a signal to damage soil quality (Li et al., 2004; Jiang et al., 2006). In most cases, soil organic C is closely associated with the level of soil organic matter. Existing studies showed that SOC concentration was similar among various farming systems, with no significant change with the farming years (less than 4 years). This demonstrated that SOC was fairly stable within a short term. In comparsion with SOC, soil total N (STN) proves to change quickly in farming system. Previous studies showed that STN was decreased by 10–18% after 3 years of winter wheat production in an alfalfa-winter wheat rotation system and a constant alfalfa planting field (Dang, 1998; Fan and Hao, 2003). In comparison with the continuous alfalfa system, alfalfa-crop rotation system significantly reduced STN and accordingly increased the SOC/STN ratio from 7.8 in year 2001 to 9.7 in year 2004 (unpublished). Comparatively, the SOC/STN ratio in the conventional system was as high as up to 11, mainly because STN level was lower than any other farming systems, while this value of SOC/STN ratio was accepted to be appropriate for sustainable soil use (Li et al., 2004; Jiang et al., 2006). In general, the SOC/STN ratio was low in alfalfa system and can be increased by alfalfa-crop rotation system. However, the way to increase the ratio was up to increasing STN. In this case, long-term practice of the two systems is not a feasible strategy to guarantee the sustainability of soil ecosystem in water-limited areas such as the semiarid region of the Loess Plateau.

As a whole, the introduction of alfalfa into cropping system has long been recognized as a means of ensuring the increase in yield of subsequent crops (Hedlin et al., 1957; Raimbault and Vyn, 1991). The core advantage is the input of N by alfalfa (Bruulsema and Christie, 1987; Hesterman et al., 1987; David et al., 1997; Li et al., 2002). At a scale of decade, alfalfa–crop rotation system can contribute to more nutrients accessible to the soil for the use of growing crops. The farming system following alfalfa rotation was found to increase water use efficiency mainly through the use of soil nutrients. In contrast, the insertion of alfalfa into a cropping sequence (alfalfa–grain rotation) can improve the aggregation of soil particles and the infiltration of water (Meek et al., 1990; Raimbault and Vyn, 1991; Angers, 1992), which in turn favor soil water conservation. Therefore, high level of soil water conservation is a critical reason to increase WUE in the alfalfa-crop rotation system (Li et al.,1999; Wang et al., 2005).

3.3. How to Reverse the Bottleneck Effect of 9-Year Alfalfa Production?

Average field productivity of perennial alfalfa kept increasing for a certain period of time (up to 9 years) and subsequently tended to decline. Comparative studies indicated that annual forage production of alfalfa varied among different climatic regions due primarily to differences in total precipitation. For example, the average annual precipitation in Changwu County, Shaanxi Proinve was approximately 550 mm, and alfalfa forage production reached the peak at the 4th - 5th year. In contrast, the peak forage production occurred at the 5th - 6th year in the mountainous area of south Ningxia Province (400 mm annual rainfall) and at the 7th - 8th year in the Beishan of Yuzhong county of Gansu (320 mm). However, the peak production year is about the same if expressed on the basis of average annual accumulative biomass (AAAB). The AAAB in all three regions was up to the peak value at the 9th year, without any exception. In other words, the 9th year was considered a turning point when the average biomass production tended to decline.

How to reverse the bottleneck effect of alfalfa forage productivity at the 9th year and thereafter? Here, we presented a case study: Field experiment was conducted in the arid and semiarid Beishan Station of Lanzhou University, China. The long-term average of annual total precipitation is around 320 mm with mean annual air temperature of 5 °C. A 9-year continuous alfalfa field was chosen for the study. There were five treatments tested, including, conventional cropping system (CS), 10-13 years of continuous alfalfa pasture (AS), the 9-year alfalfa field was followed by a 3-year of millet-wheat-potato (MWP), or of millet-potato-wheat (MPW), or millet-fallow-pea (MFPe) rotation in 2001-2003. We found that from 2001 to 2003, the total aboveground biomass was similar among the rotation cropping systems of CS, MWP, and MPW, but was lower for the AS system. Total biomass was the lowest in the millet-fallow-pea (MFPe) system (Table 3). In this study, all above-ground biomass was removed from the field. Interestingly, root biomass followed the same pattern as the aboveground biomass (Table 3).

The dynamics of soil organic C and total N were evaluated in this study. The data showed that cropping systems had little effect on soil organic C (SOC) during the period from 2001 to 2004, except for the alfalfa system (AS) that had higher SOC in 2004 (Table 4). This was probably due to the larger root system, perennial growth habit and longer growth period of alfalfa crop. As expected, prior to the introduction of alfalfa-based rotation systems in 2001, soil total N (STN) in the alfalfa system was significantly higher than that of conventional system (Table 4). From 2001 to 2004, STN concentration was reduced by 21.7% in MWP, by 18.3% in MPW and 5.9% in MFPe system, respectively. This was generally consistent with the dynamics of aboveground biomass and total N harvested with the corresponding crops (Table 3).

Table 3. Removal of above-ground biomass and soil input of root biomass under different farming systems in 2001, 2002 and 2003

Farming system	Above-ground biomass (kg ha^{-1})				Root biomass (kg ha^{-1})			
	2001	2002	2003	Total	2001	2002	2003	Total
CS	4895	6381	4669	15944	881	1021	607	2509
AS	3781	4179	2537	10496	-	-	-	-
MWP	4504	6831	5699	17034	365	1093	1026	2484
MPW	4504	5958	5093	15554	365	1072	815	2252
MFPe	4504	0	4366	8869	365	0	568	932

CS = conventional cropping system; AS = 10-13 years of alfalfa pasture; MWP = 9-year alfalfa followed by millet-wheat-potato in years 2001–2003; MPW = 9-year alfalfa followed by millet-potato-wheat in years 2001–2003; MFPe = 9-year alfalfa followed by millet-fallow-pea in years 2001–2003.

Table 4. Soil organic C and total N, and the ratio of soil organic C to total N (SOC/STN) in 2001 and 2004

Farming system	SOC (g/kg)		STN (g/kg)		SOC/STN	
	2001	2004	2001	2004	2001	2004
CS	13.20	12.79	1.22	1.25	10.81	11.07
AS	13.36	14.36	1.70	1.81	7.85	7.93
MWP	13.36	12.79	1.70	1.33	7.85	9.63
MPW	13.36	13.29	1.70	1.39	7.85	9.71
MFPe	13.36	14.07	1.70	1.60	7.85	8.88
Mean	13.33	13.46	1.60	1.48	8.44	9.44
$LSD_{0.05}$						
Land use	ns		0.38		2.53	
Year	ns		ns		ns	
Land use × Time	0.88		0.41		2.67	

For treatments: CS = conventional cropping system; AS = 10-13 years of alfalfa grassland; MWP = 9-year alfalfa followed by millet-wheat-potato in years 2001–2003; MPW = 9-year alfalfa followed by millet-potato-wheat in years 2001–2003; MFPe = 9-year alfalfa followed by millet-fallow-pea in years 2001–2003.

* $P \leq 0.05$; ns = not significant.

Interestingly, application of manure and chemical fertilizers in the conventional system did not affect STN as determined from 2001 to 2004. The SOC/STN ratio is an important indicator of soil quality. When measured in 2001 and 2004, the SOC/STN ratio was around 11 in the conventional system, around 7.9 in the alfalfa system (Table 4). In comparison, the

SOC/STN ratio was around 9.7 in the MWP and MPW systems, and around 8.9 in the MFPe system. Importantly, the increase of SOC/STN ratio of the alfalfa-based rotations was largely due to the reduction of STN, with little effect on the dynamics of SOC (Table 4).

In this study, SOC concentration did not vary much from 2001 to 2004 in all cropping systems, suggesting that SOC was a relatively stable parameter in a short term. In contrast, soil total N (STN) appeared to be a rapid variable, since it was significantly lowered in alfalfa-crop rotation systems, compared with continuous alfalfa system. This was why alfalfa-crop rotation systems increased the SOC/STN ratio from 7.8 in 2001 to 9.7 in 2004, being still lower than around 11 of the conventional system. And the value of around 11 was appropriate for soil sustainable use (Li et al., 2004; Jiang et al., 2006).

In conclusion, the SOC/STN ratio was lowered in alfalfa system but increased by crop rotations. The mechanism causing this change was associated with the reduced STN. Therefore, both farming systems may not be sustainable in the water-limited environments.

4. Innovation and Application of Alfalfa-Crop System

4.1. Issues of Dried Soil Layer in Alfalfa-Crop Rotation System

Alfalfa has a bigger root system with much higher water requirement than other crops (Saeed and Ei-Nadi, 1997; Blad and Rosenberg, 1976), and is able to take up water from deep soil layer (Wan et al., 2008). Therefore, alfalfa is likely to result in negative effect on the growth of succeeding crops due to excessive water consumption in deep soil layer (Du et al., 1999a; Li, 2002; Liu et al., 2000). An earlier study by Australian researchers showed that a certain layer of soil profile on the lucerne-based perennial prairies remained constantly drier throughout the year than that of continuous annual crop field (McCallum et al., 2001). In general, soil moisture under conventional cultivation was kept in balance, but it came to a sharp decline following years of alfalfa cultivation. This is so-called the phenomenon of dried soil layer in alfalfa-crop rotation system. According to our observations at Beishan, Yuzhong County of Gansu Province, soil moisture in the soil layer of 0-50 cm depth remained at around 10%, and declined to around 8% in the 50-100 cm soil layer in perennial alfalfa field (Wang et al., 2005; Jia et al., 2006). Soil moisture tended to decrease sharply at the 100 to 400 cm soil layer, with a measured value of 4%. In deeper soil layer of 400 to 1000 cm, soil moisture tended to maintain at a stable status of 8-10% (Wang et al., 2005; Jia et al., 2006). Similar results were also observed in other semiarid areas of the Loess Plateau.

The negative effect of alfalfa cultivation on soil moisture is obvious. The continuous cultivation of alfafa leads to the moisture decline of deep soil profile into around or below the permanent wilting point (Yang et al., 2006). Therefore, long alfalfa standing may deplete available soil water in the deep layers, which would negatively influence the production of succeeding crops (Li et al., 2002). However, alfalfa's profit strongly relies on its stand duration in view of production cost and forage output. The short stand duration is not accepted by farmers. Hence, it is crucial to balance the relationship between soil water use and economic profit, in terms of the optimal duration of an alfalfa stand. In a long run, local farmers decided the length of an alfalfa stand duration and the subsequent crop type according to their farming habits and local crop phenology. However, at the harvesting time, it was required to supplement water into the soil, which frequently resulted in less profit. Researchers had different opinions on the selection of optimal duration of an alfalfa stand in the arid and semiarid Loess Plateau region. For example, Du and Qu (1994) suggested that the stand ages of the perennial legumes, including alfalfa and erect milk vetch (*Astragalus adsurgens Pall*) should be less than 3 years in those locations with around 300 mm of annual total precipitation, since forage yield declined substantianlly after the 3rd year. Zhang et al. (2004) reported that alfalfa should be ploughed after 4–5 years of standing to obtain high forage yields in the Loess Plateau region with around 500 mm of annual total precipitation. Several studies suggested that 6- to 8-year-old alfalfa stands should be renewed in the area with annual total precipitation of 400 mm (Du et al., 1999b), and should not last up to 10 years in the areas with 450 mm of annual total precipitation (Cheng et al., 2005). In conclusion, the apt stand age of alfalfa field is expected to be 9 years or less, when the average annual accumulative forage yield is nearly maximum.

4.2. Local Solutions to Reverse the Dried Soil Layer Using Crop Rotation System: A Case Study

Two potential rotation systems have been proposed to reverse the dried soil layer problem due to long duration of alfalfa production. The experimental field was divided into two groups, one was deeply plowed, and the other was sustained with on-going alfalfa (AS) production. In the spring of the year 2001, the plowed field (October 2000) was further divided to host four treatments: 1) RS1: millet, spring wheat, potatoes and peas; 2) RS2: millet, maize, maize and spring wheat; 3) RS3: millet, potatoes, spring wheat and maize; and 4) RS4: millet, fallow, peas and potatoes. These fields were grown without any fertilizer addition during the 4-year experimental period. A conventional farming system (CS) was performed in an adjacient field where there was not history of growing alfalfa. The field was divided to grow spring wheat, peas and potatoes during the following four years. The field was applied with 1.5 t/ha of donkey manure, 90 kg/ha N fertilizer, and 30 kg/ha P_2O_5 fertilizer. The field data showed that there existed a significant difference in soil water storage as a result of long-term growing of alfalfa (Wang et al., 2008), with dried soil layer below 1 meter, while the dried soil layer can be restored by growing annual crops in a crop rotation system. For example, soil water restoration was attained ranging from 90% to 97% at a soil depth of 0–500 cm, after the plow-up work on an alfalfa field and the operation of rotating the lands with millet, maize, spring wheat, potato, and pea. On the other hand, higher total N content and soil respiration rates were observed in the alfalfa–crop rotation systems (Wang et al., 2008). The differences in yields between conventional farming system and alfalfa–crop rotation systems were not statistically significant. Particularly, including potato in the rotation system showed the highest yields. Hence, it was concluded that potato should be included in the alfalfa-based crop rotations in the Loess Plateau (Wang et al., 2008).

Table 5. Yield and water use efficiency (WUE) in conventional cropping system and rotation system from the years 2002 to 2004

Crop	Year	Treatments	Yield ($Kgha^{-1}$)	Precipitation during growth period (mm)	Water use efficiency ($kgha^{-1}mm^{-1}$)
Spring wheat	2002	CS	2397.2a‡	221.8	10.8b
	2002	RS1	2395.3aA	221.8	12.5aA
	2003	RS3	2037.3B	185.7	11.0AB
	2004	RS2	1791.5C	128.5	10.7B
Pea	2003	CS	1425.0a	185.7	6.9b
	2003	RS4	1524.0aA	185.7	8.4aA
	2004	RS1	1273.4B	128.5	7.3B
Maize	2002	RS2	3372.8A	318.2	11.2A
	2003	RS2	2586.6B	244.3	12.1A
	2004	RS3	2657.4B	178.8	10.6A
Potato	2002	RS3	3574.5A	318.2	13.8B
	2003	RS1	3191.3B	24.3	16.5A
	2004	RS4	3603.3aA	178.8	16.4aA
	2004	CS	3498.3a	178.8	13.9b

† CS, Cropping system; RS, Rotation system.

‡ Values within a column followed by the same lower-case letters: means do not differ significantly at P= 0.05 for one crop in different years after alfalfa, the same upper-case letters: means do not differ significantly at P = 0.05 for one crop in 1 year between the alfalfa-crop rotation systems and the conventional farming system.

Yield performance and water use efficiency are two critical paramters to decide the selection of cropping system in the Loess Plateau. Biomass productivity is frequently affected by the lack of water resources. Grain for food is not the mere necessity for local farmers, but farmers need to produce more crop straw biomass to feed animals, improve soil quality, for household fuel and even for construction materials. It is complex to choose the crop types and their combination pattern for a releatively optimal rotation system in terms of a specific site. The N turnover and utilization of nitrogen element is also the important issue to be taken into account for the decision on choosing a rotation pattern. As is well known, biological N_2 fixation and its related N economy in crop rotation systems have received extensive interest worldwide. Greater biomass production tends to be achieved from legume components of pastures in the cropping sequence (McCallum et al., 2000). It is a fact that a number of aspects need to be addressed when choosing the efficient crop rotation system, such as economic benefit, lifestyle, labor, custom and so on. If biomass productivity and WUE are the mere aspects for consideration, the best ensuing crop rotations in the Loess Plateau after 10 years of alfalfa would be millet–wheat–potato–pea–potato or millet–maize–maize–wheat–wheat. Obviously, the WUE and biomass production are not sufficent to influence the final decision on rotation system within a short term (generally less than five years). Ecosystem sustainability of cropland is also the primary factor to be addressed, particularly the duration of soil water restoration. In this case, we finally summarized the relatively optimal forage-crop rotation system during long-term rotation for over ten years in terms of annual precipitation in different regions of the Loess Plateau. In general, the Loess Plateau is categorized into three subregions, according to annual precipitation of 350, 450 and 550 mm. In the region with 350 mm annual precipitation, the relatively optimized rotation system is recommended to be 9 years of alfalfa plus 18 years of grain crops and then starts a new cycle. In this type of rotation system, the duration of yield peak maintenance would be 6-14 years, the duration of soil water restoration be 18 years and the total duration of rotation be 27 years. In the region with annual precipitation of 450 mm, the recommended rotation pattern is 7 years of alfalfa plus 7 years of grain crops. In this case, the duration of yield peak maintenance would be 6-10 years, the duration of soil water restoration be 7 years and the total duration of rotation be 14 years. In the relatively wet region of 550 mm annual precipitation, a rotation pattern of 6 years of alfalfa plus 5 years of grain crops is recommended (Table 6). Accordingly, the duration of yield peak maintenance would be 5-9 years, the duration of soil water restoration be 5 years and the total duration of rotation be 11 years. In conclusion, we attempted proposing such a challenging rotation system option on the basis of existing experimental data.

Table 6. Suitable grass-crop rotation pattern in the regions with different annual rainfall amount in the Loess Plateau

Rainfall amount	Years of yield peak maintenance	Years of soil water restoration	Rotation pattern	Years of rotation system cycle
Around 350 mm	6-14	18	9 years of alfalfa + 18 years of grain crops +9 years of alfalfa	27
Around 450 mm	6-10	7	7 years of alfalfa + 7years of grain crops + 7 years of alfalfa	14
Around 550 mm	5-9	5	6 years of alfalfa + 5 years of grain crops + 6 years of alfalfa	11

CONCLUSION

The Loess Plateau of northwest China is the most fragile ecological region in the world. Improving rainwater use efficiency is the key for sustainable and profitable production by small-holder farms. Since the 1990s, extensive farming practice with the aim to improve crop productivity and rainwater use efficiency has led to a general degradation in soil quality and a significant decrease in available water storage in soil profile. The farming system was generally featured by ridge-furrow planting with plastic mulching at field scale, and much attributed to massive enhancement on grain yield and water use efficiency in major staple crops such as maize. However, the overall concentration of C, N, P and the related ratios were observed to become worse in soil profile. There therefore existed a critical unsustainability in the agroecosystem. Over the last two decades, the adoption of righe-furrow planting with plastic mulching for alfalfa production appeared to be capable of restoring the soil quality and fertility on the basis of previous crop farming. However, a dried soil layer in the 1.0-4.0 m soil profile emerged after 3-5 years of the new practice, and this became a serious challenge to the agroecosystem sustainability in the Loess Plateau. To address this issue, different rotation systems were tested and compared in terms of restoring soil water storage and maintaining high crop productivity. The results showed that crop rotations, such as millet-wheat-bean-potato and millet-potato-wheat-maize-maize following a 9-year preceeding alfalfa, were among the most optimized rotation patterns in restoring soil moisture and nutrient balance. The field productivity in these systems was observed to increase to the maximum value as compared with the conventional farming system. Field productivity and water use efficiency are critical for the selection of cropping system in the Loess Plateau. It is complex to choose the crop types and their combination pattern for a releatively optimal rotation system in terms of a specific site, since a number of aspects need to be addressed when choosing the efficient crop rotation system, such as economic benefit, lifestyle, labor, custom and so on. We attempted proposing a quantitative mode in which the best ensuing crop rotations after 10 years of alfalfa would be millet–wheat–potato–pea– potato or millet–maize–maize–wheat–wheat in the Loess Plateau. The duration of yield peak maintenance would be 5-9 years, the duration of soil water restoration be 5 years and the total duration of rotation be 11 years. Therefore, it can be argued that the reintroduction of an alfalfa-based crop rotation system into the arid and semiarid Loess Plateau of northwest China would play a central role in restoring the fragile agroecosystem and advancing sustainable agriculture.

ACKNOWLEDGMENTS

The research was funded by national science and technology support program (2015BAD22B04), international cooperation program of Ministry of Science and Technology of China (2015DFG31840), the fundamental research funds for the central universities of China (lzujbky-2015-br02 and lzujbky-2015-ct02), natural science foundation of China (31570415) and the oversea master program of Ministry of Education (Ms2011LZDX059).

REFERENCES

Aubry, C., F. Papy, and A. Capillon. 1998. Modelling decision-making processes for annual crop management. *Agric Syst.* 56:45–65.

Angers, D.A. 1992. Changes in soil aggregation and organic carbon under corn and alfalfa. *Soil Sci. Soc. Am. J.* 56:1244–1249.

Bastian, B.C., A.B. Olshen, and P.E. LeBoit. 2003. Classifying melanocytic tumors based on DNA copy number changes. *Am. J. Pathol.* 163:1765–1770.

Berzsenyi, Z., B. Győrffy, and L. DangQuoc. 2000. Effect of crop rotation and fertilization on maize and wheat yields and yield stability in a long-term experiment. *Eur. J. Agron.* 13:225–244.

Blad, B.L., and N.J. Rosenberg. 1976. Evaluation of resistance and mass transport evapotranspiration models requiring canopy temperature data. *Agron. J.* 68:764–769.

Bruulsema, T.W., and B.R. Christie. 1987. Nitrogen contribution to succeeding corn from alfalfa and red clover. *Agron. J.* 79:96–100.

Buick, R.D., N.D. Stone, R.K. Scheckler, and J.W. Roach. 1992. CROPS: a whole-farm crop rotation planning system to implement sustainable agriculture. *AI Appl.* 6:29–50.

Bullock, D.G. 1992. Crop rotations. *Crit. Rev. Plant Sci.* 11:309-326.

Bullock, D.A., A.D. Bradshaw, and A.J.M. Baker. 1997. Hopes for the future: restoration ecology and conservation biology. *Science* 277: 515-522.

Cantero, M.C., G.J. O'Leary, and D.J. Connor. 1995. Stubble retention and nitrogen in a fallow-wheat rainfed cropping system. 1. Soil water and nitrogen conservation, crop growth and yield. *Soil Till. Res.* 1995:79–94.

Cao, Y.Q., Y.M. Liu, M.C. Wang, and J.Y. Li. 1994. Testing study of cultivation technique of water collection by ridge and furrow mulching rainfed farmlands. *Agric. Res. Arid Areas* 12:75-77.

Carter, M.R. 1991. The influence of tillage on the proportion of organic carbon and nitrogen in the microbial biomass of medium-textured soils in a humid climate. *Biol. Fertil. Soils* 11:135–139.

Chen, W.N., Z.B Dong, Z.S. Li, and Z.T. Yang. 1996. Wind tunnel test of the influence of moisture on the erodibility of Loessial sandy loam soils by wind. *J. Arid Environ.* 34: 391–402 (in Chinese with English abstract).

Chen, Y., and J. Liu. 2007. Design of terrace evaluation system on the Loess Plateau. *Agric. Res. Arid Areas.* 25:53–58 (in Chinese, with English abstract).

Costanza, R., R. dArge, R. de Groot, S. Farber, M. Grasso, B. Hannon, K. Limburg, S. Naeem, R.V. Oneill, J. Paruelo, R.G. Raskin, P. Sutton and M. van den Belt. 1997. The value of the world's ecosystem services and natural capital. *Nature* 387:253–260.

Dang, T.H. 1998. Influence of crop rotation on soil fertility in arid-highland of Loess Plateau. *J. Soil Eros. Soil Water Conserv.* 4:44–47 (in Chinese with English abstract).

David, J., J.K. Kelner, and M.H. Entz. 1997. The nitrogen dynamics of 1-, 2- and 3-year stands of alfalfa in a cropping system, agriculture. *Ecosyst. Environ.* 64:1–10.

Dobson, A.P., A.D. Bradshaw, and A.J.M. Baker. 1997. Hopes for the future: Restoration ecology and conservation biology. *Science* 277(5325):515-522.

Dogliotti, S., W.A.H. Rossing, and M.K. Van Ittersum. 2003. ROTAT, a tool for systematically generating crop rotations. *Eur. J. Agron.* 19:239–250.

Dregne, H.E., and W.O. Willis. 1983. Dryland Agriculture. American Society of Agronomy., Inc., Crops Sci. Soc. Am., Inc., *Soil Science Society of America,* Inc., Madison, W.I, U.S.A.

Du, G.Q., and Z.Q. Qu. 1994. The study of balancing law and renewing cycle of forage water in arid region farmland. *Pratac. Sci.* 11:25–28 (in Chinese with English abstract).

Du, S.P., L.F, Wang, and M.X. Long. 1999a. Soil water dynamics and yield response on lucerne in dry land farming in mountain region, *Northern of Ninxia. Pratac. Sci.* 18:12–15 (in Chinese with English abstract).

Du, S.P., M.X. Long, and L.F. Wang. 1999b. Study on soil moisture and yield dynamic of arid alfalfa grassland in mountain area of southern Ningxia. *Pratac. Sci.* 16:12–17 (in Chinese with English abstract).

Dwyer, L.M., D.W. Stewart, L. Carrigan, B.L. Ma, and P. Neave. 1999. Guidelines for comparisons among different corn maturity rating systems. *Agron. J.* 91:946-949.

Ellis, E.C., and N. Ramankutty. 2008. Putting people in the map: anthropogenic biomes of the world. *Front. Ecol. Environ.* 6(8):439-447.

Fan, J., and M.D. Hao. 2003. Effects of long-term rotations and fertilizations on soil microbial biomass carbon and nitrogen. *Res. Soil Water Conserv.* 10:85–87 (in Chinese with English abstract).

Gan, Y., K.H.M. Siddique, N.C. Turner, X.G. Li, J.Y. Niu, C. Yang, L. Liu, and Q. Chai. 2013. Ridge–furrow mulching systems–an innovative technique for boosting crop productivity in semiarid rain-fed environments. *Adv. Agron.* 117:429–476.

Graham, M.H., R.J. Haynesm, and J.H. Meyer. 2002. Soil organic matter content and quality: effects of fertilizer applications, burning and trash retention on a long-term sugarcane experiment in South America. *Soil Biol. Biochem.* 34:93–102.

Guo, L., J. Chua, D. Vijayakumar, K.C. Lee, K. Lim, H. Eng, F. Ghadessy, D. Coomber, and D.P. Lane. 2010. Detection of the 113p53 protein isoform: A p53-induced protein that feeds back on the p53 pathway to modulate the p53 response in zebrafish. *Cell Cycle* 9:1998-2007.

Guo, Z.P., L. Zhang, Z.Y. Ding, Z.X. Wang, and G.Y. Shi. 2010. Improving the performance of industrial ethanol-producing yeast by expressing the aspartyl protease on the cell surface. *Yeast* 27(12):1017-1027.

He, X.F., H.H. Cao, and F.M. Li. 2007. Econometric analysis of the determinants of adoption of rainwater harvesting and supplementary irrigation technology (RHSIT) in the semiarid Loess Plateau of China. Agric. *Water Manage.* 89:243–250.

Heady, E.O. 1948. The economics of rotations with farm and production policy applications. *J. Farm Econ.* 30:645–664.

Hedlin, R.A., R.E. Smith, and F.P. Leclaire. 1957. Effect of crop residues and fertilizer treatments on the yield and protein content of wheat. *Can. J. Soil Sci.* 37:34–40.

Hesterman, O.B., M.P. Russell, C.C. Sheaffer, and G.H. Heichel. 1987. Nitrogen utilization from fertilizer and legume residues in legume-corn rotations. *Agron. J.* 79:726–731.

Hu, J.H., Q.F. Han, L. Ma, and Z.K. Jia. 2002. Analysis on the industrialized production of alfalfa in themountains in South Ningxia. *Agric. Res. Arid Areas,* 20:116–119 (in Chinese with English abstract).

Jankauskas, B., and G. Jankauskiene. 2003. Erosion–preventive crop rotations for landscape ecological stability in upland regions of Lithuania. *Agric. Ecosys. Environ.* 95:129–142.

Jia, Y., F. Li, Z. Zhang, X. Wang, R. Guo, and K.H. Siddique. 2009. Productivity and water use of alfalfa and subsequent crops in the semiarid loess plateau with different stand ages of alfalfa and crop sequences. *Field Crop. Res.* 114:58-65.

Jia, Y., F.M. Li, and X.L. Wang. 2006. Soil quality responses to alfalfa watered with a field micro-catchment technique in the loess plateau of China. *Field Crop Res.* 95:64–74.

Jiang, H.M., F.M. Li, and J.P. Jiang. 2006. Soil carbon pool and effects of soil fertility in seeded alfalfa fields on the semi-arid loess plateau in China. *Soil Biol. Biochem.* 38:2350–2358.

Jun, F., G. Yu, W. Quanjiu, S.S. Malhi, and L. Yangyang. 2014. Mulching effects on water storage in soil and its depletion by alfalfa in the loess plateau of northwestern China. *Agric. Water Manag.* 138:10-16.

Leteinturier, B., J. Herman, F.D. Longueville, L. Quintin, and R. Oger. 2006. Adaptation of a crop sequence indicator based on a land parcel management system. *Agric. Ecosyst. Environ.* 112:324–334.

Li, C.H. 1989. The population pressure upon the natural resources and ecologic environment on the Loess Plateau and the ways of its elimination. *J. Nat. Resour.* 43:1294–1299 (in Chinese with English abstract).

Li, F.M., A.H. Guo, and H. Wei. 1999. Effects of clear film mulch on yield of spring wheat. *Field Crop Res.* 63:293–304.

Li, Y.Y., and M.A. Shao. 2001. Climate change vegetation evolution and lower moisture layer of soil in the Loess Plateau. *J. Arid Land Resour. Environ.* 15:72–76 (in Chinese with English abstract).

Li, Y.S. 2002. Productivity dynamics of alfalfa and its effects on water–eco-environment. *Acta Pedol. Sin.* 39:404–411 (in Chinese with English abstract).

Li, F.M., and J.Z. Xu. 2002. Rainwater-collecting eco-agriculture in semi-arid region of Loess Plateau. *Chin. J. Eco-Agric.* 10:101–103 (in Chinese with English abstract).

Li, F.M., J.Z. Xu, and G.J. Sun. 2003. Restoration of degraded ecosystem and development of water-harvesting ecological agriculture in the semi-arid Loess Plateau of China. *Acta Ecol. Sin.* 23:1901–1909 (in Chinese with English abstract).

Li, F.M., Q.H. Song, J.H. Hao, P.K. Jjemba, and Y.C. Shi. 2004. Dynamics of soil microbial biomass C and soil fertility in cropland mulched with plastic film in a semiarid agroecosystem. *Soil Biol. Biochem.* 36:1893–1902.

Liu, G. 1999. Soil conservation and sustainable agriculture on the Loess Plateau: challenges and prospects. *Ambio* 28:663–668.

Liu, X.H., M.D. Hao, and J. Fan. 2000. Effects of long-term rotation and fertilization on N supply by soil in dryland area on the Loess Plateau. *Agric. Res. Arid Areas* 18:1–7 (in Chinese with English abstract).

Liu, J., T. Dietz, S.R. Carpenter, M. Alberti, C. Folke, E. Moran, A.N. Pell, P. Deadman, T. Kratz, J. Lubchenco, E. Ostrom, Z. Ouyang, W. Provencher, C.L. Redman, S.H. Schneider, and W.W. Taylor. 2007. Complexity of coupled human and natural systems. *Science* 317:1513-1516.

Liu, Z., P. Mark, Z. David, D. Robert, H. Robert, H. Matthew, B. Henk, R.S. Sunita, R.M. Leckie, and P. Ann. 2009. Global cooling during the eocene-oligocene climate transition. *Science* 323:1187-1190.

Ma, S.L. 1991. Dryfarming. China Agriculture Publishing House, Beijing, pp. 175–185.

Maxime, F., J.M. Mollet, and F. Papy. 1995. Aide au raisonnement de l'assolement en grande culture. *Cah Agric.* 4:351–362.

Malhi, S.S., S. Brandt, and K.S. Gill. 2003. Cultivation and grassland type effects on light fraction and total organic C and N in a dark brown chernozemic soil. *Can. J. Soil Sci.* 83:145–153.

Meek, B.D., W.R. DeTar, D. Rolph, E.R Rechel, and L.M. Carter. 1990. Infiltration rate as affected by an alfalfa and no-till cotton cropping system. *Soil Sci. Soc. Am. J.* 54:505–508.

McCallum, M.H., M.B. Peoples, and D.J. Connor. 2000. Contributions of nitrogen by field pea (*Pisum sativum* L.) in a continuous cropping sequence compared with a lucerne (*Medicago sativa* L.) based pasture ley in the Victorian Wimmera. *Aust. J. Agric. Res.* 51:13–22.

McCallum, M.H., D.J. Connor, and G.J. O'Leary. 2001.Water use by lucerne and effect on crops in the Victorian Wimmera. *Aust. J. Agric. Res.* 52:193–201.

Mueller, T., L.S. Jensen, N.E. Nielsen, and J. Magid. 1998. Turnover of carbon and nitrogen in a sandy loam soil following incorporation of chopped maize plants, barley straw and blue grass in the field. *Soil Biol. Biochem*. 30:561–571.

Nannipieri, P., S. Grego, and B. Ceccanti. 1990. Ecological significance of the biological activity in soil. In: Bollag, J.M., Stotzky, G. (Eds.), *Soil viochemistry,* vol. 6. Marcel Dekker, New York, NY, U.S.A., pp. 293–355.

Navarrete, M., and M.L. Bail. 2007. SALADPLAN: a model of the decision making process in lettuce and endive cropping. *Agron. Sustain. Dev.* 27:209–221.

Nevo, A., and I. Amir. 1991. CROPLOT—an expert system for determining the suitability of crops to plots. *Agric. Syst*. 37:225–241.

Nevo, A., R. Oad, and T.H. Podmore. 1994. An integrated expert system for optimal crop planning. *Agric. Syst*. 45:73–92.

Nuthall, P. 2010. Farm Business Management: The Human Factor. CAB Int.

O'Leary, G.J., and D.J. Connor. 1997. Stubble retention and tillage in a semi-arid environment. 3. Response of wheat. *Field Crop. Res*. 54:39–50.

Powlson, D.S., P.C Brooks, and B.T. Christensen. 1987. Measurement of soil microbial biomass provides an early indication of changes in total soil organic matter due to straw incorporation. *Soil Biol. Biochem*. 19:159–164.

Raimbault, B.A., and T.J. Vyn. 1991. Crop rotation and tillage effects on corn growth and soil structural stability. *Agron. J.* 83:979–985.

Ren, J.Z. 1992. Ecological productivity of grassland farming system on the Loess Plateau of China. In: Ren, J.Z. (Ed.), Proceedings of the International Conference on Agroecosystems in the Loess Plateau. Gansu Science and Technology Press, Lanzhou, pp. 3–7.

Reynolds, F.J., D.M.S. Smith, E.F. Lambin, B.L. Turner, M. Mortimore, S.P.J. Batterbury, T.E. Downing, H. Dowlatabadi, R.J. Fernandez, J.E. Herrick, E.H. Sannwald, H. Jiang, R. Leemans, T. Lynam, F.T. Mastre, M. Ayarza, and B. Walker. 2007. Global desertification: building a science for dryland development'. *Science* 316:847-851.

Ridley, A.M., B. Christy, and P.J. Haines. 2001. Lucerne in crop rotations on the Riverine plains. I. Soil water balance. *Aust. J. Agric. Res*. 52:263–277.

Rounsevell, M.D.A., J.E. Annetts, E. Audsley, T. Mayr, and I. Reginster. 2003. Modelling the spatial distribution of agricultural land use at the regional scale. *Agric. Ecosyst. Environ.* 95:465–479.

Saeed, I.A.M., and A.H. Ei-Nadi. 1997. Irrigation effects on growth, yield, and water use efficiency of alfalfa. *Irrig. Sci*. 17:63–68.

Schwab, G.O., D.D. Fangmeier, and W.J. Elliot. 1996. In: Soil Water Manage. Sys. 4th ed. John Wiley and Sons, Inc., New York, Chrichester, Brisbane, Toronto, Singapore.

Shan, L., and G.L. Chen. 1993. The theory and practice of dryland agriculture in the Loess Plateau of China. *Sci. Press,* Beijing (in Chinese).

Shen, Y.Y., Z.B. Nan, C.Y Gao, W.D. Bellotti, and W. Chen. 2004. Spatial and temporal characteristics of soil water dynamics and crop yield response from a 4-year lucerne and winter wheat rotation in the Loess Plateau. *Acta Ecol. Sin*. 24(3):640–647 (in Chinese with English abstract).

Singh, K.P., and V. Kumar. 1981. Water use and water use efficiency of wheat and barley in relation to seeding dates, levels of irrigation and nitrogen fertilization. Agric. *Water Manage*. 3:305–316.

Sparling, G.P. 1992. Ratio of microbial biomass carbon to soil organic carbon as a sensitive indicator of changes in soil organic matter. *Aust. J. Soil Res.* 30:195–207.

Sparling, G.P. 1997. Soil microbial biomas, activity and nutrient cycling as indicators of soil health. In: Pankhurst, C., B.M.,Doube, V.V.S.R. Gupta, (Eds.), Biological Indicators of Soil Health. *CAB International*, Wallingford, U.K., pp. 97–119.

Stewart, B.A., and G. Peterson. 2015. Managing green water in dryland agriculture. *Agron. J.*, 106:1–10.

Stoate, C., N.D. Boatman, R.J. Borralho, C.R. Carvalho, G.R. Snoo, and P. Eden. 2001. Ecological impacts of arable intensification in Europe. *J. Environ. Manage.* 63:337–365.

Tansley, A.G. 1939. British ecology during the past quarter-century: The plant community and the ecosystem. *J. Ecol.* 27:513–530.

Tao, L.F., L.X. Wang, and S.F. Han. 1993. Water production potential of dry land area in north Chinese. China Weather Publishing House, Beijing, pp. 21–28.

Vandermeer, J., N.M. Van, J. Anderson, C. Ong, and I. Perfecto. 1998. Global change and multi-species agroecosystems: concepts and issues. *Agric. Ecosyst. Environ.* 67:1–22.

Vereijken, P. 1997. A methodical way of prototyping integrated and ecological arable farming systems (I/EAFS) in interaction with pilot farms. *Eur. J. Agron.* 7:235–250.

Wang, J., J.M. Cheng, and H.E Wan. 2005. Alfalfa growth and its relation with soil water status in loess hilly and gully region. *Chin. J. Appl. Ecol.* 16:435–438 (in Chinese with English abstract).

Wan, S.M., Z.K. Jia, Q.F. Han, and B.P. Yang. 2008. Dry soil layer forming and soil moisture restoration of alfalfa grassland in the semi-humid region of the Loess Plateau. *Acta Ecol. Sin.* 28:1045–1051.

Wang, Z.Q., B.Y. Liu, and B.J. Lu. 2003. A study on water restoration of dry soil layers in the semi-arid area of the Loess Plateau. *Acta Ecol. Sin.* 23:1944–1950 (in Chinese with English abstract).

Wang, X.L., F.M. Li, Y. Jia, and W.Q. Shi. 2005. Increasing potato yields with additional water and increased soil temperature. *Agric. Water Manage.* 78:181–194.

Wang, X.L., J.S. Guo, J. Yu, M.L. Feng, and Z.X. Jin. 2008. Crop yield and soil water restoration on 9-year-old alfalfa pasture in the semiarid Loess Plateau of China'. Agric. *Water Manage.* 95:190-198.

Wang, X., Q. Sun, Z. Ding, J. Ji, J. Wang, X. Kong, J. Yang, and G. Cai. 2014. Redefining the modular organization of the core Mediator complex. *Cell Res.* 24:796-808.

Wijnands, E. 1999. Crop rotation in organic farming: theory and practice. In: Designing and testing crop rotations for organic farming. Proceedings from an international workshop. *Danish Research Centre for Organic Farming,* pp. 21–35.

Xie, Z.K., Y.J. Wang, and F.M. Li. 2005. Effect of plastic mulching on soil water use and spring wheat yield in arid region of Northwest China. *Agric. Water Manage.* 75:71–78.

Yang, W.Z., and M.A. Shao. 2000. Study on soil moisture in the Loess Plateau. Sci. Press, Beijing, pp. 251–258 (in Chinese).

Yang, J.Y., W. Joël, A. Amar, C. Marion, W. Gérard, T. Bernard, and W. Christian. 2005. Expression of an uncleavable N-terminal rasgap frangment in insulin screening cells increases their resistance towards apoptotic stimuli without affecting their glucose-induced insulin secretion. *J. Biol. Chem.* 280:32835–32842.

Yang, J.J., H.K. Li, G.J. Zheng, G.H. Lu, Q.D. Shi, and T.B. Taxifulati. 2006. Preliminary research on counter measures of desiccation of deep soil layer in Loess Plateau. *J. Soil Water Conserv*. 20:186–190 (in Chinese with English abstract).

Yang, L., L. Chen, and W.I. Wei. 2015. Effects of vegetation restoration on the spatial distribution of soil moisture at the hillslope scale in semi-arid regions. *CATENA* 124:138-146.

Yongqiang, W., Z. Xinlei, H. Zhang, L. Yi, H. Haolong, D. Xiaoxi, C. Jinan, D. Jiuhong, Y.X. Yang, H. Haiying, and J. Taijiao. Coiled-coil networking shapes cell molecular machinery. *Mol. Biol. Cell.* 23(19):3911-3922.

Zhang, C.X., D.M. Hao, X.R. Wei, and X.G. Wang. 2004. Soil water distribution characteristics of alfalfa with different planting years in the gully region of the Loess Plateau. *Plant Nutr. Fertil. Sci*. 10:604–607 (in Chinese with English abstract).

Zhou, L., S. Jin, C. Liu, Y. Xiong, J. Si, X. Li, Y. Gan, and F. Li. 2012. Ridge-furrow and plastic-mulching tillage enhances maize–soil interactions: Opportunities and challenges in a semiarid agroecosystem. *Field Crop. Res.* 126:181-188.

Zhou, L.M., F.M. Li, S.L. Jin, and Y.J. Song. 2009. How double ridges and furrows mulched with plastic film affect soil water, soil temperature and yield of maize on the semiarid Loess Plateau of China. *Field Crop. Res*. 113:41–47.

In: Crop Rotations
Editor: Bao-Luo Ma
ISBN: 978-1-63484-496-3

Chapter 7

CROP ROTATION AND COVER CROP IN PEST AND DISEASE MANAGEMENT IN SUSTAINABLE AGRICULTURE

Qing Yu[1,*] and Yuejing Qiao[2]
[1]Ottawa Research and Development Centre,
Agriculture and Agri-Food Canada, Ottawa, ON, Canada
[2]College of Agronomy, Shanxi Agricultural University, Taigu,
Jinzhong, Shanxi, China

ABSTRACT

Crop rotations, and crop rotation with cover crop or with fallow, have had a very long history of use for pest and disease controls; however these practices had been neglected during the Green Revolution. Only recently have we come to the realization that the ecological and environmental problems associated with the monoculture and excessive use of chemical pesticides and fertilizers in the Green Revolution can be mitigated through well-planned crop rotation and cover crop. One of the newly found benefits of the cultural practices such as crop rotation and cover crop is the increased diversity of flura and fauna within the agroecosystem, which tends to be healthy, and resilent to pests and diseases. A successful crop rotation and cover crop for pest and disease controls starts with the understanding of the pests and diseases. This chapter reviews the latest scientific knowledge by studying the current cropping systems, and cover crops for reducing pest and disease populations in the context of integrated pest management (IPM) systems. The shift of adoption from selecting a specific (nonhost) to a general (only taxonomically distant) rotating crop or cover crop for pest and disease control is discussed.

Keywords: biodiversity, cover crop, crop rotation, diseases, insects, IPM, nematodes, sustainable agriculture

[*] Corresponding author: Qing.Yu@agr.gc.ca.

INTRODUCTION

Crop rotation is the practice of growing a series of different types of crops in the same area through time. This is in contrast to continuous monoculture, which is the practice of growing a single species repeatedly on the same land. The practice is a documented ancient practice, but it is also probably prehistoric, dating back to the early time when humans learned growing crops, the dawn of agriculture, a process usually called the Neolithic Revolution. At beginning, it was forced upon because of the temperature changes of the seasons. During the most part of human history, the practice was widely used in all civilizations of the world, until the Green Revolution in the 20th century.

The Green Revolution from the 1960s heavily relied on monoculture of high-yielding varieties, pesticides and synthetic fertilizers. Without doubt, it saved billions of people from starvation; but it came with a huge cost to the environment and biodiversity. Monoculture production systems compromise biodiversity, utilize resources inefficiently, and are susceptible to pest outbreaks (Martens et al., 2015). Many pesticides since have been banned from production and consumption because they kill wild life, pollute environments and are toxic to humans. The public was surprised to learn that several species of bee pollinators, on which we depend on for successful production of broad ranges of crops, have disappeared. And for a number of years, people read news around the world that the honey bee populations collapsed, especially in developed countries. Now we know that their disappearances are caused by loss of habitats and the toxic effect of neonicotinoids, one family of pesticides. We know that large scale of monoculture entailed loss of wild flowers that bees live on. Red-tide, the other well known phenomenon is also the result of excessive use of fertilizers in monoculture. The Green Revolution has saved millions of lives, but it isunlikely sustainable, neither environmentally nor socially.

In the 1980s, the concept of sustainable agriculture became popular (Jackson, 1980). One of the practices within sustainable agriculture is crop rotation. The two most obvious benefits of crop rotation are the replenishing of nitrogen, if legumes are included, and pest and disease control if non-host crops are included.

Green manure is created by leaving uprooted or mown crop parts to decompose on a field so that they serve as a mulch and soil amendment to improve the soil fertility. The practice was documented in ancient civilizations in China, Greece, and India for thousands of years. In China, the practice was used in Han Dynasty about 2000 years ago. In ancient Greece too, farmers ploughed broad bean plants into the soil to increase soil nitrogen. A cover crop is now a crop planted primarily to manage soil erosion. The practice has its deep root in Canada and USA to combat the Dust Bowl in the 1930s, one of the most serious soil erosion disasters. The crop usually is not harvested, but all plant parts are left on the field to be withered or tilled in soil to be decomposed as green manure. The practice has evolved and is now becoming increasingly important for sustainable agriculture. Disease and pest control aspect of the practice was observed early on in a similar way as the crop rotation practice, because in essence, the inclusion of a cover crop and the incorporation of the plant parts in soil as green manure are commonly known as the fallow cycle of crop rotation. This cover crop/green manure practice over crop rotation for disease and pest management gives farmers more freedom in selecting plant species.

Interestingly, in early 1980s, times when sustainable agriculture was getting attention, the term biodiversity became frequently and widely used in science and environmental policy documents. Its importance to mankind was manifested in adopting and signing of the Convention on Biological Diversity (CBD) in 1990.

Today few would argue against the concept that humans rely on biodiversity for its goods and services. Biodiversity is the basis for agriculture. However, all human activities reduce biodiversity, including agriculture which has reduced it since the domestication process several thousand years ago, although it was in a small scale at beginning. Only recently we have finally realized that agriculture needs to conserve biodiversity as well. The Green Revolution relied on pesticide uses, to the detriment of the environment and the biodiversity, which was realized with the publishing of Silent Spring. A total of 2 million tonnes of pesticides are consumed globally each year. Only about 2 to 5% of that land on the targeted pests; this means that over 95% ends up in the environment. Sustainable agriculture depends on biodiversity.

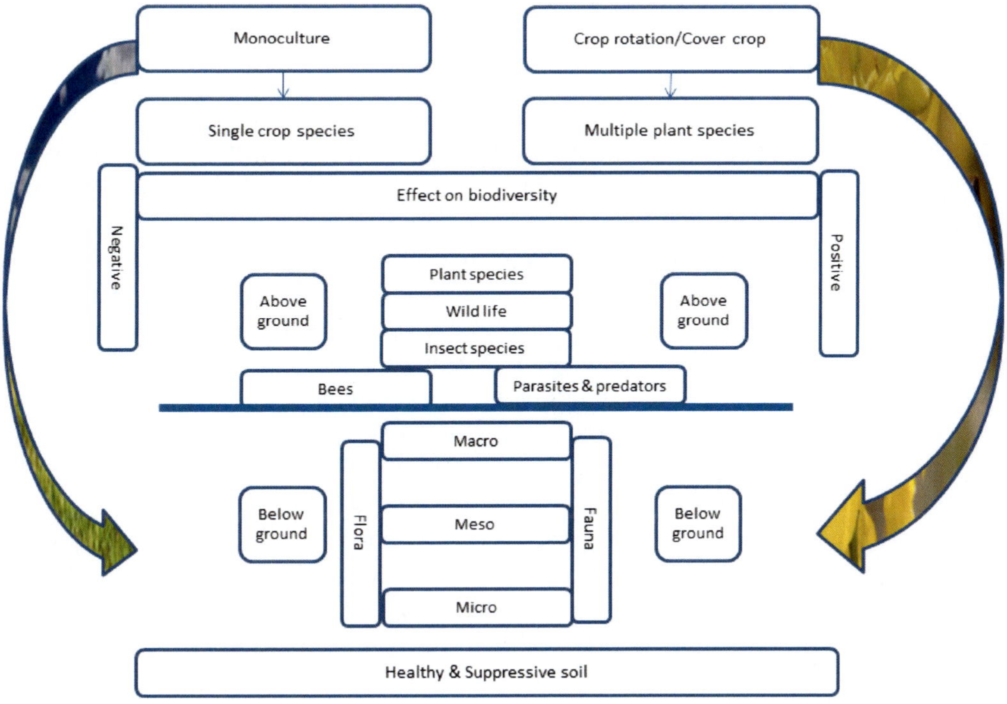

Figure 1. Impact of monoculture and crop rotation with cover crop on biodiversity.

The one cultural practice that can be and have been used to reduce pest and pathogen populations to manageable levels (Satti, 2012) while enhancing overall biodiversity and restoring soil nutrients is crop rotation, or crop rotation with cover crop (Figure 2), along with biological control (Ehler and Miller, 1978).

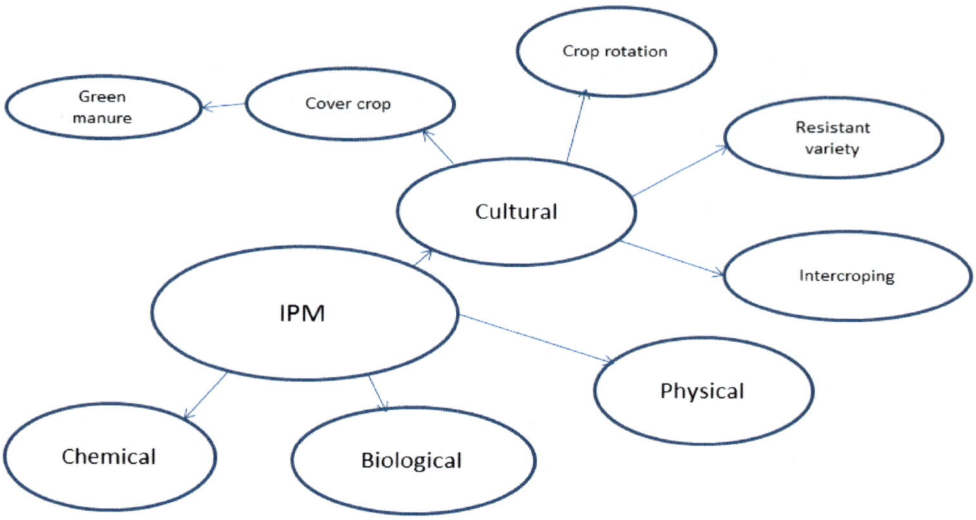

Figure 2. Roles of crop rotation and cover crop in an integrated pest management system.

An ecosystem which is species-rich is more resilient and adaptable to external stress such as pests and diseases than one in which the range of species is limited. In a system where species are limited, the loss or temporary reduction of any one could disrupt a complex food chain with serious effects on other species in that same system. Once biodiversity is sufficient, if one nutrient cycling path is affected another pathway can function and the ecosystem - and the biological species it supports –can thrive. That is exactly what happens between monoculture and multi-cropping systems. Healthy, thriving ecosystems are generally highly diverse with numerous taxa. Also, more recently numerous studies covering terrestrial and aquatic ecosystems show that high-diversity ecosystems are approximately twice as productive as monocultures of the same species, and that this difference increases through time (Cardinale et al., 2011; Hayden et al., 2014; Miyazawa et al., 2014; Tilman et al., 2014; Smith et al., 2014; Wortman et al., 2012).

There has been a shift in understanding of crop rotation and cover crop for disease and pest management from using non-host starvation to biodiversity. Crop rotation and rotating with cover crops brings biodiversity into agriculture, starting with the plants, and then followed by more diversified insect species including pollinators such as bees and parasites and predators which provide checks on pest populations above ground; below ground, crop diversity leads to increased diversity of macro to microorganisms, leading to a soil that becomes healthy and suppressive to nematodes and pathogens. Suppressive soils are soils that are unfit for certain pathogens and nematodes, and have been well identified (Hornby, 1983) which contain diversified microbial populations of different families and kingdoms (Figure 1). In general, more taxonomically distant plants mean more functional characteristic differences.

Sustainable agriculture and integrated pest management both involve many different dimensions and interacting factors. Management of diseases, pests and weeds with crop rotation and cover crops is no exception; and they are linked to other sustainable management practices such as biological control. One of the objectives of this review is to look at the practices in conjunction with the subjects covered in other chapters of this book.

SECTION I: CROP ROTATION

Rotating land out of susceptible crops can be an effective and relatively inexpensive means for managing some pests and diseases (Bullock, 1992). Using crop rotation for pest and disease management, however, requires understanding the life cycles of the pest and disease-causing organisms (bacteria, fungi, nematodes and insects). Primarily, the technique of using crop rotation for pest and disease management is to grow non-host plants until the pathogen in the soil dies or its population is reduced to a level that will result in negligible crop damage. More recently the practice has been shifted: namely if a non-host is not possible, a more taxonomically distant crop will provide a different habitat for different flaura and fauna resulting in increased biodiversity, which helps hold the pest and disease populations in check (Tilman et al., 2014). General considerations when selecting a crop rotation for pest and disease control include how long the pest and pathogen can survive in the soil; which additional plant species (including weeds and cover crops) it can infect or survive on; other ways it can survive between susceptible crops; how it can be spread or be reintroduced into a field; and methods for managing other pathogen sources (Figure 3). For example, a pathogen that can survive in the soil but can also disperse by wind may not be successfully managed by rotation if an infected planting occurs nearby or the spores can disperse long distances.

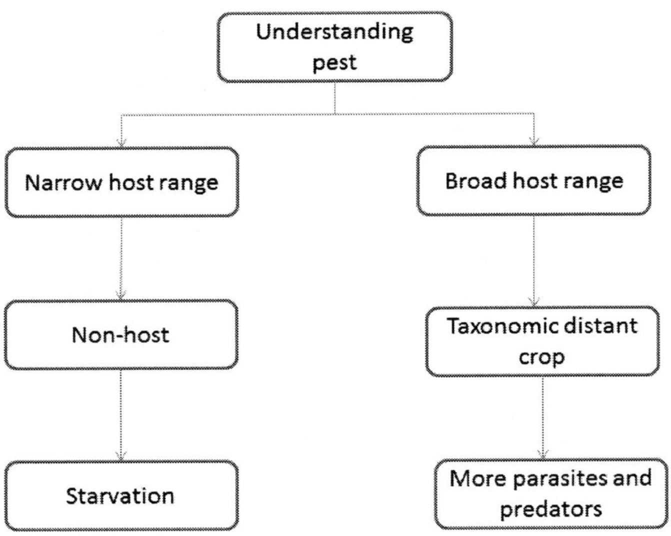

Figure 3. Decision making flowchart to select a crop in a crop rotation system for pest and disease management.

The specific in crop rotation is host or non-host selection, the general selection is a crop that is taxonomically distant from the previous one. Taxonomically distant crops present different habitats both above and below ground. Crops that are taxonomically close (i.e., same family) tend to have similar pests and diseases.

One recent study (Matthew et al., 2016) has shown that introduction of canola into corn and soybean production system in the Central and Midwestern US would benefit honey bees and wild bees in the region.

Crop Rotation and Weed Management

Crop rotation has been shown to provide the foundation for long-term weed management (Liebman and Dyckm, 1993; Liebman and Gallandt, 1997; Liebman and Staver, 2001). Planting a wide variety of crops with varied characteristics reduces the likelihood that specific weed species will become adapted to the system and become problematic. According to Liebman and Dyck (1993), "the success of rotation systems for weed suppression appears to be based on the use of crop sequences that employ varying patterns of resource competition, allelopathic interference, soil disturbance, and mechanical damage to provide an unstable and frequently inhospitable environment that prevents the proliferation of a particular weed species." Attributes to consider when selecting a rotating crop for weed control purposes are listed in Table 1 (Mohler, 2009; Mohler, 2001a, b).

Table 1. Crop characteristics in crop rotation for weed management

Characteristics	Current crop	Next crop
Phenotic	Grassy	Broadleaved
Life cycle	Annual	Perennial
Climate	Cool season	Warm season
Seeding date	Spring	Fall
Competitive	Low	High
Fertilize requirement	Low demanding	High
Taxonomic distance	Close	Distant

Crop diversification is important to consider even for solving immediate weed problems (Liebman and Staver, 2001). For example, one study found that 83 percent of farmers surveyed in Saskatchewan and Manitoba, Canada, noticed decreased weed problems such as Canada thistle (broadleaved, annual) following sod crops (grassy, perennial), and most indicated that the effect lasted more than one year (Entz et al., 1995).

Crop Rotation for Insect and Disease Control

Insect Control

There are four criteria defining successful insect and disease management by crop rotation: 1) The source of the pest inoculum must be from the field itself; 2) the host range of the pest needs to be fairly narrow or at least not include crops that are reasonable alternatives for a given area; 3) the pest must be incapable of long periods of existence without a living host; and 4) the insect pest must be relatively immobile (Deana et al., 2005).

Because of the mobility of most insect pests, crop rotation has not been thought effective in controlling insect pests. However there are a few important pests that have been successfully controlled by crop rotation. The classical example is the Colorado potato beetle, *Leptinotarsa decemlineata*, which overwinters adjacent to field edges and infests nearby fields in the spring, primarily by walking. Spring infestations can be delayed or reduced by moving potatoes to a new field distant from the previous year's potatoes (Sexson and Wyman, 2005; Hunt and Vernon, 2001).

Another example is the corn root worms consisting western corn rootworm (*Diabrotica virgifera virgifera*) and northern corn rootworm (*Diabrotica barberi* L.) which can be managed with crop rotation taking advantage of their overwintering preference on corn roots. Annual rotation of corn with soybean in USA and Canada has largely controlled corn root worms. However there are reports of adaption of the pests to this practice (Smith and McSorley. 2000).

Much needs to be learned about whether diversified habitat from a distantly related rotation crop provides these parasites and predators with alternative sources of food, shelter and breeding sites. Nevertheless, including taxonomical distant crops in a crop rotation system increases diversity, which inevitably results in a healthy ecosystem.

Table 2. Cover crops that attract beneficial insects and mites (adopted from Wang, 2012)

Beneficial arthropoda	Plant
Various parasitoids and predators	Fennel, dill, coriander (cilantro), parsley, carrot, wild carrot (Queen Anne's-lace), angelica, yarrow (milfoil), sow thistle, dandelion, zinnia, tansy, marigold, sunflowers
Predatory wasps, hoverflies	sweet alyssum, buckwheat, mustard, Cuban oregano, sage, salvia, lavender, oregano, thyme, marjoram, perilla
Lady beetles	Dill, marigold, Mexican tea, morning glory, oleander, yarrow.
Lacewing	Carrot, oleander, red cosmos, wild lettuce, tansy
Minute pirate bug	Carrot, Mexican tea, oleander, sunn hemp, cowpea
Ground beetles	Low-growing plants: thyme, rosemary, mint or mulches
Spider	Marigold, yellow-sweet clover, white clover

Disease Control

Rotation can effectively suppress a crop disease when the target pathogen is capable of surviving in the soil or on crop debris for no more than a few years. Some fungal and bacterial pathogens can survive in soil only in crop debris, and these are the most suitable pathogens to target for management with crop rotation because they cannot survive once the debris has decomposed (Davis and Nunez, 1999). Pathogens that survive on soil organic matter only for a few years can also be managed with crop rotation (Dillard and Cobb, 1998). These short-term residents of the soil are called soil invaders or soil transients. Pathogens in this group vary in the length of time they can survive, and thus in the length of rotation needed.

Survival time partly reflects the type of plant host tissue infected. For example, the barley scald pathogen primarily infects leaves and leaf sheaths, which decompose fairly rapidly. In contrast, the net blotch pathogen infecting barley stems, including the nodes, which are more resistant to decay; consequently, a longer rotation is needed. In addition, infected seed, and also wind-dispersed spores for the net blotch pathogen, are additional sources of these pathogens that need to be managed to ensure successful control through rotation.

Choice of crop in a rotation may also garner microbial benefits beyond those normally associated with pathogen host range and saprophytic pathogen survival (Curl, 1963). For example, analyses of microbial populations in plant tissues and soils when clover preceded potato in a rotation revealed that 25 bacterial species were common to both clover and potatoes and represented 73% of culturable bacteria recovered from clover roots and potato tubers (Sturz et al., 1998). Of the bacteria tested, 74% showed in vitro antibiosis to *Rhizoctonia solani* Kühn (Sturz et al., 1998). More bacteria inhibitory to *R. solani* were found within plant tissues than in root zone soil (Sturz et al., 1998). Rhizosphere-inhabiting bacteria can be crop specific (Glandorf et al., 1993) and lants under monoculture have been shown to support and respond to populations of rhizosphere microorganisms antagonistic to their pathogens (Cook et al., 1995). Seed source has also been shown to influence *Rhizoctonia* disease severity, presumably due to differing loads of microbial antagonists (Jager and Velvis, 1983). Mechanisms by which endophytes can act as biocontrol agents include production of antibiotic agents (Lambert et al., 1987), siderophore production (Kloepper et al., 1980), nutrient competition (Kloepper et al., 1980), niche exclusion (Cook and Baker, 1983), and induction of systematic acquired host resistance (Chen et al., 1995). Bacterial endophytes can thus play a role in pathogen suppression (Chen et al., 1993; Sturz et al., 1998; and Sturz et al., 2000), and complementary crop sequences can encourage beneficial allelopathy (Sturz et al., 1998).

Nematode Control

Because nematodes are not able to move large distances by themselves, crop rotation has been tested and applied widely for reducing nematode populations; especially for the species which have narrow host ranges. The best known example is the rotating corn with soybean to control soybean cyst nematode (*Heterodera glycines*) in Canada and USA (Epps, 1960; Epps and Chambers, 1958, 1965; Ross, 1960; Sasser and Grover, 1991; Sasser and Uzzell, 1963a, b).

Crop rotation can also reduce the negative impact of nematodes on crop production by suppressing population levels of most plant parasitic nematodes from a more diversified community of free living nematodes brought about by diversified cropping system. For some nematode species and crops (e.g., ring nematode *Helicotylenchus dihystera)*, short rotations such as maize-soybean are effective, whereas in other situations (e.g., *Criconemoides ornatus*), short rotations are not effective and the host crop must not be present for many years. Crop rotation in the Southeast USA has proven to be an effective method for reducing the severity of root-knot nematodes and cyst nematodes (*Heterodera glycines*) in soybean (Ross, 1962). The reduction of nematode pressure may account for most of the rotation benefits for soybean in the Southeast USA. Another example is that using a 1-year rotation with barley, clean fallow, or a resistant processing tomato cultivar has been shown to be effective in controlling the root knot nematode (*Meloidogyne incognita*).

Suppressive soils which have microbial antagonistic to plant parasitic nematodes have been assessed, and polyculture of multiple crops provides a more conducive environment for those antagonistic microbial communities to flourish. However, very few studies are available demonstrating that crop rotation enhances microbial suppressiveness to nematodes.

Crop rotation cannot solve all weed, pest and disease problems in crops. But it is an essential practice to be included in any integrated pest and disease management systems. The selection of crops in any crop rotation system is limited by economic and climatic factors. Crop rotation with a cover crop offers more freedom in the selection.

SECTION II: COVER CROP

In addition to helping to control pests and disease, cover crops can improve soil health; producing organic matter; reducing compaction and erosion,; scavenging nutrients, and regulating soil temperature (Snapp et al., 2005). For many years, cereal rye was the only cover crop that farmers considered planting in North America. Now farmers are planting several different species of cover crops on their lands: Brassica mustard and tillage radish; cereals like triticale, and oat; grasses like annual rye grass; and sorghum sudangrass, and legumes like clovers;vetch and peas. Figure 4 lists potential cover crop species currently applicable to farmers in North America.

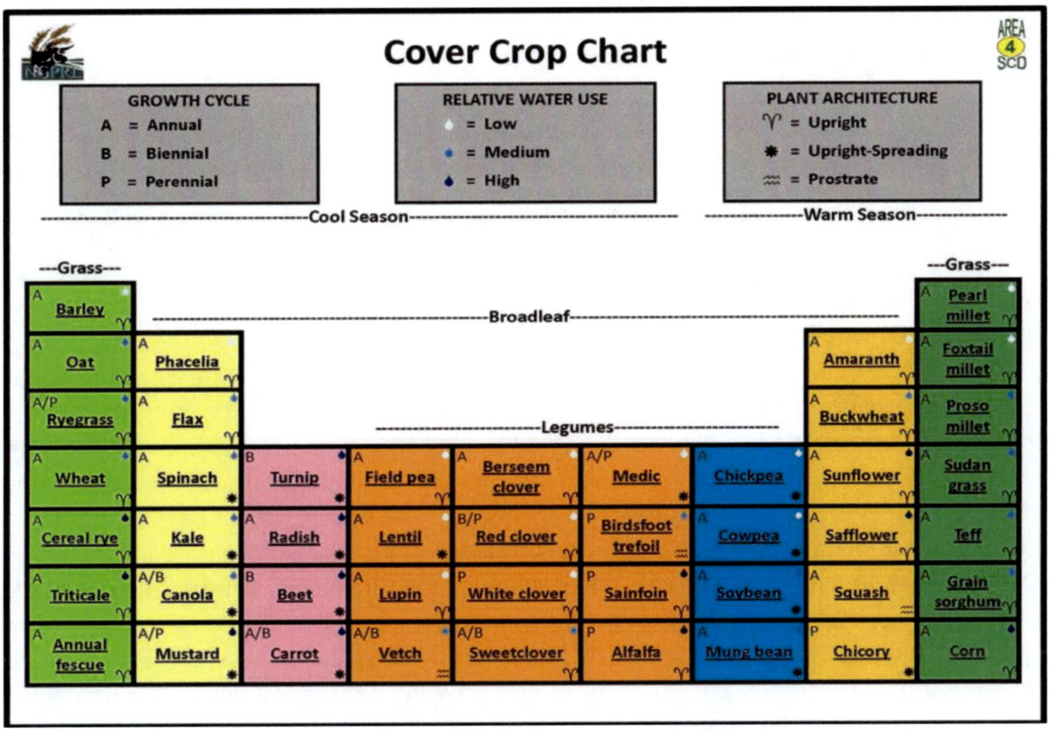

Figure 4. Cover crop table for North American farmers.

Cover crops are poised to play increasingly important roles on North American farms. In addition to slowing erosion, improving soil structure and providing fertility, we are learning how cover crops help farmers to manage pests. With limited tillage and careful attention to cultivar choice, placement and timing, cover crops can reduce infestations by insects, diseases, nematodes and weeds. Pest-fighting cover crop systems help minimize reliance on pesticides, and as a result cut costs, reduce chemical exposure, protect the environment, and increase consumer confidence in the food quality.

Farmers and researchers are using cover crops to design new strategies that preserve a farm's natural resources while remaining profitable. Key to this approach is to see a farm as an "agro-ecosystem"— a dynamic relationship of the mineral, biological, weather and human resources involved in producing crops or livestock. The goal is to develop agricultural practices that are environmentally sound, economically feasible and socially acceptable.

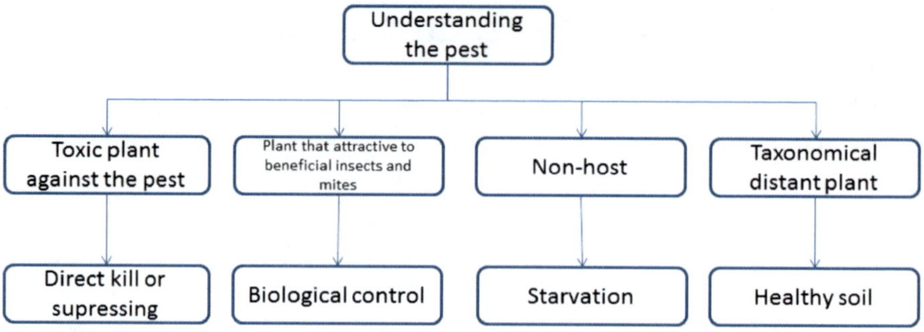

Figure 5. Decision making flowchart for a cover crop for pest and disease management.

Cover Crops for Weed Management

Cover crops suppress weeds either by smothering growth of established weeds or creating an environment that interferes with weed emergence and establishment (Teasdale, 1998). Cover crop species that emerge and grow rapidly are most effective in smothering weeds. Vigorous species that produce high biomass yields are often the most effective competitors that deprive weeds of light, water, and nutrients. Winter annual cover crops are particularly adapted to developing a dense canopy in early spring that prevents establishment and growth of annual weeds later in season.

After a cover crop is killed, residue can inhibit the establishment of weeds through a number of mechanisms (Teasdale, 1998). If the residue remains on the soil surface, it can eliminate environmental cues required for weed seed germination such as light or alternating temperature. Residue can also act as a physical barrier that impedes the emergence of weed seedlings after germination. Phytotoxic (allelopathic) compounds released by residue can inhibit germination and growth of weeds. In contrast, cover crop residue may encourage weed emergence in selected circumstances, either when residue maintains soil moisture at a level suitable for germination during droughty periods or when germination-stimulating compounds such as nitrate are released. Weed suppression by cover crop residue is generally observed early but not for the duration of the growing season. Cover crops need to be integrated with other techniques to obtain optimal weed control.

Cover Crop for Pest and Disease Control

Insect Control

Cover crops have been shown to affect insect, pathogen, and nematode pests (Phatak, 1992). Bugg and Wilson (1989) found that generalist predators may be important in the biological control of insects that attack warm-season vegetable crops (Table 2). They observed that during periods when pests are scarce or absent, several important predators subsisted on nectar, pollen, and alternative prey offered by cover crops. Bugg et al. (1990) have shown that the flower bug (*Orius incidiosus*), big-eyed bugs (*Geocoris* spp.), and various lady beetles (Coleoptera: coccinellidae) attained high densities in various vetches, clovers, and certain Cruciferae. These predators subsisted and reproduced on nectar, pollen, thrips, and aphids and were established before the arrival of pests (Bugg et al., 1987; Ehler and Miller, 1978; Murdoch et al., 1985; Tamaki, 1981; Tedders, 1983). In these studies, narrow strips of hairy vetch and crimson clover were maintained along the borders of tomato fields. Insect predators used the cover crops as host plants and then moved into the tomato crop.

Many plants produce repellents, poisons, antifeedants, growth regulators, and antibiotics to reduce exploitation by herbivores and pathogens. Because insects comprise more than 90% of all planetary species and are clearly the dominant herbivores, plants have evolved an extensive variety of defensive strategies targeted to limit their exploitation by pests. The earliest insecticides used by man were simple powders or aqueous extracts of poisonous plants. Hundreds of toxic plants have been employed by primitive cultures for purposes of insect control and most of these have been evaluated as prospective insecticides (Bowers, 1993). There are too many to list and the following are a few widely known plants derived chemicals that have been used in agriculture:pyrethrins from chrysanthemum, nicotine from tobacco, and ryanodine from *Ryania speciose* are a few examples.

Disease Control

When sweet corn was used as cover crop following potato in Idaho, USA and when residues were turned under, populations of *Verticillium dahliae*, the cause of *Verticillium* wilt, in the soil declined (Stark, 1995). Recently, priority has been given to developing biological approaches to control phytopathogenic nematodes in vegetable and strawberry systems as alternatives to methyl bromide.

There is a long list of plant derived compounds that are suppressive to the growth and reproduction of fungi, but a few have been used as cover crops for disease management. One example, in an Idaho study, *Verticillium* wilt of potato was reduced by 24 to 29% following sudangrass green manure. Yield of U.S. No.1 potatoes increased by 24 to 38% compared with potatoes following barley or fallow in an Oregon study (Wiggins and Kinkel. 2005). It is known that sudangrass, a member of Poaceae family produce volatile biocidal compounds such as hydrogen cyanide (Stapleton et al., 2009).

Nematode Control

Early nematode control using cover crop by utilizing the bioactive compounds which are plant secondary metabolites of low-molecular weight compounds, and they are regarded as not essential for sustaining life, but crucial for the survival of the producing organism (Hadacek, 2002). Traditional medicines in ancient civilizations such as Chinese and Indian make good use of these bioactive compounds. Some compounds are insecticidal, antibiotic, antifungal, herbicidal and nemicidal. Between 50, 000 to 100, 000 structures have been identified (Yazdani et al., 2011). Table 3 lists some known plants that produce nematicidal compounds. Much more remain to be discovered.

Of course, non-hosts have been widely tested as well. Rapeseed and sudangrass green manures grown prior to potatoes at Prosser, Washington State, USA provided 72% and 86% control of the root-knot nematodes on potatoes, respectively (Stark, 1995). Cereals such as rye, bahiagrass, and barley, are non-hosts to *Meloidogyne incognita*, the dominant species of root-knot nematode attacking fruit and vegetable crops in warm-temperate and subtropical regions (Minton, 1986). Other non-host cover crops include several lines of cowpeas (Roberts et al., 1996), marigold (*Tagetes patula*), hairy indigo (*Indigofera hirsuta*), sunn hemp

(*Crotalaria junda*), velvetbean (*Mucuna deeringiana*), and castorbean (Leonard, 1991). It is likely those crop rotations that utilize non-host cover crops and nematode-resistant horticultural and field crops in well-planned production rotations will reduce the need for chemical nematicides (Fortnum and Currin, 1993).

Table 3. Rotation/cover crops suppressive to plant parasitic nematodes

Common name	Scientific name	Active ingredient	References
Vetch	Vicia spp.	Vicianin (Cyanogen)	Minton et al., 1965, 1966 Minton, 1992
Grassy species			
Sudan grass	Sorghum vulgare var. sudanense	Dhurrin(Cyanogen)	Mojtahedi et al., 1993; Good et al., 1965
Sorghum	S. bicolor X S. vulgare var. sudanense	Cyanogen	Kinloch and Dunavin, 1993
Millet		Cyanogen	Jagdale et al. 2000.
Sunn hemp	Crotalaria juncea	unknown	Sipes and Arakaki, 1997, Robinson et al., 1998 McSorley et al., 1999 Wang et al., 2001, 2003
Castor bean	Ricinus communis	Ricin	Rodriguez-Kabana et al., 1991
Marigold	Tagetes spp	Essential oil	Uhlenbroek and Bijloo, 1958, 1959, Good et al., 1965; Jagdale et al.1999
Sesame	Sesamum indicum	Essential oil	Rodríguez-Kábana et al., 1994
Neem	Azadirachta indica	Azadirachtin	Schmutterer, 1990
Velvet bean	Mucuna deeringiana	Cyanogen	Rodriguez-Kabana et al., 1992, Weaver et al., 1993 Taylor and Rodriguez-Kabana, 1999, Vargas-Ayala and Rodriguez-Kabana, 2001
Mustard	Brassica spp.	Isothiocyanate	Yu et al. 2005, 2007, 2009
Tobacco	*Nicotiana tabacum*	Nicotin	Yu and Potter 2008

In crop rotation with cover crop, Rodriguez-Kabana et al. (1992a, b) reported that velvet bean (*Mucuna deeringianna*) was not a host to *Meloidogyne* spp. and therefore could be used to manage root-knot nematode problems in several arable crops. Also, the suppressing effect of *M. deeringianna* on *Meloidogyne* spp. in rotation with vegetable and other crops has been well documented (McSorley et al., 1994; McSorley and Dickson, 1995). While crops like soybean, peanut and okra may be susceptible to root-knot nematodes, a number of other crops (sorghum, corn, sudangrass, horsebean and summer hemp) may suppress some root-knot nematode species (Rodriguez-Kabana et al., 1987, 1992; Kinloch and Dunivan, 1993; McSorley and Gallaher, 1991; Sipes and Arakaki, 1997; Wang et al., 2006; Weaver et al., 1993). A list of plants which have been known to produce compounds toxic to nematodes is listed in Table 3.

Many studies have reported the ecological effects of biomass amendment to soil (including cover crops) on free-living nematode communities (McSorley and Frederick, 1999; Porazinska et al., 1999; Ferris and Matute, 2003; Forge et al., 2003; Ferris et al., 2004; Wang et al., 2004, 2006; Fiscus and Neher, 2002; Ferris and Bongers, 2006). A few papers have reported on the free-living nematode community after brassicaceous or rye cover crop incorporation (Lundquist et al., 1999; Georgieva et al., 2002, 2005a, 2005b), but no published studies were found that described in detail the effects of brassicaceous cover crops on the free-living nematode community.

CONCLUSION

For sustainable agriculture, biodiversity is key. Crop rotation and cover crop provide opportunities to increase plant and soil biodiversity. More biodiversified soils are more likely to be suppressive to pests and diseases. To get the most benefit from crop rotations and cover crops, one needs to make them an integral part of a cropping system that satisfies the social, economic, and environmental requirments. Because of the complexity of the practices, it is important to remember that there are no single crop roataion and cover crop systems that will meet all the possible objectives in a region. One working for insect control might not work for disease management. Nevertheless, choosing a taxonomically more distant crop or a cover crop will bring more biodiversity into the cropping system. Sustainable agriculture increases biodiversity.

To meet the growing food demand, and the expected dietary changes, agriculture is at a critical juncture. Some are the same pressures spawned the birth and development of the Green Revolution. Some of the practices developed during that period will stay. Breeding high yielding and disease resistance varieties will likely intensify, while some practices will have to be adapted to the new reality, and some will be introduced for sustainable agriculture.

REFERENCES

Bowers. W.S. 1993. Insecticidal compounds from plants. Pp 227-235. In H.N. Nigg, and D. Seigler (eds.) Phytochemical resources for medicine and agriculture, Plenum press, New York.

Bugg, R.L., L.T. andWilson, L. T. (1989). *Ammi visnaga* (L.) Lamarck (Apiaceae): Associated beneficial insects and implications for biological control with emphasis on the bell-pepper ecosystem. *Biol. Agr. Hort.* 6: 241–268.

Bugg, R.L., F.L. Wackers, K.E.Brunson, S.C. Phatak,and J.D.Dutcher.1990. Tarnished plant bug, *Lygus lineolaris* (Palisot de Beauvois) (Hemiptera: Miridae), on selected cool season leguminous cover crops: Abundance, survival and performance. *J. Entomol Sci.* 25:463–474.

Bugg, R.L., L.E. Ehler, and L.T. Wilson. 1987. Effect of common knotweed (*Polygonum aveculare*) on abundance and efficiency of insect predators of crop pests. *Hilgardia* 55: 1–52.

Bullock, D.G. 1992. Crop rotation. *Critical Reviews in Plant Sciences* 1: 309-326.

Cardinale, B. J., K.L. Matulich, D.U. Hooper, J.E. Byrnes, E. Duffy, L. Gamfeldt, and A. Gonzalez. 2011. The functional role of producer diversity in ecosystems. *Am. J. Bot.* 98: 572–592.

Chen, C., E.M. Bauske, G. Musson, R. Rodríguez-Kábana, and J.W. Kloepper. 1995. Biological control of *Fusarium* wilt on cotton by use of endophytic bacteria. *Biol. Contr.* 5: 83–9.

Chen, C., G. Musson, E. Bauske, and J.W. Kloepper. 1993. Potential of endophytic bacteria for biological control of Fusarium wilt of cotton. Phytopathol. 83: 1404

Cook, R.J., L.S. Thomashow, D.M. Weller, D. Fujimoto, M. Mazzola, G. Bangera, and D. Kim. 1995. Molecular mechanisms of defense by rhizobacteria against root disease, *Proc. Natl. Acad. Sci.* 92:4197–420.

Cook, R.J., and K.F. Baker. 1983. The nature and practice of biological control of plant pathogens. Pp 453. *American Phytopathological Society*, St. Paul, MN.

Curl, E.A. 1963. Control of plant diseases by crop rotation. *Bot. Rev.* 29: 413–479.

Davis, R.M. and Nunez, J. J. (1999). Influence of crop rotation on the incidence of *Pythium*- and *Rhizoctonia*-induced carrot root dieback. *Plant Dis.* 83, 146–148.

Dillard, H.R., and A.C. Cobb. 1998. Survival of *Colletotrichum coccodes* in infected tomato tissue and in soil. *Plant Dis.* 82: 235–238.

Ehler, L.E., and J.C Miller. 1978. Biological control in temporary agroecosystems. *Entomophaga* 23: 207–212.

Entz, M.H., W.J. Bullied, and F. Katepa-Mupondwa. 1995. Rotational benefits of forage crops in Canadian prairie cropping systems. *J. Prod. Agr.* 8:521–29.

Epps, J.M., and A.Y Chambers. 1958. New host records for *Heterodera glycines*: including one host in the Labiatae. *Plant Dis. Rep.* 42: 194.

Epps, J.M. 1960. Evaluation of crops rotation and soil fumigation for controlling the soybean cyst nematode. *Phytopathol.* 50: 635 (Abstr.).

Epps, J.M., andA. Chambers. 1965. Population dynamics of *Heterodera glycines* under various cropping sequences in field bins. *Phytopathol.* 55:100-103.

Ferris, H., and T. Bongers. 2006. Nematode indicators of organic enrichment. *J. Nematol.* 38: 3–12.

Ferris, H., and M.M. Matute. 2003. Structural and functional succession in the nematode fauna of a soil food web. *Appl. Soil Ecol.* 23: 93–110.

Ferris, H., R.C. Venette, and K.M. Scow. 2004. Soil management to enhance bacterivore and fungivore nematode populations and their nitrogen mineralisation function. *Appl. Soil Ecol.* 25:19–35.

Fiscus, D.A., and D.A Neher. 2002. Distinguishing sensitivity of free-living soil nematode genera to physical and chemical disturbances. *Ecol. Appl.* 12: 565–575.

Forge, T.A., E. Hogue, G. Neilsen, and D. Neilsen. 2003. Effects of organic mulches on soil microfauna in the root zone of apple: Implications for nutrient fluxes and functional diversity of the soil food web. *Appl. Soil Ecol.* 22:39–54.

Fortnum, B.A., and R.E. Currin. 1993. Crop rotation and nematicide effects on the frequency of *Meloidogyne* spp. in a mixed population. *Phytopathol.* 83:350–355.

Georgieva, S., S. Christensen, H. Petersen, P. Gjelstrup, and K. Thorup-Kristensen. 2005a. Early decomposer assemblages of soil organisms in litterbags with vetch and rye roots. *Soil Biol. Biochem.* 37: 1145–1155.

Georgieva, S., S. Christensen, and K. Stevnbak. 2005b. Nematode succession and microfauna—microorganism interactions during root residue decomposition. *Soil Biol. Biochem.* 37:1763–1774.

Georgieva, S., S.P. McGrath, D.J. Hooper, and B.S. Chambers. 2002. Nematode communities under stress: the long-term effects of heavy metals in soil treated with sewage sludge. *Appl. Soil Ecol.* 20: 27–42.

Glandorf, D.C.M., L.G.L. Peters, I. van der Sluis, P.A.H.M.Bakker, and B. Schippers. 1993. Crop specificity of rhizosphere pseudomonads and the involvement of root agglutins. *Soil Biol. Biochem.* 25:981–989.

Good, J.M., N.A. Minton, and C.A. Jaworski. 1965. Relative susceptibility of selected cover crops and coastal bermudagrass to plant nematodes. *Phytopathol.* 55:1026-1030.

Hadacek, F. 2002. Secondary metabolites as plant traits: current assessment and future perspectives. *Crit. Rev. Plant Sci.* 21:273-322.

Harborne, J.B., and H. Baxter. 1993. Phytochemical dictionary. A handbook of bioactive compounds from plants. CABI international, England.

Hayden, Z.D., M. Ngouajio, and D.C. Brainard. 2014. Rye–vetch mixture proportion tradeoffs: cover crop productivity, nitrogen accumulation, and weed suppression. *Agron. J.* 106:904.

Hornby, D. 1983. Suppressive Soils. *Ann. Rev. Phytopathol.*, 21:65-85.

Hunt, D.W.A., and R.S. Vernon. 2001. Portable trench barrier for protecting edges of tomato fields from Colorado potato beetle (Coleoptera: Chrysomelidae). *J. Econ. Entomol.* 94: 204–207.

Jackson, W. 1980. New Roots for Agriculture. University of Nebraska Press.

Jagdale, G.B., B. Ball-Coelho, J. Potter, J. Brandle, and R.C. Roy 2000. Rotation crop effects on *Pratylenchus penetrans* and subsequent crop yields. *Can. J. Plant Sci.* 80:543-549.

Jagdale, G.B., B. Reynolds, B. Ball-Coelho, and J. Potter. 1999. Nematicidal activity of marigold plant parts against root-lesion nematodes (*Pratylenchus penetrans*). *J. Nematol.* 31:546–547 (Abstr.).

Jager, G., and H. Velvis. 1983. Suppression of *Rhizoctonia solani* in potato fields. II. Effect of origin and degree of infection with *Rhizoctonia solani* of seed potatoes on subsequent infestation and on formation of sclerotia. *Neth. J. Plant Pathol.* 89:141–152.

Stapleton, J., G. Charles, J. Summers, P. Mitchell, and T.S. Prather 2009. Deleterious activity of cultivated grasses (Poaceae) and residues on soilborne fungal, nematode and weed pests. *Phytoparasit.* 38: 61–69.

Kinloch, R.A., and L.S. Dunivan. 1993. Summer cropping effects on the abundance of *Meloidogyne arenaria* race 2 and subsequent soybean yield. Suppl. *J. Nematol.* 25: 806-808.

Kloepper, J.W., J. Leong, M. Tientze, and M.N. Schroth. 1980. Enhanced plant growth by siderophores produced by plant growth promoting rhizobacteria. *Nature* 286:885–886.

Lambert, B., F. Leyns, L. Van Rooyen, F. Gosselé, Y. Papon, Y., and J. Swings. 1987. Rhizobacteria of maize and their fungal activities. *Appl. Environ. Microbiol.* 53:1866–1871.

Leonard, D.1991. Nematodes. Horticulture, *LXIX*, 3:66–73.

Liebman, M., and C.P. Staver. 2001. Crop diversification for weed management. Pp 322-374 In M. Liebman, C. L. Mohler and C. P. Staver (eds.) Ecological management of agricultural weeds. Cambridge University Press, Cambridge, UK.

Liebman, M., and E. Dyck. 1993. Crop rotation and intercropping strategies for weed management. *Ecol. Appl.*, 3:92-122.

Liebman, M., and E.R. Gallandt. 1997. Many little hammers: Ecological management of crop-weed interactions. Pp 291–343. *In* L. E. Jackson (ed.) Agricultural Ecology Academic Press, San Diego, CA.

Lundquist, E.J., L.E. Jackson, K.M. Scow, and C. Hsu. 1999. Changes in microbial biomass and community composition, and soil carbon and nitrogen pools after incorporation of rye into three California agricultural soils. *Soil Biol. Biochem.* 31:221–236.

Martens, T.J.R., M.H. Entz, and M. D. Wonneck. 2015. Review: Redesigning Canadian prairie cropping systems for profitability, sustainability, and resilience. Can. J. Plant Sci. 95:1049–1072

Matthew, D.T., C.A. Eberle, F. Forcella, R. Gesch, S. Weyers, and J.G. Lundgren. 2016. Nectar production in oilseeds: Food for pollinators in an agricultural landscape. *Crop Sci.* (*in press*).

Minton, N.A. 1986. Impact of conservation tillage on nematode populations, *J. Nematol.* 18:135–140.

Minton, N.A. 1992. Nematode management in minimum-till soybean with resistant cultivars, rye rotation, and aldicarb. *Nematropica* 22:21–28.

Minton, N.A., E.D. Donnelly, and R.L. Shepherd. 1965. Vetches and nematodes. *Highlights of Agri. Res.* 12: 7.

Minton, N.A., E.D. Donnelly, and R.L. Shepherd. 1966. Reaction of *Vicia* species and F5 hybrids from *V. sativa* × *V. angustifolia* to five root knot nematode species. *Phytopathol.* 56:102–107.

Miyazawa, K., M. Takeda, T. Murakami, and T. Murayama. 2014. Dual and triple intercropping: potential benefits for annual green manure production. *Plant Prod. Sci.* 17:194–201.

McSorley, R., and J.J. Frederick. 1999. Nematode population fluctuations during decomposition of specific organic amendments. *J. Nematol.* 31:37–44.

McSorley, R., D.W. Dickson, J.A. de Brito, T.E. Hewlett, and J.J. Frederick. 1994. Effects of tropical rotation crops on *Meloidogyne areneria* population densities and vegetable yields in microplots. *J. Nematol.* 26:175-181.

McSorley, R., and D.W. Dickson. 1995. The effect of tropical rotation crops on *Meloidogyne incognita* and other plant-parasitic nematodes. *J. Nematol.* 27:535-544.

McSorley, R., and R.N. Gallaher. 1991. Nematode population changes and forage yields of six corn and sorghum cultivars. Suppl. *J. Nematol.* 23:673-677.

McSorley, R., M. Ozores-Hampton, P.A. Stansly, and J.M.Conner. 1999. Nematode management, soil fertility, and yield in organic vegetable production. *Nematropica* 29:205–213.

Mohler, C.L. 2009. The role of crop rotation in weed management. In Crop Rotation on Organic Farms. A Planning Manual. Pp 44-46. *In* C. L. Mohler and S. E. Johnson, (eds.) Natural Resource, Agriculture, and Engineering Service (NRAES), Ithaca, NY.

Mohler, C.L. 2001a. Weed life history: Identifying vulnerabilities. Pp 40-98. *In* M. Liebman, C.L. Mohler, and C.P. Staver, (eds.) Ecological Management of Agricultural Weeds, Cambridge University Press, New York.

Mohler, C.L. 2001b. Enhancing the competitive ability of crops. Pp 269-321. *In* M. Liebman, C. L. Mohler, and C. P. Staver, (eds.) Ecological Management of Agricultural Weeds, Cambridge University Press, New York.

Mojahedi, H., G.S. Santo, and R.E. Inghram. 1993. Suppression of *Meloidogyne chitwoodi* with sudangrass cultivars as green manure. *J. Nematol.* 25:303-311.

Murdoch, W.W., J. Chesson, and P.L. Chesson. 1985. Biological control in theory and practice, *Amer. Nat.* 125:344–266.

Phatak, S.C. 1992. An integrated sustainable vegetable production system. *HortScience* 27:738–741.

Porazinska, D.L., L.W. Duncan, R. McSorley, and J.H. Graham. 1999. Nematode communities as indicators of status and processes of a soil ecosystem influenced by agricultural management practices. *App. Soil Ecol.* 13:69–86.

Roberts, P.A., W.C. Matthews, and J.D. Ehlers. 1996. New resistance to virulent root-knot nematodes linked to the Rk locus of cowpea. *Crop Sci.* 36:889–894.

Robinson, A.F., C.G. Cook, and A.C. Bridges. 1998. Comparative reproduction of *Rotylenchulus reniformis* and *Meloidogyne incognita* race 3 on kenaf and sunn hemp grown in rotation with cotton. *Nematropica* 28:143 (Abstr.).

Rodríguez-Kábana, R., D.G. Robertson, C.F. Weaver, and L. Wells. 1991. Rotations of bahiagrass and castorbean with peanut for the management of *Meloidogyne arenaria.* Suppl. *J. Nematol.* 23:658-661.

Rodríguez-Kábana, R., J.W. Kloepper, D.G. Robertson, and L.W. Wells. 1992a. Velvet bean for the management of rootknot and southern blight in peanut. *Nematropica* 22:75-80.

Rodriguez-Kabana R., J. Pinochet, D.G. Robertson, and L.W. Wells. 1992b. Crop rotation studies with velvet bean (*Mucuna deeringiana*) for the management of *Meloidogyne* spp. Suppl. *J. Nematol.* 24 (4S):662-668.

Rodríguez-Kábana, R., N. Kokalis-Burelle, D.G. Robertson, and L.W. Wells. 1994. Evaluation of sesame for control of *Meloidogyne arenaria* and *Sclerotium rolfsii* in peanut. *Nematropica* 24:55-61.

Rodriguez-Kabana, R., H. Ivey, and P.A. Backman. 1987. Peanut-cotton rotations for the management of *Meloidogyne arenaria*. *J. Nematol.* 19:484-486.

Ross, J.P. 1960. Soybean cyst nematode control by crop rotation. *Phytopathol.* 50:652 (Abstr.).

Ross, J.P. 1962. Crop rotation effects on the soybean cyst nematode population and soybean yields. *Phytopathol.* 52:815-818.

Sasser, J.N., and G. Uzzell, Jr. 1963a. Control of the soybean cyst nematode by crop rotation. *Phytopathol.* 53:625 (Abstr.).

Sasser, J.N., and G. Uzzell, Jr. 1963b. Influence of nonhost crops alone or in combination with a nematicide on the longevity of the soybean cyst nematode in soil. *Phytopathol.* 53:625 (Abstr.).

Sasser, J.N., and U.J. Grover. 1991. Control of the soybean cyst nematode by crop rotation in combination with a nematicide. *J. Nematol.* 23:344-347.

Satti, A. 2012. Combating agricultureal pests and diseases through cultural means. *The Experiment* 5:304-314.

Schmutterer, H. 1990. Properties and potential of natural pesticides from the neem tree, *Azadirachta indica*. *Ann. Rev. Entomol.* 35:271-397.

Sexson, D.L., and J.A. Wyman. 2005. Effect of crop rotation distance on populations of Colorado potato beetle (Coleoptera: Chrysomelidae): development of area wide Colorado potato beetle pest management strategies. *J. Econ. Entomol.* 98:716-724.

Sipes, B.S., and A.S. Arakaki. 1997. Root-knot nematode management in dryland taro with tropical cover crops. Suppl. *J. Nematol.* 29:721–724.

Smith, H.A., and R. McSorley. 2000. Intercropping and pest management: A review of major concepts. *Am. Ento.* 46:154–61.

Smith, R.G., L.W. Atwood, and N.D. Warren. 2014. Increased productivity of a cover crop mixture is not associated with enhanced agroecosystem services. *PLoS ONE* 9:e97351.

Snapp, S.S., S.M. Swinton, R. Labarta, D. Mutch, J.R. Black, R. Leep, J. Nyiraneza, and K. O'Neil. 2005. Evaluating cover crops for benefits, costs and performance within cropping System niches. *Agron. J.* 97:322-332.

Stark, J.C. 1995. Development of sustainable potato production systems for the Pacific NW. SARE Project report #LW91-029.Western Region SARE. Logan, Utah. http://www.sare.org/MySare/ProjectReport.aspx? do= viewProj&pn=LW91-029.

Sturz, A.V., B.P. Christie, and B.G. Matheson. 1998. Associations of bacterial endophyte populations from red clover and potato crops with potential for beneficial allelopathy. *Can. J. Microbiol.* 44: 162–167.

Sturz, A.V., B.P. Christie, and J. Nowak. 2000. Bacterial endophytes: potential role in developing sustainable systems of crop production. *Crit. Rev. Plant Sci.* 19:1–30.

Tamaki, G. 1981. Biological control of potato pests. Pp 178–192. *In* J. H. Lashomb and R. Casagrande (eds.) Advances in Potato Pest Management. Hutchinson Ross, Stroudsberg, PA.

Tedders, W.L. 1983. Insect management in deciduous orchard ecosystems: Habitat manipulation, Environ. *Mgmt.* 7:29–34.

Tilman, D., F. Isbell, and J.M. Cowles. 2014. Biodiversity and ecosystem functioning. *Ann. Rev. Ecol. Evol. Syst.* 45:471-493.

Teasdale, J.R. 1998. Cover crops, smother plants, and weed management. Pp 247-270. *In* J. L. Hatfield, D.D. Buhler, and B.A. Stewart, (eds.), Integrated Weed and Soil Management, Ann Arbor Press, Chelsea, MI.

Taylor, C.R., and R. Rodriguez-Kabana. 1999. Population dynamics and crop yield effects of nematodes and white mold in peanut, cotton and velvetbean. *Agric. Syst.* 59:177-191.

Uhlenbroek, J.H., and J.D. Bijloo. 1958. Investigations on nematicides. 1958. I. Isolation and structure of a nematicidal principle occurring in Tagetes roots. *Recuceil des Travaux Chimiques des Pays-Bas* 77 :1004-1008.

Uhlenbroek, J.H., and J.D. Bijloo. 1959. Investigations on nematicides. II. Structure of a second nematicidal principle isolated from Tagetes roots. *Recuceil des Travaux Chimiques des Pays-Bas* 78 :382-390.

Vargas-Ayala, R., and R. Rodriguez-Kabana. 2001. Bioremediative management of soybean nematode population densities in crop rotations with velvetbean, cowpea, and winter crops. *Nematropica* 31:37-46.

Wang, K.H., B.S. Sipes, and D.P. Schmitt. 2001. Suppression of *Rotylenchulus reniformis* by *Crotalaria juncea*, *Brassica napus*, and *Target erecta*. *Nematropica* 31:237–251.

Wang, K.H., R. McSorley, and R.N. Gallaher. 2003. Effect of *Crotalaria juncea* amendment on nematode communities in soil with different agricultural histories. *J. Nematol.* 35:294–301.

Wang, K.H., R. McSorley, and N. Kokalis-Burelle. 2006. Effects of cover cropping, solarization, and soil fumigation on nematode communities. *Plant Soil* 286:229–243.

Wang, K.H., R. McSorley, A.J. Marshall, and R.N. Gallaher. 2004. Nematode community changes associated with decomposition of *Crotalaria juncea* amendment in litterbags. *Appl. Soil Ecol.* 27:31–45.

Wang, K.H. 2012. Cover crops as insectary plants to enhance above and below ground beneficial organisms. *Hānai'Ai / The Food Provider*, March-April-May.

Weaver, D.B., R. Rodriguez-Kabana, and E.L. Carden. 1993. Velvet bean in rotation with soybean for management of *Heterodera glycines* and *Meloidogyne arenaria*. Suppl. *J. Nematol.* 25:809-813.

Wiggins, B.E., andL.L. Kinkel. 2005. Green manures and crop sequences influence potato diseases and pathogen inhibitory activity of indigenous streptomycetes. *Phytopathol.* 95:178–85.

Wortman, S.E., C.A. Francis, and J.L. Lindquist. 2012. Cover crop mixtures for the western corn belt: Opportunities for increased productivity and stability. *Agron. J.,* 104:699-705.

Yazdani, D., Y.H. Tan, M.A. Zainal Abidin, and I.B. Jaganath, 2011. A review on bioactive compounds isolated from plants against plant pathogenic fungi. *J. Medi Plants Res.* 5:6584-6589.

Yu, Q., and J.W. Potter. 1988. Population development of *Pratylenchus penetrans* on sweet corn under controlled conditions. *Can. J. Plant Sci.* 78:703-706.

Yu, Q., R. Tsao, M. Chiba, and J.W. Potter. 2005. Selective nematicidal activity of allyl isothiocyanate. *J. Food Agric. Environ.* 3:218-221.

Yu, Q., R. Tsao, M. Chiba, and J.W. Potter. 2007. Elucidation of the nematicidal activity of bran and seed meal of Oriental mustard (*Brassica juncea* L.) under controlled conditions. *J. Food Agric. Environ.* 5:374-379.

Yu, Q., and J.W. Potter. 2008. Selective nematicidal activity of allyl isothiocyanate. *J. Food Agric. Environ.* 6:132-136.

In: Crop Rotations
Editor: Bao-Luo Ma

ISBN: 978-1-63484-496-3

Chapter 8

CARBON FOOTPRINTS IN CROP ROTATION SYSTEMS

***Chang B. Liang*[1]*, Aruna W. Herath*[2] *and Bao-Luo Ma*[2,*]**
[1]Environment Canada, Pollutant Inventories and Reporting Division,
Gatineau, QC, Canada
[2]Ottawa Research and Development Centre,
Agriculture and Agri–Food Canada, Ottawa, ON, Canada

ABSTRACT

Global climate is rapidly changing due to anthropogenic greenhouse gas (GHG) emissions, causing substantial risks to agricultural production systems associated with frequent occurrence of catastrophic weather events. Since agriculture itself is one of the major contributors to GHG emissions, producers, researchers and policy makers strive to develop effective crop management practices to minimize GHGs while maximizing farmer's net returns. Thus, quantification of GHGs with diverse cropping systems is essential to mitigate GHGs from agriculture and, in turn, develop more sustainable practices. Carbon footprint is a measure of the intensity of GHGs and productivity of different agricultural practices. Because of the easy conveyance of information to the general public about the GHG intensity of a variety of products and diverse activities, carbon footprint, as a new quantitative indicator, has attracted the attention of scientists and policy-makers and gained public acceptance. Although the scientific literature on carbon footprints targeting GHGs from farming practices is still sparse, the accumulated convincing evidence indicates that a significant part of the GHGs related to agriculture can be mitigated through improved agronomic practices, including the adoption of diversified cropping systems with well-defined crop sequences including cereal, oilseed and legume crops in site-specific rotation systems. Effective crop rotation systems have been shown to increase crop productivity with an efficient use of resources by individual crops, as well as improved soil carbon storage and reduced carbon footprints. This chapter comprises an overview of GHGs in relation to different crop management practices, followed by the concept and general principle of estimating carbon footprints of agricultural products. It also reviews the available scientific literature on calculations

[*] Corresponding author: baoluo.ma@agr.gc.ca.

of carbon footprints, its application, boundaries and challenges, and effective measures to reduce GHGs in agriculture, particularly focusing on diverse crop rotation systems.

1. INTRODUCTION

Climate change as a result of global warming is recognized as one of the most serious threats to the resilience of the earth's ecosystem, and hence to human civilization. The effects of climate change and global warming are evidently manifested through shifting weather patterns, receding ice caps, a rise in sea level, erratic distribution of precipitation (frequent occurrence of floods and droughts), increased frequency and intensity of heat waves, and serious ecological imbalances (IPCC, 2007, 2013; Oreskes, 2004). All of such changes are largely associated with the increasing emissions of anthropogenic greenhouse gases (GHGs) into the atmosphere (Ruddiman, 2003; IPCC, 2006). Greenhouse gases act as a radiation blanket over the Earth's surface, absorbing longwave infrared radiation in the planetary atmosphere, thereby trapping heat and warming the lower atmosphere and the ocean, potentially contributing to global warming. This phenomenon is often termed as the 'greenhouse effect'. Since the late 19th century, the mean annual global temperature has risen by 0.74 ± 0.18°C, and from 1951 to 2010, the mean global surface temperature has increased in the range of 0.5°C to 1.3°C (IPCC, 2013). The enhanced greenhouse effect has also generated a series of other feedback effects within the interactive climate system. For example, a warmer climate could cause more water vapors in the atmosphere and in turn, such higher concentration of atmospheric water vapor will further aggravate the GHG effect. Hence, it is imperative to control the build-up of GHGs in the atmosphere to avoid global warming and serious ecological and economic threats and imbalances caused in natural ecosystems by changing climate.

Key anthropogenic GHGs associated with global warming are carbon dioxide (CO_2), methane (CH_4) and nitrous oxide (N_2O). As shown from some geological and ice core studies, the rates of increment in GHG concentrations are extraordinarily high, far exceeding the natural range (IPCC, 2007). Since the beginning of industrialization, global CO_2, CH_4 and N_2O levels have increased by 40%, 150% and 20%, respectively (IPCC, 2013). Such increases are largely caused by anthropogenic activities including fossil fuel combustion, land-use changes, increasingly intensive agriculture, and an expanding global human population (Reay and Hogan, 2008). However, not all of the three GHGs have equal culpability in causing atmospheric warming. Their relative contributions to warming depend on radiative forcing by a given gas molecule (the effect from the addition of a unit of gas on the radiation balance of the Earth), the average duration of time that such a gas molecule stays in the atmosphere, and the total quantity of gas emitted. Considering all of these factors together, the average warming that specific gas can cause is referred to as 'global warming potential' (GWP), which is expressed relative to CO_2 and the unit of GWP is CO_2-equivalent (CO_2eq). For example, GWP of N_2O is 298 times greater than that of CO_2; CH_4, 25 times greater (IPCC, 2006).

Globally, more than 75% of anthropogenic CO_2 emissions are from fossil fuel combustion (Snyder et al., 2009). The remaining share originates from the land use changes including deforestation, biomass burning, and conversion of natural lands to agricultural ecosystems (Lal, 2004; Snyder et al., 2009). The largest sources of CH_4 emissions are the enteric fermentation of ruminant livestock and waste decomposition, including animal waste, crop residues and landfills. In addition, fossil fuel mining and rice cultivation also generate significant amounts of CH_4 emissions (US EPA, 2015). Most of the N_2O emissions in the atmosphere are associated with anthropogenic activities with agricultural soils, where N fertilizers and animal manures are converted to N_2O by soil microorganisms. In addition, N_2O is also released to the atmosphere with fossil fuel combustion and biomass decomposition. Even a small quantity of N_2O can cause a significant radiative forcing, as N_2O has a very high GWP. On the flip side, given the high GWP of N_2O, even a small reduction in N_2O flux can have a relatively large remedial impact on the overall GWP.

2. Agriculture and Anthropogenic Greenhouse Gases

Agriculture and agri-food system is a significant contributor of anthropogenic GHG emissions. Most of the practices involve crop production, food processing and the marketing of food products to the consumers; all of these activities generate GHGs contributing to global climate change (Dyer et al., 2010). Globally, the agricultural sector has been estimated to account for nearly 13.5% of total GHG emissions (Montzka et al., 2011).

In general, GHG fluxes from agricultural systems are complex, given that the production dynamics of CO_2, CH_4 and N_2O are affected by soil, environment and management factors. Studies have found that soil and crop management practices, including crop type, fallow frequency, crop residue management, soil amendments, crop rotations, tillage, irrigation, drainage, cover crops, and fertilization can play a major role in regulating GHGs (Gregorich et al., 1997; Paustian et al., 1997; Collins et al., 1999). However, the effects of these practices on GHGs are highly variable depending on the cultivation and environmental conditions.

Researchers and producers are encouraged to develop and adopt effective farming practices to reduce GHG emissions from all agricultural activities. For this, the quantification of GHG contribution from various agricultural activities and from various farm products is required. Such quantification can be used to identify important sources of emissions and develop more sustainable agricultural management practices to reduce GHG emissions. As a result, the term 'carbon footprint' has been proposed. Carbon footprint is the quantitative expression of GHG intensity of a diverse set of activities and products. According to Gan et al. (2011a), carbon footprint relevant to agricultural products can be defined as the total amount of GHG emissions associated with a food product or a service, expressed in CO_2eq. Although most of the scientific analyses of carbon footprints are being conducted mainly for consumer products and industrial processes, carbon footprint also has a potential for assessing and comparing different agricultural practices and products (Hillier et al., 2009). These are used by farmers to develop management strategies to lower carbon footprints of the products produced from their farms; and identify opportunities to improve production efficiencies.

2.1. Anthropogenic Greenhouse Gases, Sources and Its Contributions

2.1.1. Carbon Dioxide

Carbon dioxide is the most abundant GHG in the atmosphere after water vapour (Kiehl and Trenberth, 1997). The major anthropogenic sources of CO_2 are combustion of fossil fuel (burning of coal, oil and natural gas) and land use changes from natural to intensive agriculture (including clear-cutting trees and burning of wood). Since the industrial revolution, fossil fuel usage and deforestation have increased; the atmospheric CO_2 concentration increased from 278 ppm in 1750 to 390 ppm in 2011 (IPCC, 2013) with an unprecedented increasing rate of 2.1 ppm yr^{-1} from 2003 to 2012 (NOAA, 2013).

Carbon dioxide is cycled largely through agricultural cropping systems; crop plants consume large amounts of atmospheric CO_2 through photosynthesis to make food, feed and fiber. These products eventually convert back to CO_2 when they are consumed by animals and people, or decomposed after they die. The net emission of CO_2 in agriculture is relatively smaller than the total CO_2 volume cycling in agriculture. This net emission is mostly owing to fuel and energy use in the on-farm operations and in the manufacturing, and transportation of agricultural products.

Economic activities related to agriculture consume fossil fuels starting from the manufacturing of machineries, equipment, fertilizers and other chemical inputs, as well as during the operations of such machinery in land preparation and cultivations, harvesting, applications of fertilizers and pesticides and grain handling. In addition, agriculture-related CO_2 emissions are generated from the oxidation of soil organic C (West and Marland, 2002). Furthermore, soil CO_2 is produced as a result of biological and chemical activities, such as decomposition of crop residues by heterotrophic micro-organisms, plant roots (Hanson et al., 2000; Mosier et al., 2006; Smith et al., 2011) and soil microbial respiration (Luo and Zhou, 2006). Soil factors including temperature and moisture, cropping system and N availability can influence soil microbial activity (Al-Kaisi et al., 2008) and hence affect the process of decomposition of soil organic matter and CO_2 production.

2.1.2. Methane

Methane is the main hydrocarbon in the atmosphere, and it has increased by a factor of 2.5 from 722 ± 25 ppb in 1750 to 1803 ± 2 ppb in 2011 (IPCC, 2013). Global warming potential of CH_4 is 25 times greater than that of CO_2 for a 100-year time horizon, even though CH_4 has a brief life time (10-12 years) (IPCC, 2007). Methane is produced in soil, both naturally and anthropogenically as part of the biological processes in low oxygen environments. Of the total CH_4 emissions in the world, between 70% and 80% of atmospheric CH_4 is of biological origin (LeMer and Roger, 2001). According to IPCC (2013), approximately 55% of the total CH_4 is contributed by natural and cultivated submerged soils, while upland soils are responsible for only about 6%. The main anthropogenic sources of CH_4 are from livestock production, including enteric fermentation from ruminants and animal waste storage. The U.S. EPA (2007) estimated that about 28% of CH_4 emissions were owing to livestock products, second only to landfill emissions. The other major sources of CH_4 emissions are from waste decomposition of animal, crop residue and landfills, rice cultivation, and fossil-fuel mining (Schlesinger, 1997; NOAA, 2013).

Methane emissions in soil are regulated by several environmental factors, including soil moisture, temperature, pH, and soil management practices (Mapanda et al., 2011; Sistani et al., 2011; Ma et al., 2007). Generally, the soil is both a producer (source) and a consumer (sink) for atmospheric CH_4 (Smith et al., 2011). The net balance of CH_4 flux depends on two processes in the soil, i.e., methanogenesis (production by methanogenic bacteria under anaerobic conditions) and methanotrophy (consumption by methanotrophic bacteria, mainly under aerobic conditions) (LeMer and Roger, 2001). If the balance between methanogenesis and methanotrophy is positive, the soil is a CH_4 source, and if the balance is negative, the soil is a CH_4 sink.

2.1.3. Nitrous Oxide

Nitrous oxide is the most significant GHG emissions from agricultural practices (Janzen et al., 2006). During the last three decades, atmospheric N_2O has increased at a rate of about 0.7 ppb or 0.26% yr^{-1} (Smith et al., 2010; IPCC, 2013), largely owing to the increased use of N fertilizer (Schwenke et al., 2015). Bouwman (1990) estimated that about 70% of the N_2O emitted was derived from the soil, and of that, agricultural activities contributed about 4.2-7 Tg N yr^{-1} (Del Grosso et al., 2008). Rochette et al. (2008b) reported that agriculture accounts for approximately 72% of Canadian anthropogenic N_2O emissions, and globally 50% of the total N_2O emissions originated from agriculture. In 2013, direct soil N_2O emissions from synthetic N fertilizers and animal manure N applied as fertilizers accounted for 10.4 Mt CO_2eq for Canada (Environment Canada, 2015), and 50.7 Mt CO_2eq for the U.S.A. (U.S. EPA 2015), respectively (Table 1).

Table 1. Estimates of N_2O emissions from cropland in the U.S. and Canada in 2013

Region	Area Mha	Fertilizer N applied Mt	Animal manure N applied Mt	Emissions Mt CO_2eq
Canada	50	2.5	0.34	10.4
U.S.	145	11.7	2.7	50.7

Sources: Environment Canada (2015) and U.S. EPA (2015).

Studies show that the major source of N_2O emissions from agricultural soils is the N fertilizer inputs (Eichner, 1990; Matthews, 1994; Mosier et al., 1998; Maggiotto et al., 2000; Bouwman et al., 2002; Rochette et al., 2008b; Ma et al., 2012). In addition, organic N inputs such as farm manure (Velthof et al., 2003), decomposition of crop residues (legumes and non-legumes), and mineralization of native soil organic matter from forest and grassland conversion (Rochette et al., 2008b) also contribute significantly to agriculture-related N_2O emissions. In addition, crop management practices and soil factors can influence N_2O emissions from agricultural soils (Parkin and Kaspar, 2006).

The production of N_2O in a soil is a function of nitrification and denitrification mediated by soil microorganisms (Robertson and Groffman, 2007). Soil factors including available C, O_2, pH, temperature and moisture, regulate nitrification and denitrification processes, hence N_2O production (Robertson and Groffman, 2007). These controlling factors are interrelated and complex. For example, the amount of N_2O production from soil depends on soil O_2 concentrations, which is influenced by the soil moisture content. Soil moisture is influenced by soil temperature that determines the rate of nitrification and denitrification by microorganisms. Other factors, including soil texture, the amount of ammonium available for nitrification, and the amount of nitrate available for denitrification also influence N_2O emissions (Firestone, 1982).

Generally, N_2O fluxes from soils are highly sporadic. The peak emissions of N_2O are found after wetting of a dry soil (JØrgensen et al., 1998), following the application of N fertilizers (Maggiotto et al., 2000; Ma et al., 2010b) and also during the events of spring thawing of frozen soils (Bremner et al., 1980; Burton and Beauchamp, 1994; Wagner-Riddle and Thurtell, 1998). Emissions of N_2O associated with freezing and thawing appear to be a result of the accumulation of organic N in frozen soil, followed by N mineralization as the soil thaws, and finally nitrification and subsequent denitrification leading to N_2O emissions during the post-thaw period (Wagner-Riddle et al., 2008). Others have found substantial N_2O emissions from grasslands in response to soil freezing and thawing processes (Velthof et al., 1996; Kammann et al., 1998; Williams et al., 1999). In addition, in a crop rotation (corn–soybean–wheat) study in Ontario, Canada, Wagner-Riddle et al. (2007) observed that N_2O emissions during the non-growing-season (November–April, mostly during the spring thaw period) amounted to a 30 - 90% of the total N_2O emissions. These findings indicate that soil N_2O emissions are sporadic within the growing season as well as beyond the growing season, especially in locations where soils undergo freeze–thaw cycles, hence the need for year-round N_2O measurements.

2.2. Impact of Agricultural Management Practices on Greenhouse Gas Emissions

Inappropriate agricultural management practices may create diverse negative impacts on the environment, including GHG emissions and water pollution. The magnitude of GHGs is highly variable among different environments, land use, soil types, and crop management practices. For example, Kim and Neff (2009) analysed crop production systems in eight major corn-producing counties across the U.S. Corn Belt, and they reported that GHG emissions varied from 254 to 825 kg CO_2eq Mg^{-1} of grain, mainly owing to the variations in agronomic practices such as N application rate, tillage and irrigation practices, in addition to weather (rainfall and temperature) and soil conditions. Comparing two corn farms in Ontario, for crop years from 2006 to 2008, Fast (2008) found that GHGs are 52% lower in Southern than in Eastern Ontario (145 and 305 kg CO_2eq Mg^{-1} of grain, respectively). This was attributed to higher grain yields, lower N fertilizer rates, use of no-till and animal manure in the southern compared to the eastern Ontario farm. From a long-term rotation study, Ma et al. (2012) estimated the GHG emissions of corn production under different rotation systems with various N fertilizer rates in eastern Ontario. Across 18 years of the study, the estimated GHGs ranged from 287 to 354 kg CO_2eq Mg^{-1} of grain with rotational corn receiving 150 kg N ha^{-1}. Other farming operations, including planting operations, pesticide application, crop harvest, storage and shipping also contributed considerably to the GHG emissions (Gan et al., 2011a). Hence, the absolute amounts of GHG emissions depend on the crop inputs especially N fertilization, cropping system, farm operations such as tillage, crop residue management, use of agro-chemicals and environmental conditions.

2.2.1. Nitrogen Fertilizer Applications

Nitrogen fertilization is identified as one of the main sources of anthropogenic N_2O emissions in the atmosphere (Cole et al., 1997). The amount of N_2O emissions is related to the quantity of N applied to the soil (Dyer et al., 2010), and varies widely on a site-specific basis (Snyder et al., 2009). Therefore, the coefficients of fertilizer N induced N_2O emissions can range between 100% and 300% (Thornton and Valente, 1996). In general, N_2O emissions increase with increasing rates of N (Gregorich et al., 2005; McSwiney and Robertson, 2005; Ma et al., 2010b; Zebarth et al., 2008). However, the relationship between N rate and N_2O emissions is not linear and has a large scattering along the regression line (Gregorich et al., 2005; Roelandt et al., 2005). As demonstrated by McSwiney and Robertson (2005), N_2O flux is low to moderate until the N input is within the crop's N needs and N_2O flux is nearly doubled when that limit is exceeded.

Nitrogen fertilizer is often used inefficiently in crop production (Cassman et al., 2003). Only part of the applied N is incorporated into the soil organic matter, and inorganic N pools which are not taken up by the crop may be susceptible to losses from the soil. Various studies have consistently shown that excess N in the soil can greatly increase N_2O emissions (Conrad et al., 1983; Maggiotto et al., 2000; Ma et al., 2010b). Applying fertilizer N in excess of crop N requirements leads to elevated mineral N (NH_4^+-N and NO_3^--N) levels in the soil (Andraski et al., 2000; Wagner-Riddle et al., 2007; Ma et al., 2010a). Due to surplus mineral N, soil microbial transformation of N is increased and in turn as a by-product N_2O fluxes increase (Cole et al., 1997; McSwiney and Robertson, 2005). Hence, excess N would lead to higher N_2O emissions during the growing season (Conrad et al., 1983; Bouwman, 1990; Maggiotto et al., 2000) and even after harvesting the crop (Gregorich et al., 2005; McSwiney and Robertson, 2005; Wagner-Riddle et al., 2007).

2.2.2. Crop Residues as a Source of Nitrous Oxide

Although N fertilization is the largest contributor of N_2O emissions, crop residues are estimated to account for 24% of N_2O emissions in Canada (Rochette et al., 2008a). Crop residues are a source of organic C for soil microorganisms and a source of plant nutrients, especially N. For example, crop residues could obtain a credit of 8–10 kg N ha^{-1} fertilizer N equivalent for oilseed rape grown in Denmark (Thomsen and Christensen, 1996). In western Canada, a credit of 28 kg N ha^{-1} was given to crop production following a preceding pea (Beckie and Brandt, 1997). Since crop residues can serve as an important N source for nitrification and denitrification, retaining crop residues on the soil contribute significant N_2O emissions to the atmosphere (Singh et al., 2015).

The magnitude of GHG contributions from decomposing crop residues depends on factors intrinsic to the crop as well as other environmental factors. The net productivity of the crop (Forster et al., 2007), and quality of the crop residue (N concentrations) (Janzen et al., 2006; Gan et al., 2011b) are intrinsic to the crop. Agricultural management practices such as tillage, crop rotation, N inputs, weather (Gan et al., 2009) and growing conditions (soil temperature and moisture) (Flynn et al., 2005; Novoa and Tejeda, 2006; Merrill et al., 2007) are extrinsic factors. For example, Gan et al. (2012b) reported that the decomposition of barley crop residues directly and indirectly contributed to GHG emissions, averaging 173 kg CO_2eq ha^{-1}, i.e., 19% of the total emissions, and decomposition of wheat crop residues could emit 53 kg CO_2eq ha^{-1}, or on average, 9.3% of the total emissions (Gan et al., 2012a). In comparison, the decomposition of durum wheat straw and roots contributed about 25% of the crop's carbon footprint (Gan et al., 2011c), and oilseed straw and roots, about 10% of its total carbon footprint (Gan et al., 2011a). The difference in carbon footprint between oilseeds and cereals is owing to the greater amounts of straw and roots in cereals than in oilseeds under similar growing conditions.

2.2.3. Cropping Systems and Crop Types

The GHG emissions from agriculture largely depend on the crop species and the cropping system. Dyer et al. (2010) estimated GHG emissions for major field crops grown in Canada and reported that canola production had the largest GHG emissions per unit of grain dry matter. Similarly, Gan et al. (2011a) estimated the GHG emissions from different crop species grown on the Brown and Dark Brown soil zones across the Canadian semiarid prairies and concluded that canola had the largest GHG emissions, averaging 1105 kg CO_2eq ha^{-1}, followed by spring wheat at 943 kg CO_2eq ha^{-1} and flaxseed at 636 kg CO_2eq ha^{-1}. Under the same growing conditions, N-fixing pulse crops such as chickpea, lentil, and dry pea had the lowest emissions of 339 kg CO_2eq ha^{-1}, i.e., 65% lower than the emissions from canola and spring wheat.

Gan et al. (2003, 2010) have shown that using improved management practices such as proper crop rotation can increase crop yields without increasing production inputs. In general, diversified cropping systems in a well-defined crop sequence generate lower GHG emissions than a monoculture cropping system. For example, in a study conducted in southern Saskatchewan, durum wheat grown in diversified cropping systems (including a pulse crop, such as chickpea, dry pea, or lentil as preceding crop) had 46% lower carbon footprint than monoculture wheat systems (Gan et al., 2011c), which was associated with the reduced production inputs and the increased yields of durum wheat in diversified cropping systems. Durum wheat following legumes benefits from increased availability of soil N that may reduce external N inputs to the crop, leading to lower carbon footprint. Carbon footprint was also noted to be 19% lower when durum wheat is grown following oilseed crop, compared to following a cereal. Adviento-Borbe et al. (2007) reported a 47% reduction in CO_2eq m^{-2} yr^{-1} in a maize-soybean crop rotation after changing from continuous maize culture.

2.2.4. Tillage Practices

Tillage can influence GHG emissions mainly through its impact on crop residue, soil moisture and soil organic C level. With tillage, crop residues moved downwards and mixed to the deeper soil layers, altering the vertical distribution of soil organic matter, and leading to the elevated decomposition of soil organic matter. Some studies have reported that reduced organic matter levels with tillage and therefore reducing tillage intensity can inhibit the loss of organic matter from soil (Havlin et al., 1990). Tillage brings crop residues closer to the decomposing microorganisms, and creates more favourable micro-environmental (temperature and moisture) conditions (Douglas et al., 1980; Christensen, 1986). Soil tilling could generally accelerate soil drying and heating/cooling as it disturbs the soil surface (Lichter et al., 2008; Ussiri and Lal, 2009). Moreover, the mechanical action of tillage and conventional ploughing breaks up soil aggregates and may alter the soil structure, thereby enhancing the decomposition of physically protected soil organic matter (Roberts and Chan, 1990; Bronick and Lal, 2004). In contrast, no-till soils are characterized by the accumulation of crop residues on the soil surface, thus high soil moisture and reduced soil temperature, which in turn may restrict the decomposition of soil organic matter. Hence, no-till systems physically protect soil organic matter and stabilize soil aggregates, leading to lower mineralization rates (Lichter et al., 2008) and a reduction in GHG emissions (Lal, 2003). On the other hand, leaving crop residues on the soil surface in a no-till system can increase C and N contents and hence higher denitrification rates, particularly in the surface soil layer, compared to conventionally tilled soils (Staley et al., 1990; McKenney et al., 1993), leading to higher CO_2 emissions with no-till soil compared to conventional tillage (Oorts et al., 2007). Almaraz et al. (2009) reported an elevated GHG emissions from a study conducted in fine-textured soil conditions of Quebec, when changing from a conventional tillage to no-till system. In their study, similar cumulative CO_2 emissions were found in both systems, but there were higher cumulative N_2O emissions under no-till than under conventional tillage system.

Generally, emission of N_2O is the result of many interacting and interdependent processes in the soil. For example, increased soil organic C and organic N together with lower soil temperature in a no-till system compared to conventional tillage, may reduce soil N_2O emissions. On the other hand, high soil organic matter content, high moisture level and high mineral N content may increase emissions of N_2O from soils of no-till system. These contradictory results of N_2O emissions under no-till conditions could be a compound effect of different soil factors, such as soil moisture, temperature, organic C and mineral N contents in no-till soils. The effect of tillage on GHG emissions is inconclusive in the literature. For example, several studies reported a nil effect of tillage systems on N_2O emissions in the soil (Jantalia et al., 2008), and also on the overall GHG emissions (Venterea et al., 2005; Grandy et al., 2006; Parkin and Kaspar, 2006). Similarly, Gregorich et al. (2005) did not find a consistent relationship between tillage and N_2O emissions, with both increased and decreased N_2O emissions when comparing no-till to conventional tillage. In addition, increased N_2O emissions from no-till compared to conventional tilled soils have been reported in a noteworthy number of studies (Aulakh et al., 1984; Dendoncker et al., 2004; Freibauer et al., 2004; Mackenzie et al., 1997; McKenney et al., 1993; Robertson et al., 2000; Steinbach and Alvarez, 2006). Increased GHG emissions were also noted in studies on changing conventional tillage to conservation tillage (Blanco-Conqui and Lal, 2008; Gregorich et al., 2005; Lal, 2003). In a review paper, Smith and Conen (2004) reported the generalized trend of more N_2O emissions under no-till than under conventional tillage systems, indicating the less ability of no-till systems to mitigate GHGs.

2.2.5. Use of Agrochemicals (Herbicides and Fungicides)

The use of herbicides and fungicides in field crop production is increasing worldwide. The value of the global herbicide market has grown by 39% from 2002 to 2011, and is expected to grow another 11% by 2016 (Gianessi, 2013). The commonly used herbicides and fungicides in field crop production in Canada include boscalid, bromoxynil, glyphosate, imazamox, imazethapyr, pyraxlostrobin, and sethoxydim. Although these herbicides are used to act on target weeds or fungus diseases of the cropland, they may also have the potential to influence soil microbial and enzymatic activities (Seghers et al., 2005; Chen et al., 2009) that may alter the processes of GHG emissions. However, there is a lack of convincing literature on this research area.

Some herbicides could inhibit N_2O or CH_4 emissions. For example, in a laboratory incubation study with two types of soils that amended with two types of organic matter, Kyaw and Toyota (2007) found that the application of both glyphosate and propanil suppressed cumulative N_2O production. Similarly, a decreased rate of N_2O emissions over short time periods was reported in a laboratory study on the application of prosulfuron herbicide to fertilized soil (Kinney et al., 2005). Application of herbicide butachlor inhibited CH_4 emissions as much as by 20% in a direct-seeded and flooded rice field (Mohanty et al., 2001, 2004). These studies further demonstrated that even at very low concentrations of butachlor application, CH_4 production and oxidation were inhibited. Evidence suggests that prosulfuron degradation in grassland soils stimulated soil microbial activity that is responsible for the gas flux. Das et al. (2011) demonstrated that although separate applications of the herbicides bensulfuron-methyl and pretilachlor resulted in reduced N_2O and CH_4 emissions, the combined application of these two herbicides resulted in increased N_2O and CH_4 emissions from a flooded rice field. Accordingly, the potential effects of widely used herbicides and fungicides on GHG emissions are not consistent and remain unclear.

2.3. Impacts of Environmental Factors on Greenhouse Gas Emissions

The emission of GHGs from agricultural soils can vary between ecological regions and within a specific agro-ecosystem. Environmental factors, especially temperature and rainfall influence GHG emissions from agricultural soils, because soil moisture level strongly regulates soil aeration, nutrient availability, microbial activity and soil temperature. For example, increased soil temperatures coupled with high moisture levels during cooler months will promote denitrification process and increase soil N_2O production. In a compilation of Canadian studies, Helgason et al. (2005) reported an increased trend of N_2O emissions in humid regions (e.g., Eastern Canada), and reduced rate of GHG emissions in arid and semiarid regions (e.g., Western Prairies) under a no-till system. The N_2O emissions were reported to be correlated with precipitation in addition to soil and crop management practices. Almaraz et al. (2009) evaluated the effects of two tillage systems and N fertilization regimes on CO_2 and N_2O fluxes in Quebec, and reported higher N_2O fluxes during the spring that were associated with precipitation events, and higher CO_2 fluxes in mid-season that were associated with the temperature.

Rochette et al. (2008a) estimated that N_2O emissions from fine-textured soils were 50% greater than emissions from coarse- and medium-textured soils in eastern Canada. In the Quebec and Ontario mixed wood region, emissions during winter and spring thaw are reported to correspond to 40% of emissions during the snow-free season. They also observed N_2O emissions were 10% greater in eastern and 20% lower in western Canada in no-till soil systems compared with conventional tillage, indicating regional, climatic, and land use impacts on N_2O emissions from soil.

3. Greenhouse Gas Emissions and Carbon Footprints

3.1. Concept of Carbon Footprint

Carbon footprint is a term originated from the first academic publication developed by Rees in 1992 (Rees, 1992). It discussed "ecological footprint" and refers to the biologically productive land and sea area required to sustain a given human population expressed as global hectares. According to this concept, carbon footprint was defined as the land area that will assimilate CO_2 during the lifetime of a person or total global population. The calculation of carbon footprint as a part of the ecological footprint was tedious and complex. No methodology has been standardized for ecological footprint calculation and its scientific analyses. Carbon footprints are much more specific than ecological footprints, as they measure direct emissions of gases that are related with climate change, and are associated with human production or consumption activities. Wiedmann and Minx (2008) defined carbon footprint as "a measure of the exclusive total amount of CO_2 emissions that is directly and indirectly caused by an activity or is accumulated over the life stages of a product." This definition only highlights the GHG emissions as CO_2. However, in agricultural systems, GHG emissions are mainly associated with N_2O and CH_4, rather than with CO_2 emissions (Janzen et al., 2006). Therefore, Gan et al. (2011a) defined the carbon footprint relevant to agricultural products as "the total amount of GHG emissions associated with a food product or a service, expressed in CO_2 equivalents (CO_2eq)." With this definition, all GHGs are converted into CO_2 equivalent, and the definition is hence proposed as "The quantity of GHGs expressed in terms of CO_2eq, emitted into the atmosphere by an individual, organization, process, product or event from within a specified boundary" (Pandey and Agrawal, 2014).

3.2. Importance of Carbon Footprint

Since carbon footprint is a quantitative expression of GHG intensity for a diverse set of activities and products, it helps to manage GHG emissions and to evaluate the GHG mitigation measures (Carbon Trust, 2007). For example, with the quantification of the GHG emissions, the important sources of emissions can be identified, and the areas of GHG reductions can be prioritized. This provides the opportunity for environmental efficiencies and cost reduction.

Besides policy concerns, carbon footprint has an enormous importance for business purposes. Currently, consumers are willing to pay for measures leading to reductions in GHG emissions. Most consumers prefer to buy products that display the information about their carbon footprints, and are also willing to pay more for the products with a relatively low carbon footprint. Therefore, some international food companies have proposed that suppliers provide the amount of CO_2eq emissions released in the production of that particular food item on product label (Gan et al., 2011a). In addition, farmers are willing to adopt improved GHG mitigation strategies on their farms to minimize the GHG emissions of the production system. However, the GHG emissions of the agriculture sector are still relatively unknown and such emissions depend on different geographical regions, environmental conditions, and management practices (Seip, 2011). The boundaries of agriculture must be expanded to include all relevant emissions of GHGs. Only then will carbon footprinting be a useful tool in agricultural systems by providing a detailed map of different sources of GHGs and identifying the points where environmental efficiencies can be improved. This will also facilitate a comparison of different management options and their environmental cost-benefit analyses.

Hence, with this growing awareness on carbon footprinting and environmental quality, standard guidelines for estimating carbon footprint for agricultural products and activities are needed for the effective application of this tool in the quantification of the GHG intensity. Currently, scientific literature is growing with more and better case studies of carbon footprinting, thus adding to development of such standard methods.

4. Estimating Greenhouse Gas Emissions, Boundaries and Calculating Carbon Footprints of Field Crop Production

4.1. Estimating Greenhouse Gas Emissions

Greenhouse gas emissions from agricultural systems mostly include emissions from the application of synthetic N fertilizers, crop residue decomposition and various farm operations including tillage practices, planting crops, spraying herbicides and fungicides, and harvesting crops (Ma et al., 2012). Such estimates should also include GHG emissions from manufacturing, storage and transportation of synthetic fertilizers and agrochemicals (herbicides and pesticides) to the farm gate. The effects of these factors on GHG emissions are highly variable and inconsistent, depending on the production system and the environmental conditions. Therefore, the selection of a set of GHGs for footprint calculation depends on the need for carbon footprinting and the type of activity.

4.2. Boundaries

The boundary defines the extent of processes or activities that are included in the carbon footprint calculation. Setting the boundary is critical as it determines which activities are included and the level up to which carbon footprints are to be calculated (Pandey and Agrawal, 2014), and to ensure that it does not ignore important sources of environmental effects across the life cycle of the product. The boundaries vary with the objectives of the study and the characteristics of the entity for which the footprint is calculated.

Generally, the boundary lines of the system are set as the product's life cycle starting from the production of inputs (fertilizers, pesticides, and seeds), delivery of inputs to the farm gates, and the harvesting of the crop and storage of products on the farm. In agriculture, the carbon footprint of the crop production includes the activities related to the cultivation of the crop up to the final harvest and readiness for use as raw material. To calculate the carbon footprint of an agri-food product, the boundary is usually focused on the cultivation process up to the farm gate for comparing different agricultural production practices and efficiencies of different management systems. Extending the boundary beyond these activities to other adjacent processes, such as transportation of products to the market, their distribution, and food preparation techniques and preferences, which are more sensitive to local and personal conditions (Pandey and Agrawal, 2014), may complicate the protocol for carbon footprint calculations.

Most protocols define carbon footprint inventories under a "three-tier" approach to facilitate convenient accounting (WRI/WBCSD, 2004; Carbon Trust, 2007; BSI, 2011). The scope of these protocols is diverse. For the carbon footprint calculation in agriculture, the standard three-tier approach can be followed and this approach would maintain uniformity among different footprint studies. One of examples on common farm activities and their classification into different tiers is given in Pandey and Agrawal (2014).

The emissions of GHGs related to agricultural activities can be estimated by direct measurements or by using emission factors with model approaches. Although direct measurements give near accurate estimates with known uncertainties and clearly prescribe globally acceptable protocols, their applications may be limited (WRI/WBCSD, 2004). For those situations, estimations can be carried out indirectly through emission factors and models. The emission factors or models are developed for a particular region or a sector and they provide fairly accurate estimates. Generally, customized tools depend on combinations of direct measurements, emission factors, and models are popular and practicable.

4.3. Calculation of Product Carbon Footprint

The carbon footprint of a crop product is estimated by using the sum of the GHG emissions from all the activities covered within the boundary lines. Some studies have been conducted to evaluate the magnitude of these emissions from agricultural activities for Canadian environmental conditions (Gregorich et al., 2005; Rochette et al., 2008b). Common activities included in the boundaries are N fertilization, decomposition of crop residues, manufacturing of N and P fertilizers, production of agro-chemicals, and various field operations including tillage, planting/seeding, spraying pesticides, and harvesting (Ma et al., 2012).

Both synthetic and organic N applications provide N source for nitrification and denitrification, contributing directly and indirectly to N_2O emissions. However, the amounts of direct and indirect emissions vary with the quantity of N applied along with the environmental conditions (Gregorich et al., 2005; Dyer et al., 2010). Using a large number of observations on measured N_2O fluxes from Canadian farmland, Rochette et al. (2008b) developed a simple and reliable model to determine the N_2O emission factors based on a growing season moisture deficit; a linear function of the ratio of growing season precipitation (P) to potential evapotranspiration (PE) as follows:

$$EF = 0.022P/PE - 0.0048$$

where EF is the emission factor with a unit of kg N_2O-N kg^{-1} of N and P/PE is the ratio of precipitation to potential evapotranspiration during the growing season based on actual weather data.

Soil mineral N, particularly nitrates in the rooting zone, has a tendency to leach out (Campbell et al., 2004) or can undergo further transformations to be emitted as N_2O. To estimate N_2O emissions from nitrate leaching, a fraction of N as total input N needs to be determined. Similarly, a method was developed to estimate the fraction of N that can be leached ($FRAC_{LEACH}$) based on P/PE (Rochette et al., 2008b):

$$FRAC_{LEACH} = 0.3247P/PE - 0.0247$$

Both these equations showed predominant impacts of weather conditions on N_2O emission factors and leaching. Hence, using the method developed by the IPCC adopted for Canadian conditions (IPCC, 2006), emissions of N_2O from synthetic N applications can be estimated. Generally, for synthetic fertilizer applied in crop production, a portion of N is volatilized and emitted to the atmosphere (Ma et al., 2010a). Therefore, the IPCC default volatilization factor of NH_3 and NO_x ($FRAC_{GASM} = 0.1$) can be used to represent the emission factor associated with the NH_3 volatilization (IPCC, 2006):

$$CO_2eq_{SNF\text{-}N2O} = Q_{SNF} \times \{(FRAC_{GASM} \times EF_{VD}) + EF + (FRAC_{LEACH} \times EF_{LEACH})\} \times 44/28 \times 298$$

where $CO_2eq_{SNF\text{-}N2O}$ is the total emissions from the synthetic N fertilizer application (kg CO_2eq ha^{-1}), Q_{SNF} is the quantity of synthetic N fertilizer applied (kg N ha^{-1}), $FRAC_{GASM}$ is the fraction of synthetic N fertilizer that volatilizes as NH_3^- and NO_x^-N ($FRAC_{GASM} = 0.1$ kg N kg^{-1} N) (IPCC, 2006), EF_{VD} is the N_2O emission factor for volatilized NH_3^- and NO_x^-N ($EF_{VD} = 0.01$ kg N_2O-N kg^{-1} N) (IPCC, 2006). EF_{LEACH} is the N_2O emission factor for nitrate leaching ($EF_{LEACH} = 0.0075$ kg N_2O-N kg^{-1} N) (IPCC, 2006), 44/28 is the conversion coefficient from N_2O-N to N, and 298 is the global warming potential of N_2O over 100 years (IPCC, 2006).

Urea is commonly used as an N source in field crop production, and during urea hydrolysis, the C contained in urea is released as CO_2 (IPCC, 2006). The emissions of CO_2 from urea-based N fertilizer can be calculated as:

$$CO_2eq_{SNF\text{-}CO2} = Q_{SNF\text{-}UREA} \times 12/28 \times 44/12$$

where $CO_2eq_{SNF\text{-}CO2}$ is the emissions of CO_2 from the urea application (kg CO_2eq ha^{-1}), $Q_{SNF\text{-}UREA}$ is the quantity of urea fertilizer applied (kg N ha^{-1}), 12/28 is the ratio of C to N in urea, and 44/12 is the conversion factor of C to CO_2.

When a field crop is harvested, a portion of crop residue is left on the soil surface to decompose. The remaining crop residues (straw and roots) act as an additional source of N for nitrification and denitrification that contribute to N_2O emissions directly and indirectly. The quantity of N in crop residues (Q_{CRD}) can be calculated using the aboveground and belowground crop residue biomass values, multiplied by its respective N concentration. Similar to synthetic N fertilization, emissions from crop residue decomposition can be estimated using the following equation (Gan et al., 2011a, 2011c):

$$CO_2eq_{CRD} = Q_{CRD} \times \{EF + (FRAC_{LEACH} \times EF_{LEACH})\} \times 44/28 \times 298$$

The Haber–Bosch process that converts N_2 together with H_2 gases into ammonia (NH_3) is energy and emission intensive (Gan et al., 2011a). Through an extensive literature review on emissions from manufacturing fertilizers, Lal (2004) reported an average emission factor of 4.8 kg CO_2eq kg^{-1} N and 0.73 kg CO_2eq kg^{-1} P_2O_5 from production, transportation, storage, and transfer of N and P fertilizers to farm gates.

Although herbicides and fungicides applications are common agronomy practices in agriculture, the emission factors for individual pesticides are not readily available. Therefore the emissions during processes of manufacture, transportation, storage, and field application are assumed to be similar among pesticides within a similar category. Based on the active ingredient of fungicides or herbicides products, an average emission factor of 23.1 kg CO_2eq ha^{-1} for herbicides and 14.3 kg CO_2eq ha^{-1} for fungicides are considered for the footprint calculations (Lal, 2004; Gan et al., 2012a, 2012b). However, the absolute values of emissions from individual fungicides and herbicides can vary due to differences in manufacturing of each product.

The emissions related to various farm activities can be estimated using a factor of 14 kg CO_2eq ha^{-1} for no-till planting, 5 kg CO_2eq ha^{-1} for spraying of herbicides and fungicides, and 37 kg CO_2eq ha^{-1} for harvesting crops (Lal, 2004; Gan et al., 2011a). Based on these estimations, the total GHG emissions of crop production can be calculated as the emissions per unit of areas, expressed as kg CO_2eq ha^{-1}. The carbon footprint of crop production can also be calculated as the emissions per kg of grain produced under the specific growing conditions, expressed as kg CO_2eq kg^{-1} of grain:

$$\text{Carbon footprint} = \text{Total GHG emissions (kg } CO_2\text{eq ha}^{-1}\text{)/grain yield (kg ha}^{-1}\text{)}$$

Soil C sequestration, a climate change mitigation strategy for agriculture, can increase or decrease as a result of crop rotations. To date very few researchers have included changes in soil organic C in its assessment of C footprints (i.e., Gan et al., 2012a, 2014). Goglio et al. (2015) have provided a recent review on accounting for soil C changes in agricultural life cycle assessment. More detail on calculating C footprints including the changes in soil organic carbon is given by Gan et al. (2012a).

4.4. Limitations

The sources of uncertainties and limitations need to be pointed out when reporting carbon footprints of a product. One of the important limitations in agricultural footprint calculation is that activity-specific emission factors are not readily available for each and every crop management practice. Besides, the lack of advanced scientific tools required for a full life cycle assessment to estimate carbon footprint for a whole production-marketing chain or under various choices of boundaries is another significant limitation. Agricultural crop production is largely influenced by climatic factors, particularly rainfall and temperature. Therefore, though demanding, long-term monitoring and calculations are needed to generate accurate estimations of carbon footprint. The lack of sector- and region-specific emission factors for important agricultural inputs aggravates the uncertainty of the estimates. Moreover, standard methodologies are required to address the GHG emissions associated with farm equipment and other relevant activities, in addition to an immediate need for uniformities of techniques that are used in GHG estimation. Further, these standard methods must address how to deal with alternative scenarios and with land use changes.

5. Management Options and Strategies for Reducing Carbon Footprints in Field Crop Production

Although the agricultural sector contributes significant amounts of GHGs to the atmosphere, there are opportunities with the possibility of reducing GHG emissions at every stage of agricultural system. However, all three GHGs need to be considered when assessing the options to mitigate GHG emissions in agriculture. This is because management practices that reduce emissions of one GHG can lead to increases in other GHGs. For example, N fertilizer increases plant productivity with C fixation through photosynthesis and also C sequestration in the soil, leading to reduce CO_2 emissions from the soil. However, applying more N than crop requirement results in high soil mineral N levels and hence N_2O emissions (Ma et al., 2010b). Therefore effective agricultural management practices with potential mitigation opportunities need to be identified for sustainable agriculture. Some strategies identified that have potential to lower the carbon footprint of field crops are as follows.

5.1. Integrating Improved Farming Practices

Integrating improved farming practices have been shown to reduce carbon footprint effectively. Gan et al. (2014) found that integrating farming practices of fertilizing crops based on soil tests, reducing summer fallow frequencies and rotating cereals with grain legumes lowered total GHG emissions by an average of 256 kg CO_2eq ha^{-1} per year, and for each kg of wheat grain produced, a net 0.027–0.377 kg CO_2eq is sequestered into the soil.

Soil tests are important to determine the soil residual nutrients and potential mineralized N from soil organic matter that may be available for the next growing season (Ma and Wu, 2008). Fertilizing crops based on soil tests is an effective method of reducing carbon footprint, because applying N according to the available soil N reserves would lead to concurrent improvements in crop N use efficiency and grain yield (Ma et al., 2005, 2015; Ma and Biswas, 2015; Wang et al., 2014, 2015; Wu and Ma, 2015), and reduce emissions of N_2O from surplus N (Cassman et al., 2002; Fageria and Baligar, 2005; Wu et al., 2008). Reducing summer-fallow frequencies with the adoption of more intensified crop rotation systems would sequester greater amounts of CO_2 from the atmosphere to offset C emissions from crop production inputs (Lal, 2004a). Crop rotations with inclusion of grain legumes to fix atmospheric N_2 into plant-available N are found to be greatly effective for C sequestration (Nishimura et al., 2008), in addition to reduce the synthetic N inputs in crop production. Gan et al. (2011a) reported that, among these integrated practices, the choice of cropping systems had the highest impact on carbon footprint of wheat production in the Canadian Prairies, with a lentil-wheat rotation system having the lowest per-area GHG emissions (−552 kg CO_2eq ha^{-1}) and the most negative per-yield carbon footprint (−0.377 kg CO_2eq kg^{-1}) of grain.

5.2. Improving Nutrient Use Efficiency

Nitrogen is a key input for high yields in most of the non-legume field crops (Ma and Biswas, 2015). However, N applied as synthetic fertilizers and organic manures is not always efficiently used by crops, and the recovery of fertilizer N is generally less than 50% (Cassman et al., 2002; Krupnik et al., 2004; Bundy and Andraski, 2005; Ma and Dwyer, 1998; Ma et al., 2015). This low recovery is associated with N losses from NO_3^- leaching, NH_3 volatilization, surface runoff, and denitirification. Besides, farmers usually apply more N than crop requirement that often results in relatively high NH_4^+-N and NO_3^--N concentrations near the surface of the soil (Ma et al., 2010a, 2010b). The low recovery of fertilizer N and accumulation of excess N in soil may lead to increase GHG emissions (Raun and Johnson, 1999; Cassman et al., 2002; Fageria and Baligar, 2005). Therefore, the efficient use of fertilizer N by crops is necessary to minimize GHG emissions from soil (Subedi and Ma, 2005).

Adopting improved N management practices can increase the efficiency of fertilizer N through increasing crop N uptake and crop N utilization (Ma and Biswas, 2015; Ma et al., 2015), while reducing leaching losses and direct and indirect emissions of N_2O from the soil. One of the improved management practices includes 4Rs nutrient concept, meaning applying the right source of N (both organic and inorganic) at the right rate, at the right time and in the right place (Roberts, 2008). The 4Rs nutrient concept provides a framework to achieving cropping system goals, including increased crop production, increased farmer's profitability, and enhanced environmental protection with improved sustainability. Examples of these management practices are optimizing N fertilizer rates by using soil tests and accounting for N credits from previously-grown legumes and applied manures. Applying N according to available soil N reserves and synchronizing the soil N supply with crop N demands (Ma et al., 2005, 2014) could reduce emissions of N_2O largely from surplus N (Cassman et al., 2002; Fageria and Baligar, 2005; Wagner-Riddle et al., 2007). Traditional uniform N application results in over- or under-application of N in various parts of the crop field due to in-field variability. The ability to variably apply adjusted levels of N fertilizer corresponding to site-specific field conditions has been shown to increase N use efficiency, grain yields, crop quality, and net dollar returns, while decreasing nutrient overload to the environment (Ma et al., 2014). These practices focus on ensuring adequately available N when required by plants and to prevent exceeding plant N demand (Crews and Peoples, 2005). Gan et al. (2014) demonstrated the relationship between precipitation, N input, and grain yield and carbon emissions of spring wheat in southwestern Saskatchewan, Canada (Figure 1).

5.3. Improving Crop Residue Management

Improved management of crop residues can increase crop productivity, while reducing the carbon footprints of crop products. However, the availability of crop residues varies with crop species, crop rotation, tillage, nutrient inputs, and environmental conditions (Gan et al., 2009). These factors alone or in combination influence the soil organic C storage (Campbell et al., 2007), hence CO_2 emissions.

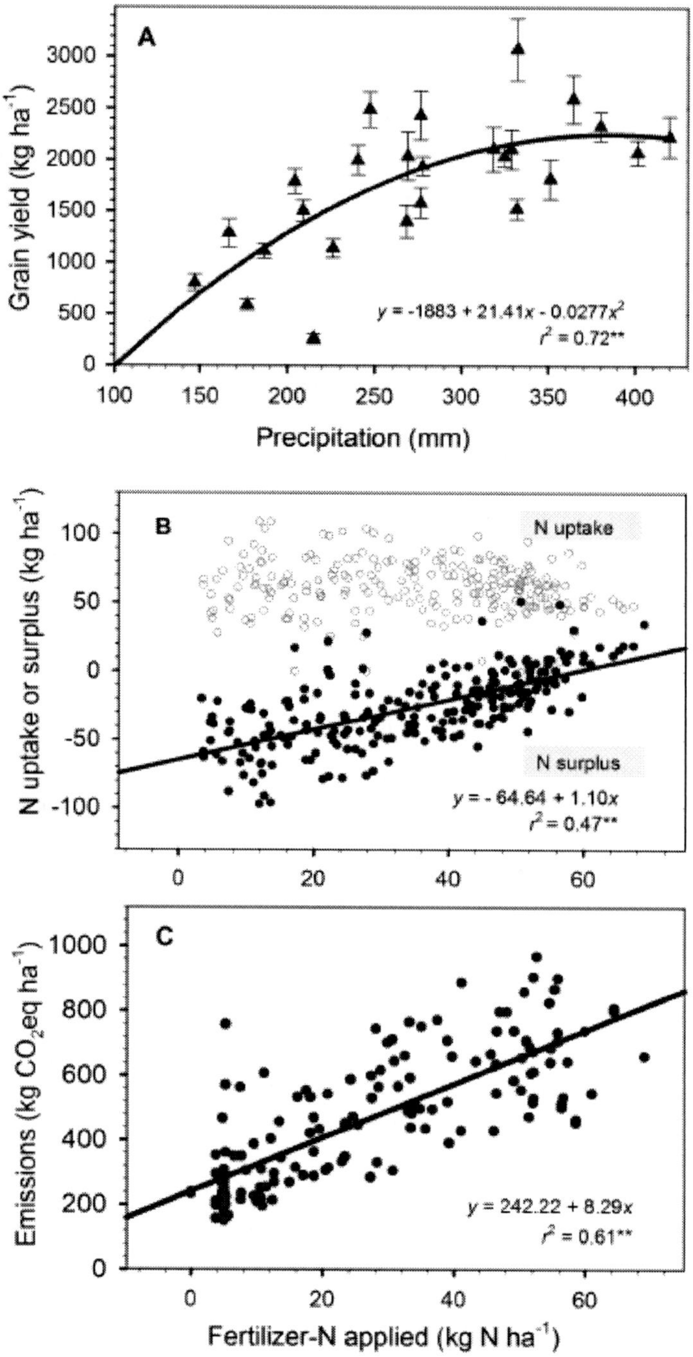

Figure 1. Effects of precipitation and N input on wheat grain yield and carbon emissions. (A) Wheat grain yield is a quadratic function of the growing-season precipitation during the 25-year period (the line bars are the standard error of the means); (B) increasing fertilizer-N input in wheat production increases the amount of N-surplus in the soil (solid circles) in a linear relationship, while N input has no impact on the total N uptake in the aboveground plant parts (open circles); and (C) increasing fertilizer-N input increases carbon emissions with each kg of N increase giving rise to the emission of 8.29 kg CO_2 equivalents. Adopted from Gan et al. (2014).

Numerous factors influence how much change in soil organic C can occur under typical agricultural management practices. Some of these factors include the C input from primary production (Liang et al., 1997), tillage intensity (McConkey et al., 2003; VandenBygaat et al., 2008), frequency of summerfallow (VandenBygaat et al., 2003), crop rotations (Gan et al., 2014; VandenBygaat et al., 2003) and application of animal manure (VandenBygaat et al., 2003). For example, Liang and MacKenzie (1992) observed that an 18% increment of soil C levels during a 6-year period at high levels of N fertilization. The net increase in soil C from crop residues depends on the magnitude of the crop biomass and the proportion that is stabilized and retained in the soil (Liang et al., 1997). Therefore, agricultural practices that influence crop residue decomposition, such as fertilization and tillage practices will also influence the amount of residue C retained and stored in the soil.

When crop residues are retained on the soil surface, it increases the C sequestration of atmospheric CO_2 into the soil (VandenBygaart et al., 2003), hence reduces CO_2 emissions (West and Marland, 2002). Moreover, the remaining crop residues can improve soil aggregate stability (Liu et al., 2005), efficiency of capturing rainfall, and in turn water holding capacity of the soil (Campbell et al., 1995). It may also improve the biodiversity in both above and below ground (Swift et al., 1996), enhance plant-mycorrhizae associations (McGonigle and Miller, 1993) and lower soil N_2O emissions (Ussiri et al., 2009). When crop residues are incorporated deeper into the sub-soil horizons, C placed beneath the plow layer will decompose very slowly due to reduced exposure to oxygen and other environmental elements.

5.4. Diversifying Cropping Systems to Reduce Carbon Footprints

Diversification of cropping systems, particularly with the inclusion of grain legumes in cereal-based crop rotations enhances plant-available N in soil. Such legume fixed N_2 can reduce inorganic N fertilizer requirement for cereal crop production (Ma et al., 2003) and also increase energy use efficiency (Zentner et al., 2004), decrease pest infestation (Krupinsky et al., 2002), and lower carbon footprints significantly (Gan et al., 2011a; Ma et al., 2012).

Diverse cropping systems generally increased crop productivity (Tanaka et al., 2007) because resources can be used by individual crops more efficiently (Robertson et al., 2000; Zentner et al., 2004; Adviento-Borbe et al., 2007). It also reduces pest infestation (Krupinsky et al., 2002) and improves water use efficiency of the crop (Miller et al., 2003). Furthermore, diverse cropping systems can generate higher inputs of crop residual C, leading to an increase in soil C storage (Lal, 2004; Gan et al., 2011a), hence reducing CO_2 emissions and lowering carbon footprint effectively (Gan et al., 2014).

Moreover, adoption of diverse cropping systems where various crop species (oilseed, legume, and cereal crops) are arranged in well-defined crop sequences in rotation systems has great environmental advantages over conventional monoculture farming systems. Gan et al. (2011c) reported that durum wheat grown in diversified cropping systems had lower carbon footprints than monoculture wheat systems in southern Saskatchewan. Drinkwater et al. (1998) demonstrated that legume-based cropping systems generally reduced organic C and N losses from soil compared with cereal-based cropping systems.

Another advantage with diversifying cropping systems is that weeds are subjected to a wider spectrum of herbicides. The changes from one crop species to another in crop rotation could generate microenvironments that may not favor the establishment and proliferation of any one specific weed species. Therefore, crop diversification leads to reduced weed abundance (Westerman et al., 2005) and herbicide inputs (Harker et al., 2009), while increasing crop productivity (Menalled et al., 2001) and yield stability, in addition to reducing the carbon footprint of the product.

CONCLUSION

Sustainability of a crop production system depends on crop yield as well as the carbon footprint of the crop product. Nitrogen inputs are essential for higher and consistent crop yields that are needed to meet the increasing global demand for food, feed, fibre and fuel. However, more than 75% of the total carbon footprints from agricultural sector are from N inputs that are used in crop production systems. It is a significant research challenge to adapt agricultural farming systems for producing high-quality and affordable food in adequate quantities in such a fashion that minimize potentially negative environmental impacts, especially from the GHG emissions. The appropriate strategy to manage GHG emissions must involve ecologically intensive crop management practices that enhance nutrient use efficiency while continuing to achieve gains in crop productivity.

There are notable opportunities for reducing GHG emissions and carbon footprints in agriculture. Some of these opportunities consisting of integrating improved farming practices including soil tests, use of improved cultivars, reducing summer fallow frequencies and adoption of diverse cropping systems where various crop species (oilseeds, legumes, and cereals) are arranged in well-defined crop sequences in rotation systems. However, numerous challenges in exploiting such opportunities are identified and those are needed to be overcome.

Although scientists and policy makers promote the use of carbon footprints as a management tool toward responding to global warming, its use in the agricultural sector is still limited owing to some significant challenges. For example, the activity-specific emission factors are not readily available for every crop management practice. Moreover, the lack of sector- and region-specific emission factors for important agricultural inputs worsens the precision of the estimates of carbon footprints. Furthermore, standard methodologies are required to address the GHG emissions associated with farm equipment and other relevant activities, in addition to an immediate need for uniformities of techniques that are used in GHG estimations. These standard methods must address how to deal with alternative scenarios and with land use changes as well. Such methods have yet to be developed. Furthermore, the advanced scientific tools required for full life cycle assessment to estimate carbon footprint for a whole production-marketing chain or under various choices of boundaries are not available.

Many agricultural mitigation opportunities have both co-benefits (improving efficiency, reducing cost and less footprints in the environment) and trade-offs involving potential negative effects. Balancing the co-benefits with potential negative effects is necessary for successful implementation. Moreover, due to widespread differences in agricultural activities all over the world, it is essential to have proper guidelines, particularly on setting the boundaries which are essential for meaningful comparison of GHG emissions and effectiveness of remedial actions.

Carbon footprint is considered a new farm management indicator and can be the focal point in the evaluation of environmental legislative actions, giving the basis to assess how damaging or beneficial a particular industry is to the environment. However, there is a knowledge gap in the subject and more research is needed to estimate carbon footprints of different agricultural products using standard methodology.

ACKNOWLEDGMENTS

This work was financially supported, in part, by the Canola Council of Canada Canola/Flax Cluster program (CL08) of Agriculture and Agri-Food Canada (AAFC) Growing Forward II. This is a joint contribution between Environment Canada and Agriculture and Agri-Food Canada. AAFC-ECORC contribution No. 15-083.

REFERENCES

Adviento-Borbe, M. A. A., M.L. Haddix, D. L. Binder, D. T. Walters, and A. Dobermann. 2007. Soil greenhouse gas fluxes and global warming potential in four high-yielding maize systems. *Glob. Chang. Biol.* 13:1972-1988.

Al-Kaisi, M. M., M. L. Kruse, and J. E. Sawyer. 2008. Effect of nitrogen fertilizer application on growing season soil carbon dioxide emission in a corn-soybean rotation. *J. Environ. Qual.* 37:325-332.

Almaraz, J. J., F. Mabood, X. Zhou, C. Madramootoo, P. Rochette, B.L. Ma, and D.L. Smith. 2009. Carbon dioxide and nitrous oxide fluxes in corn grown under two tillage systems in Southwestern Quebec. *Soil Sci. Soc.* Am. J. 73:113-119.

Andraski, T. W., L. G., Bundy, and K. R. Brye. 2000. Crop management and corn nitrogen rate effects on nitrate leaching. *J. Environ. Qual.* 29:1095-1103.

Aulakh, M.S., D.A. Rennie, and E.A. Paul. 1984. Gaseous nitrogen losses from soils under zero-till as compared with conventional till management systems. *J. Environ. Qual.* 113:130-136.

Beckie, H.J., and S.A. Brandt. 1997. Nitrogen contribution of field pea in annual cropping systems. 1. Nitrogen residual effect. *Can. J. Plant Sci.* 77:311-322.

Blanco-Conqui, H., and R. Lal. 2008. No-tillage and soil profile carbon sequestration: An on-farm assessment. *Soil Sci. Soc. Am. J.* 72:693-701.

Bouwman, A. F. 1990. Exchange of greenhouse gases between terrestrial ecosystem and atmosphere. pp 61-127. *In*: Bouwan, A.F. (ed.) *Soils and the Greenhouse Effect.* John Wiley and Sons, Chichester, West Sussex, UK.

Bouwman, A. F., and L. J. M. Boumans. 2002. Emissions of N_2O and NO from fertilized fields: Summary of available measurement data. *Glob. Biogeochem. Cycl.* 16:1-13.

Bremner, J. M., S. G. Robbins, and A. M. Blackmer. 1980. Seasonal variability in emission of nitrous oxide from soil. *Geophysical Research Letters* 7:641-644.

Bronick, C. J., and R. Lal. 2004. *Soil structure and management: A review*. Geoderma 124:3–22.

BSI. 2011. Publicly Available Specification 2050. Specification for the assessment of the life cycle greenhouse gas emissions of goods and services. British Standards Institution. Available online http://www.ghgprotocol.org/files/ghgp/public/GHG%20Protocol%20PAS%202050%20Factsheet.pdf.

Bundy, L. G., and T. W. Andraski. 2005. Recovery of fertilizer nitrogen in crop residues and cover crops on an irrigated sandy soil. *Soil Sci. Soc. Am. J.* 69:640-648.

Burton, D. L., and E. G. Beauchamp. 1994. Profile of nitrous oxide and carbon dioxide concentrations in a soil subject to freezing. *Soil Sci. Soc. Am. J.* 58:115-122.

Campbell, C. A., B. G. McConkey, R. P. Zentner, F. B. Dyck, F. Selles, and D. Curtin. 1995. Carbon sequestration in a Brown Chernozem as affected by tillage and rotation. *Can. J. Soil Sci.* 75:449-458.

Campbell, C. A., R. P. Zentner, F. Selles, V.O. Biederbeck, B.G. McConkey, R. Lemke, and Y.T. Gan. 2004. Cropping frequency effects on yield of grain, straw, plant N, N balance and annual production of spring wheat in the semiarid prairie. *Can. J. Plant Sci.* 84:487–501.

Campbell, C. A., A. J. VandenBygaart, B. Grant, R.P. Zentner, B.G. McConkey, R. Lemke, E.G. Gregorich, and M.R. Fernandez. 2007. Quantifying carbon sequestration in a conventionally tilled crop rotation study in southwestern Saskatchewan. *Can. J. Soil Sci.* 87:23–38.

Carbon Trust. 2007. *Carbon Footprint Measurement Methodology*, Version 1.1. The Carbon Trust, London, UK.

Cassman, K. G., A. Dobermann, and D. T. Walters. 2002. Agroecosystems, nitrogen-use efficiency, and nitrogen management. *Ambio* 31(2):132–140.

Cassman, K. G., A. Dobermann, D. T. Walters, and H. Yang. 2003. Meeting cereal demand while protecting natural resources and improving environmental quality. *Ann. Rev. Environ. Res.* 28:315–358.

Chen, H., J. Yao, F. Wang, M. M. F. Choi, E. Bramanti, and G. Zaray. 2009. Study on the toxic effects of diphenol compounds on soil microbial activity by a combination of methods. *J. Hazardous Materials* 167:846-851.

Christensen, B. T. 1986. Barley straw decomposition under field conditions: Effect of placement and initial nitrogen content on weight loss and nitrogen dynamics, *Soil Biol. Biochem.* 18:523-529.

Cole, C. V., J. Duxbury, J. Freney, O. Heinemeyer, K. Minami, A. Mosier, K. Paustian, N. Rosenberg, N. Sampson, D. Sauerbeck, and Q. Zhao. 1997. Global estimates of potential mitigation of greenhouse gas emissions by agriculture. *Nutr. Cycl. Agroecosyst.* 49:221-228.

Collins, H. P., R. L. Blevins, L. G. Bundy, D. R. Christensen, W. A. Dick, D. R. Huggins, and E.A. Paul. 1999. Soil carbon dynamics in corn-based agroecosystems: Results from carbon-13 natural abundance. *Soil Sci. Soc.* Am. J. 63:584–591.

Conrad, R., W. Seiler, and G. Bunse. 1983. Factors influencing the loss of fertilizer nitrogen into the atmosphere as N_2O. *J. Geophysical Res.* 88:6709-6718.

Crews, T. E., and M. B. Peoples. 2005. Can the synchrony of nitrogen supply and crop demand be improved in legume and fertilizer-based agroecosystems? A review. *Nutr. Cycl. Agroecosyst.* 72:101-120.

Das, S., A. Ghosh, and T. K. Adhya. 2011. Nitrous oxide and methane emission from a flooded rice field as influenced by separate and combined application of herbicides bensulfuron methyl and pretilachlor. *Chemosphere* 84:54–62.

Del Grosso, S. J., A. D. Halvorson, and W. J. Parton. 2008. Testing DAYCENT model simulations of corn yields and nitrous oxide emissions in irrigated tillage systems in Colorado. *J. Environ. Qual.* 37:1383-1389.

Dendoncker, N., B. Van Wesemael, M. D. A. Rounsevell, C. Roelandt, and S. Lettens. 2004. Belgium's CO_2 mitigation potential under improved cropland management. *Agric. Ecosys. Environ.* 103:101-116.

Douglas, C. L., R. R. Allmaras, P.E. Rasmusse, R.E. Ramig, and N.C. Roager. 1980. Wheat straw composition and placement effects on decomposition in dryland agriculture of the Pacific Northwest. *Soil Sci. Soc. Am. J.* 44:833-837.

Drinkwater, L. E., P. Wagoner, and M. Sarrantonio. 1998. Legume-based cropping systems have reduced carbon and nitrogen losses. *Nature* 396:262–265.

Dyer, J. A., X. P. C. Vergé, R. L. Desjardins, D. E. Worth, and B. G. McConkey. 2010. The impact of increased biodiesel production on the greenhouse gas emissions from field crops in Canada. *Energy for sustainable development* 14(2):73-82.

Eichner, M.J. 1990. Nitrous oxide emissions from fertilized soils: Summary of available data. *J. Environ. Qual.* 19:272–280.

Environment Canada. 2015. National Inventory Report: 1990-2013 - Greenhouse gas sources and sinks in Canada. The Canadian Government's Submission to the United Nations Framework Convention on Climate Change, 351 St-Joseph, Gatineau, Quebec, Canada (available online: http://unfccc.int/national_reports/annex_i_ghg_inventories/national_inventories_submissions/items/8812.php).

Fageria, N., and V. Baligar. 2005. Enhancing nitrogen use efficiency in crop plants. *Adv. Agron.* 88:97-185.

Fast, S. 2008. A life cycle assessment (LCA) of the potential for biofuels to reduce greenhouse gas emissions and fossil fuel consumption: Comparative assessment of two regions in Ontario, Canada. MA thesis, Department of Geography and Environmental Studies, Carleton University, Ottawa, ON.

Firestone, M.K. 1982. Biological denitrification. pp289–326. *In*: Stevenson, F.J. (ed.) *Nitrogen in agricultural soils*. Agronomy monographs 22. ASA-CSSA-SSSA, Madison, WI.

Flynn, H.C., J. Smith, K.A. Smith, J. Wright, P. Smith, and J. Massheder. 2005. Climate- and crop-responsive emission factors significantly alter estimates of current and future nitrous oxide emissions from fertilizer use. *Glob. Chang. Biol.* 11:1522–1536.

Freibauer, A., M. D. A. Rounsevell, P. Smith, and J. Verhagen. 2004. Carbon sequestration in the agricultural soils of Europe. *Geoderma* 122:1-23.

Gan, Y. T., C. A. Campbell, H. H. Jansen, R. Lemke, L. P. Liu, P. Basnyat, and C.L. McDonald. 2009. Carbon input to soil by oilseed and pulse crops in semiarid environment. *Agric Ecosys. Environ.* 132:290–297.

Gan, Y. T., R. Kutcher, F. Menalled, L. Lafond, and S. Brandt. 2010. Crop diversification and intensification with broadleaf crops in cereal-based cropping systems in the Northern Great Plains of North America. *In*: S.S. Malhi, Y.T. Gan, J.J. Schoenau, R.L. Lemke, and M.A. Liebig (eds.) *Recent Trends in Soil Science and Agronomy Research in the*

Northern Great Plains of North America. Research Signpost, Trivandrum, Kerala, India, pp 427.

Gan, Y. T., B. C. Liang, C. Hamel, H. Cutforth, and H. Wang. 2011a. Strategies for reducing the carbon footprint of field crops for semiarid areas. A review. *Agron. Sustain. Develop.* 31:643-656.

Gan, Y. T., B. C. Liang, L. P. Liu, X. Y. Wang, and C. L. McDonald. 2011b. C:N ratios and carbon distribution profile across rooting zones in oilseed and pulse crops. *Crop Pasture Sci.* 62:496-503.

Gan, Y. T., B. C. Liang, X. Y. Wang, and B. G. McConkey. 2011c. Lowering carbon footprint of durum wheat through diversifying cropping systems. *Field Crops Res.* 122:199–206.

Gan, Y. T., B. C. Liang, C. A. Campbell, R. P. Zentner, R. L. Lemke, H. Wang, and C. Yang. 2012a. Carbon footprint of spring wheat in response to fallow frequency and soil carbon changes over 25 years on the semiarid Canadian prairie. *Eur. J. Agron.* 43:175–184.

Gan, Y. T., B. C. Liang, W. May, S. S. Malhi, J. Niu, and X. Wang. 2012b. Carbon footprint of spring barley in relation to preceding oilseeds and N fertilization. *The Int. J. Life Cycle Assess.* 17: 635-645.

Gan, Y. T., B. C. Liang, Q. Chai, R. L. Lemke, C. A. Campbell, and R. P. Zentner. 2014. Improving farming practices reduces the carbon footprint of spring wheat production, *Nature Commun.* 11(5):1-13.

Gan, Y. T., P. R. Miller, B. G. McConkey, R. P. Zentner, F.C. Stevenson, and C.L. McDonald. 2003. Influence of diverse cropping sequences on durum wheat yield and protein in the semiarid northern Great Plains. *Agron J.* 95:245–252.

Goglio, P., W. N. Smith, B. B. Grant, R. L. Desjardins, B. G. McConkey, C. A. Campbell, and T. Nemecek. 2015. Accounting for soil carbon changes in agricultural life cycle assessment (LCA): A review. *J. Cleaner Production* (*In press*).

Gianessi, L. P. 2013. The increasing importance of herbicides in worldwide crop production. *Pest Manag. Sci.* 69:1099–1105.

Grandy, S., T. D. Loecke. S. Parr, and G. P. Robertson. 2006. Long-term trends in nitrous oxide emissions, soil nitrogen, and crop yields of till and no-till cropping systems. *J. Environ. Qual.* 35:1487-1495.

Gregorich, E. G., M. R. Carter, J. W. Doran, C. E. Pankhurst, and L. M. Dwyer. 1997. Biological attributes of soil quality. pp.81–113. *In*: Gregorich, E. G., and M. R. Carter (eds.) *Soil quality for crop production and ecosystem health*. Elsevier Science Publisher, Amsterdam, Netherlands.

Gregorich, E. G., P. Rochette, A. J. VandenBygaart, and D. A. Angers. 2005. Greenhouse gas contributions of agricultural soils and potential mitigation practices in Eastern Canada. *Soil Till. Res.* 83:53–72.

Hanson, P. J., N. T. Edwards, C. T. Garten, and J. A. Andrews. 2000. Separating soil and root microbial contributions to soil respiration: A review of methods and observations. *Biogeochem.* 48:115-146.

Harker, K. N., J. T. O'Donovan, R. B. Irvine, T. K. Turkington, and G. W. Clayton. 2009. Integrating cropping systems with cultural techniques augments wild oat (*Avena fatua*) management in barley. *Weed Sci.* 57:326–337.

Havlin, J. L., D. E. Kissel, L. D. Maddux, M. M. Claassen, and J. H. Long. 1990. Crop rotation and tillage effects on soil organic carbon and nitrogen. *Soil Sci. Soc. Am. J.* 54:448-452.

Helgason, B. L., H. H. Janzen, M. H. Chantigny, C. F. Drury, B. H. Ellert, E.G. Gregorich, and C. Wagner-Riddle. 2005. Toward improved coefficients for predicting direct N_2O emissions from soil in Canadian agroecosystems. *Nutr. Cycl. Agroecosyst.* 72(1):87-99.

Hillier, J., C. Hawes, G. Squire, A. Hilton, and S. Wale. 2009. The carbon footprint of food crop production. *Int. J. Life Cycle Assess*. 7:107–118.

IPCC. 2006. IPCC Guidelines for National Greenhouse Gas Inventories. Volume 4: Agriculture, Forestry and Other Land Use. *Intergovernmental Panel on Climate Change.* Paris, France.

IPCC. 2007. Climate Change: The physical science basis. In: Solomon, S., D. Qin, M. Manning, Z. Chen, M. Marquis, K.B. Averyt, M. Tignor, and H.L. Miller (eds.). Contribution of working group I to the fourth assessment report of the intergovernmental panel on climate change. Cambridge University Press, Cambridge, U.K.

IPCC. 2013. Summary for Policymakers. *In*: Climate Change: The physical science basis. Contribution of working group 1 to the fifth assessment report of the Intergovernmental Panel on Climate change. Cambridge University Press, Cambridge, U.K. and New York, NY, USA.

Jantalia, C. P., H. P. Dos Santos, S. Urquiaga, R. M. Boddey, and B. J. R. Alves. 2008. Fluxes of nitrous oxide from soil under different crop rotations and tillage systems in the South of Brazil. *Nutr. Cycl. Agroecosyst.* 82:161–173.

Janzen, H. H., D. A. Angers, and M. Boehm. 2006. A proposed approach to estimate and reduce net greenhouse gas emissions from whole farms. *Can. J. Soil Sci.* 86:401–418.

JØrgensen, R. N., B. J. JØrgensen, and N.E. Nielsen. 1998. N_2O emission immediately after rainfall in a dry stubble field. *Soil Biol. Biochem.* 30:545-546.

Kammann, C., L. Grunhage, C. Muller, S. Jacobi, and H.J. Jager. 1998. Seasonal variability and mitigation options for N_2O emissions from differently managed grasslands. *Environ. Pollut.* 102:179–186.

Kiehl, J. T., and K. E. Trenberth. 1997. Earth's annual global mean energy budget. *Bull. Am. Meteorol. Soc.* 78:197-208.

Kim, B., and R. Neff. 2009. Measurement and communication of greenhouse gas emissions from U.S. food consumption via carbon calculators. *Ecol. Econ.* 69:186–196.

Kinney, C. A., K. W. Mandernack, and A. R. Mosier. 2005. Laboratory investigations into the effects of the pesticides mancozeb, chlorothalonil, and prosulfuron on nitrous oxide and nitric oxide production in fertilized soil. *Soil Biology and Biochemistry.* 37:837-850.

Krupinsky, J. M., K. L. Bailey, M. P. McMullen, B. D. Gossen, and T. K. Turkington. 2002. Managing plant diseases risk in diversified cropping systems. *Agron. J.* 94:198–209.

Krupnik, T. J., J. Six, J. K. Ladha, M. J. Paine, and C. van Kessel. 2004. An assessment of fertilizer nitrogen recovery by grain crops. pp 193–207. *In*: Mosier, A. R., J. K. Syers, and J. R. Freney (eds.) *Agriculture and the nitrogen cycle. The Scientific Committee on Problems of the Environment* (SCOPE). Island Press, Covelo, California, USA.

Kyaw, K. M., and K. Toyota. 2007. Suppression of nitrous oxide production by the herbicides glyphosate and propanil in soils supplied with organic matter. *Soil Sci. Plant Nutr.* 53(4): 441-447.

Lal, R. 2003. Global potential of carbon sequestration to mitigate the greenhouse effect. *Crit. Rev. Plant Sci.* 22:151–184.

Lal, R. 2004. Carbon emission from farm operations. *Environ. Int.* 30:981–990.

LeMer, J. L., and P. Roger. 2001. Production, oxidation, emission and consumption of methane by soils: A review. *Eur. J. Soil Biol.* 37:25–50.

Liang, B. C., and A. F. Mackenzie. 1992. Changes in soil organic carbon and nitrogen after six years of corn production. *Soil Sci.* 153:307-313.

Liang, B. C., A.F. MacKenzie, M. Schnitzer, C.M. Monreal, P. R. Voroney, and R. P. Beyaert. 1997. Management-induced change in labile soil organic matter under continuous corn in eastern Canadian soils. *Biol. Fert. Soils* 26:88-94.

Lichter, K., B. Govaerts, J. Six, K. D. Sayre, J. Deckers, and L. Dendooven. 2008. Aggregation and C and N contents of soil organic matter fractions in the permanent raised-bed planting system in the Highlands of Central Mexico. *Plant Soil* 305:237–252.

Liu, A., B. L. Ma, and A. A. Bomke. 2005. Effects of cover crops on soil aggregate stability, total organic C and polysaccharides. *Soil Sci. Soc. Am. J.* 69:2041-2048.

Luo, Y., and X. Zhou. 2006. *Soil respiration and the environment.* Academic Press. New York, NY.

Ma, B. L., and D. K. Biswas. 2015. Precision nitrogen management for sustainable corn production. pp. 33-62. *In*: E. Lichtfouse and A. Goyal (eds.) *Sustainable Agriculture Reviews volume 16 – Cereals*, Chapter 2. Springer Publisher.

Ma, B. L., D. K. Biswas, A. W. Herath, J. K. Whalen, S. Q. Ruan, C. Caldwell, H. Earl, A. Vanasse, P. Scott, and D.L. Smith. 2015. Growth, yield and yield components of canola as affected by nitrogen, sulphur and boron fertilization. *J. Plant Nutr. Soil Sci.* 178:658-670.

Ma, B. L., and L. M. Dwyer. 1998. Nitrogen uptake and utilization of two contrasting maize hybrids differing in leaf senescence. *Plant Soil* 199:283-291.

Ma, B. L., B. C. Liang, D. K. Biswas, M. J. Morrison, and N. B. McLaughin. 2012. The carbon footprint of maize production as affected by nitrogen fertilizer and maize-legume rotations. *Nutr. Cycl. Agroecosyst.* 94:15-31.

Ma, B. L., K. D. Subedi, and C. Costa. 2005. Comparison of crop-based indicators with soil nitrate test for corn nitrogen management. *Agron. J.* 97:462-471.

Ma, B. L. and T. Y. Wu. 2008. Plant available N in the soil: Relationships between preplant and presidedress nitrate tests for corn production. *J. Plant Nutr. Soil Sci.* 171: 458-465.

Ma, B. L., T. Y. Wu, and J. Shang. 2014. On-farm comparison of variable rates of nitrogen with uniform application to maize on canopy reflectance, soil nitrate and grain yield. *J. Plant Nutr. Soil Sci.* 177: 216–226.

Ma, B. L., T. Y. Wu, N. Tremblay, W. Deen, N. B. McLaughlin, M. J. Morrison, and G. Stewart. 2010a. Rate and timing effects of fertilizer nitrogen application to corn on ammonia volatilization in cool and humid regions. *Agron. J.* 102:134-144.

Ma, B. L., T. Y. Wu, N. Tremblay, W. Deen, M. J. Morrison, N. B. McLaughlin, E. G. Gregorich, and G. Stewart. 2010b. Nitrous oxide fluxes from corn fields: On-farm assessment of the amount and timing of nitrogen fertilizer. *Glob. Chang. Biol.* 16:156–170.

Ma, B. L., J. Ying, L. M. Dwyer, E.G. Gregorich, and M.J. Morrison. 2003. Crop rotation and soil nitrogen amendment effects on maize production in Eastern Canada. *Can. J. Soil Sci.* 83:483-495.

Ma, J., X. L. Li, H. Xu, Y. Han, Z. C. Cai, and K. Yagi. 2007. Effects of nitrogen fertilizer and wheat straw application on CH_4 and N_2O emissions from a paddy rice field. *Aust. J. Soil Res.* 45:359–367.

MacKenzie, A. F., M. X. Fan, and F. Cadrin. 1997. Nitrous oxide emission as affected as affected by tillage, corn- soybean alfalfa rotations, and nitrogen fertilization. *Can. J. Soil Sci.* 77:145-152.

Maggiotto, S. R., J. A. Webb, C. Wagner-Riddle, and G.W. Thurtell. 2000. Nitrous and nitrogen oxide emissions from turfgrass receiving different forms of nitrogen fertilizer. *J. Environ. Qual.* 29:621–630.

Mapanda, F., W. Menas, J. Nyamangara, and R. M. Rees. 2011. Effects of organic and mineral fertilizer nitrogen on greenhouse gas emissions and plant-captured carbon under maize cropping in Zimbabwe. *Plant Soil* 343:67-81.

Matthews, E. 1994. Nitrogenous fertilizers: Global distribution of consumption and associated emissions of nitrous oxide and ammonia, *Glob. Biogeochem. Cycl.* 84:411–439.

McConkey, B. G., B. C. Liang, C. A. Campbell, D. Curtin, A. Moulin, S.A. Brandt, and G.P. Lafond. 2003. Crop rotation and tillage impact on carbon sequestration in Canadian prairie soils. *Soil Till. Res.* 74:81-90.

McGonigle, T. P., and M.H. Miller. 1993. Mycorrhizal development and phosphorus absorption in maize under conventional and reduced tillage. *Soil Sci. Soc. Am. J.* 57:1002–1006.

McKenny, D. J., S. W. McKenny, C. F. Wang, and W. I. Drury. 1993. Findlay denitrification and mineralization in soil amended with legume, grass and corn crop residues. *Soil Sci. Soc. Am. J.* 57:1013–1020.

McSwiney, C. P., and G. P. Robertson. 2005. Nonlinear response of N_2O flux to incremental fertilizer addition in a continuous maize (*Zea mays* L.) cropping system. *Glob. Chang. Biol.* 11:1712–1719.

Menalled, F. D., K. I. Gross, and M. Hammond. 2001. Weed aboveground and seedbank community responses to agricultural management systems. *Ecol. Appl.* 11:1586-1601.

Merrill, S. D., D. L.Tanaka, J. M. Krupinsky, M. A. Liebig, and J. D. Hanson. 2007. Soil water depletion and recharge under ten crop species and applications to the principles of dynamic cropping systems. *Agron. J.* 99:931-938.

Miller, P. R., Y. T. Gan, B. G. McConkey, and C. L. McDonald. 2003. Pulse crops for the Northern Great Plains: I Grain productivity and residual effects on soil water and nitrogen. *Agron. J.* 95:972-979.

Mohanty, S. R., K. Bharati, B. T. S. Moorthy, B. Ramakrishnan, V. R. Rao, N. Sethunathan, and T. K. Adhya. 2001. Effect of the herbicide butachlor on methane emission and ebullition flux from a direct-seeded flooded rice field. *Biol. Fertil. Soils* 33:175–180.

Mohanty, S. R., D. R. Nayak, Y. J. Babu, and T. K. Adhya. 2004. Butachlor inhibits production and oxidation of methane in tropical rice soils under flooded condition. *Microbiol. Res.* 159:193–201.

Montzka, S. A., E. J. Dlugokencky, and J. H. Butler. 2011. Non-CO_2 greenhouse gases and climate change. *Nature* 476(7358):43-50.

Mosier, A., C. Kroeze, C. Nevison, O. Oenema, S. Seitzinger, and O. van Cleemput. 1998. Closing the global N_2O budget: Nitrous oxide emissions through the agricultural nitrogen cycle - OECD/IPCC/IEA phase II development of IPCC guidelines for national greenhouse gas inventory methodology. *Nutr. Cycl. Agroecosys.* 52:225-248.

Mosier, A. R., A. D. Halvorson, C. A. Reule, and X. J. Liu. 2006. Net global warming potential and greenhouse gas intensity in irrigated cropping systems in northeastern Colorado. *J. Environ. Qual.* 35:1584–1598.

Nishimura, S., S. Yonemura, S. Sawamoto, Y. Shirato, H. Akiyama, S. Sudo, and K. Yagi. 2008. Effect of land use change from paddy rice cultivation to upland crop cultivation on soil carbon budget of a cropland in Japan. *Agric. Ecosys. Environ.* 125:9–20.

NOAA National Climate Data Center. 2013. State of the Climate: Global analysis for March 2013, published online April 2013, retrieved on 18 March 2013 from http://www.ncdc.noaa.gov/sotc/global/.

Novoa, R. S. A., and H. R. Tejeda. 2006. Corn nitrogen requirements, utilization, and fertilization. Pp567-604. In: Stevenson, F.J. (ed.) *Nitrogen in Agricultural Soils.* Agronomy monograph 22. ASA. Madison. WI.

Oorts, K., R. Merckx, E. Gréhan, J. Labreuche, and B. Nicolardot. 2007. Determinants of annual fluxes of CO_2 and N_2O in long-term no-tillage and conventional tillage systems in northern France. *Soil Till. Res.* 95:133-148.

Oreskes, N. 2004. The scientific consensus on climate change. *Science* 306(5702):1686-1686.

Pandey, D., M. Agrawal, and J.S. Pandey. 2011. Carbon footprints: Current methods of estimation. *Environ. Monit. Assess.* 178:135–160.

Pandey, D., and M. Agrawal. 2014. Carbon footprint estimation in the agriculture sector. Pp25-47. *In*: Muthu, S.S. (ed.) Assessment of Carbon Footprint in Different Industrial Sectors, volume 1. Springer, Singapore.

Parkin, T. B., and T. C. Kaspar. 2006. Nitrous oxide emissions from corn–soybean systems in the Midwest. *J. Environ. Qual.* 35:1496–1506.

Paustian, K., H. P. Collins, and E.A. Paul. 1997. Management controls on soil carbon. pp. 39-41. *In*: Paul, E.A., K. Paustian, E.T. Elliot, and C.V. Cole (eds.) *Soil Organic Matter in Temperate Agroecosystems*. CRC Press, Inc. NY, New York.

Raun, W. R., and G. V. Johnson. 1999. Improving nitrogen use efficiency for cereal production. *Agron. J.* 91:357-363.

Reay, D., and C.M. Hogan. 2008. Greenhouse gas. *The Encyclopedia of Earth* Available at www.eoearth. org/view/article/153147/.

Rees, W. E. 1992. Ecological footprints and appropriated carrying capacity: What urban economics leaves out. Environ. *Urbanisation* 4:121-130.

Roberts, W. P., and K. Y. Chan. 1990. Tillage-induced increases in carbon dioxide loss from soil. *Soil Till. Res.* 17:143-151.

Roberts, T. L. 2008. Improving nutrient use efficiency. Turkish *J. Agric. Forest.* 32(3):177.

Robertson, G. P., and P. M. Groffman. 2007. Nitrogen transformation. Pp341-364. *In*: Paul, E.A. (ed.) *Soil Microbiology, Bio-chemistry, and Ecology.* Springer, NY, New York.

Robertson, G. P., E. A. Paul, and R. R. Harwood. 2000. Greenhouse gases in intensive agriculture: contributions of individual gases to the radiative forcing of the atmosphere. *Science* 289:1922–1925.

Rochette, P., D. E. Worth, R. L. Lemke, B.G. McConkey, D. J. Pennock, C. Wagner-Riddle, and R. L. Desjardins. 2008a. Estimation of N_2O emissions from agricultural soils in Canada. I. Development of a country-specific methodology. *Can. J. Soil Sci.* 88:641-654.

Rochette, P., D. E. Worth, E. C. Huffman, J. A. Brierley, B. G. McConkey, J. Yang, J. J. Hutchinson, R. L. Desjardins, R. Lemke, and S. Gameda. 2008b. Estimation of N_2O emissions from agricultural soils in Canada. II. 1990-2005 inventory. *Can. J. Soil Sci.* 88:655-669.

Roelandt, C., B. van Wesemael, and M. Rounsevell. 2005. Estimating annual N_2O emissions from agricultural soils in temperature climates, *Glob. Change Biol.* 11:1701-1711.

Ruddiman, W. F. 2003. The anthropogenic greenhouse era began thousands of years ago. *Clim. Change* 61:261–293.

Schlesinger, W. H. 1997. *Biogeochemistry: An analysis of global change*. Academic Press, San Diego, CA. pp 588.

Schwenke, G. D., D. F. Herridge, C. Scheer, D. W. Rowlings, B. M. Haigh, and K. G. McMullen. 2015. Soil N_2O emissions under N_2-fixing legumes and N-fertilised canola: A reappraisal of emissions factor calculations. *Agric. Eco. Environ.* 202: 232-242.

Seghers, D., S. D. Siciliano, E. M. Top, and W. Verstraete. 2005. Combined effect of fertilizer and herbicide applications on the abundance, community structure and performance of the soil methanotrophic community. *Soil Biol. Biochem.* 37:187–193.

Seip, H. M. 2011. Carbon footprints: relative drop for farming emissions. *Nature* 480:321.

Singh, P., J. Heikkinen, E. Ketoja, V. Nuutinen, A. Palojärvi, J. Sheehy, and K. Regina. 2015. Tillage and crop residue management methods had minor effects on the stock and stabilization of topsoil carbon in a 30-year field experiment. *Sci. Total Environ.* 518:337-344.

Sistani, K. R., M. Jn-Baptiste, N. Lovanh, and K. I Cook. 2011. Atmospheric emissions of nitrous oxide, methane, and carbon dioxide from different nitrogen fertilizers. *J. Environ. Qual.* 40:1797–1805.

Smith, K. A., and F. Conen. 2004. Impacts of land management on fluxes of trace greenhouse gases. *Soil Use Manage.* 20:255–263.

Smith, K., P. Crutzen, A. Mosier, and W. Winiwarter. 2010. The global nitrous oxide budget: A reassessment. pp. 63-84. *In*: Smith, K. (ed.) *Nitrous oxide and climate change*. Earth Scan, London.

Smith, D. R., G. Hernandez-Ramirez, S. D. Armstrong, D. L. Bucholtz, and D.E. Stott. 2011. Fertilizer and tillage management impacts on non-carbon dioxide greenhouse gas emissions. *Soil Sci. Soc. Am. J.* 75:1070-1082.

Snyder, C. S., T. W. Bruulsema, T. L. Jensen, and P. E. Fixen. 2009. Review of greenhouse gas emissions from crop production systems and fertilizer management effects. *Agric. Eco. Environ.* 133:247-266.

Staley, T. E., W. H. Caskey, and D.G. Boyer. 1990. Soil denitrification and nitrification potentials during the growing season relative to tillage. *Soil Sci. Soc. Am. J.* 54:1602–1608.

Steinbach, H. S., and R. Alvarez. 2006. Changes in soil organic carbon contents and nitrous oxide emissions after introduction of no-till in pampean agroecosystems. *J. Environ. Qual.* 35:3–13.

Subedi, K. D., and B. L. Ma. 2005. Effects of N-deficiency and timing of N supply on the recovery and distribution of labeled ^{15}N in contrasting maize hybrids. Plant Soil 273:189-202.

Swift, M. J., J. Vandermeer, P. S. Ramakrishnana, J. M. Anderson, C. K. Ong, and B.A. Hawkins. 1996. Biodiversity and agroecosystem function. pp 261-290. *In*: Mooney, H.A.,

J. H. Cushman, E. Medina, O. E. Sala, and E. D. Schulze (eds.) *Functional Roles in Biodiversity: A Global Perspective.* John wiley and Sons, New York, NY.

Tanaka, D. L., J. M. Krupinsky, S. D. Merrill, M. A. Liebig, and J. D. Hanson. 2007. Dynamic cropping systems for sustainable crop production in the northern Great Plains. *Agron. J.* 99:904–911.

Thomsen, K., and B. T. Christensen. 1996. Availability to subsequent crops and leaching of nitrogen in ^{15}N-labelled sugar-beet tops and oilseed rape residues. *J. Agric. Sci.* 126:191-199.

Thornton, F. C., and R. J. Valente. 1996. Soil emissions of nitric oxide and nitrous oxide from no-till corn. *Soil Sci. Soc. Am. J.* 60:1127–1133.

U.S. EPA. 2007. Inventory of U.S. greenhouse gas emissions and sinks: 1990-2005. EPA 430-R-07-002 U.S. Environmental Protection Agency, Office of Atmospheric Programs, Washington, DC.

U.S. EPA. 2015. Inventory of U.S. greenhouse gas emissions and sinks: 1990–2013. United States Environmental Protection Agency. Washington, DC (available online http://unfccc.int/national_reports/annex_i_ghg_inventories/national_inventories_submissions/items/8812.php).

Ussiri, D. A. N., and R. Lal. 2009. Long-term tillage effects on soil carbon storage and carbon dioxide emissions in continuous corn cropping system from an alfisol in Ohio. *Soil Till. Res.* 104:39–47.

Ussiri, D. A. N., R. Lal, and M. K. Jarecki. 2009. Nitrous oxide and methane emissions from long-term tillage under a continuous corn cropping system in Ohio. *Soil Till. Res.*104:247–255.

VandenBygaart, A. J., E. G. Gregorich, and D. A. Angers. 2003. Influence of agricultural management on soil organic carbon: A compendium and assessment of Canadian studies. *Can. J. Soil. Sci.* 83:363-380.

VandenBygaart, A.J., B.G. McConkey, D.A. Angers, W.N. Smith, H.de Gooijer, M. Bentham, and T. Martin. 2008. Soil carbon change factors for the Canadian agriculture national greenhouse gas inventory. *Can. J. Soil Sci.* 88:671-680.

Velthof, G. L., S. C. Jarvis, A. Stein, A.G. Allen, and O. Oenema. 1996. Spatial variability of nitrous oxide fluxes in mown and grazed grasslands on a poorly drained clay soil. *Soil Biol. Biochem.* 28:1215–1225.

Velthof, G. L., P. J. Kuikman, and O. Oenema. 2003. Nitrous oxide emission from animal manures applied to soil under controlled conditions. *Biol. Fertil. Soils* 37:221-230.

Venterea, R. T., M. Burger, and K. A. Spokas. 2005. Nitrogen oxide and methane emissions under varying tillage and fertilizer management. *J. Environ. Qual.* 34:1467–1477.

Wagner-Riddle, C., and G. W. Thurtell. 1998. Nitrous oxide emissions from agricultural fields during winter and spring thaw as affected by management practices. *Nutr. Cycl. Agroecosys.* 52: 151-163.

Wagner-Riddle, C., N. L. Furon, I. L. Mclaughlin, J. Barbeau, S. Jayasundara, G. Parkin, P. von Bertoldi, and J. Warland. 2007. Intensive measurement of nitrous oxide emissions from a corn- soybean-wheat rotation under two contrasting management systems over 5 years. *Glob. Change Biol.* 13:1722-1736.

Wagner-Riddle, C., Q. C. Hu, E. van Bochove, and S. Jayasundara. 2008. Linking nitrous oxide flux during spring thaw to nitrate denitrification in the soil profile. *Soil Sci. Soc. Am. J.* 72:908–916.

Wang, Z., B. L. Ma, R. Gao, and J. Sun. 2015. Effects of different management systems on root distribution of maize. *Can. J. Plant Sci.* 95:21-28.

Wang, Z., R. Gao, and B. L. Ma. 2014. Concurrent improvement in maize yield and nitrogen use efficiency with integrated agronomic management strategies. *Agron. J.* 106:1243–1250.

West, T. O., and G. Marland. 2002. A synthesis of carbon sequestration, carbon emissions, and net carbon flux in agriculture: comparing tillage practices in the United States. *Agric. Ecosyst. Environ.* 91:217–232.

Westerman, P. R., M. Liebman, F. D. Menalled, A. H. Heggenstaller, R. G. Hartzler, and P. M. Dixon. 2005. Are many little hammers effective? Velvetleaf (Abutilon theophrasti) population dynamics in two and four-year crop rotation systems. *Weed Sci.* 53:382-392.

Wiedmann, T., and J. Minx. 2008. A definition of carbon footprint. Pp1-11. *In*: Pertsova, C. C. (ed.) *Ecological economics research trends.* Nova Science Publishers, Hauppauge, NY, USA. https://www.novapublishers.com/catalog/product_info.php?products_id=5999.

Williams, D. L., P. Ineson, and P. A. Coward. 1999. Temporal variations in nitrous oxide fluxes from urine-affected grasslands. Soil Biol. Biochem. 31:779–788.

WRI/WBCSD. 2004. The greenhouse gas protocol: A corporate accounting and reporting standard. Revised Edition. World Business Council for Sustainable Development (WBCSD), Geneva, and World Resource Institute USA. Available online http://www.ghgprotocol.org/files/ghgp/public/ghg-protocol-revised.pdf.

Wu, T. Y., B. L. Ma, and B. C. Liang. 2008. Quantification of seasonal soil nitrogen mineralization for corn production in eastern Canada. *Nutr. Cycl. Agroecosyst.* 81:279-290.

Wu, W., and B. L. Ma. 2015. Integrated nutrient management (INM) for sustaining crop productivity and reducing environmental impact: A review. *Sci. Total Environ.* 512:415-427.

Zebarth, B. J., P. Rochette, D. L. Burton, and M. Price. 2008. Effect of fertilizer nitrogen management on N2O emissions in commercial corn fields. *Can. J. Soil Sci.* 88:189-195.

Zentner, R. P., G. P. Lafond, D. A. Derksen, C. N. Nagy, D. D. Wall, and W. E. May. 2004. Effects of tillage method and crop rotations on non-renewable energy use efficiency for a thin Black Chernozem in the Canadian Prairies. *Soil Till. Res.* 77:125–136.

EDITOR CONTACT INFORMATION

Dr. Bao-Luo Ma, PhD
Senior Research Scientist
Ottawa Research and Development Centre,
Agriculture and Agri-Food Canada
Eastern Cereal and Oilseed Research Centre
960 Carling Avenue, Ottawa, ON, Canada K1A 0C6
Tel: 613-759-1521; Fax: 613-759-1701
E-mail: baoluo.ma@agr.gc.ca

INDEX

E

F

G

I

L

M

N

O

P

Q

R

S

T

W

Y